COMPRAIL

Railway Engineering, Design and Operation

COMPRAIL

Railway Engineering, Design and Operation

Editors

C.A. Brebbia
Wessex Institute, UK

J.M. Mera
Polytechnic University of Madrid, Spain

N. Tomii
Chiba Institute of Technology, Japan

P. Tzieropoulos
Ecole Polytechnique Federale de Lausanne, Switzerland

WIT*PRESS* Southampton, Boston

Editors:

C.A. Brebbia
Wessex Institute, UK

J.M. Mera
Polytechnic University of Madrid, Spain

N. Tomii
Chiba Institute of Technology, Japan

P. Tzieropoulos
Ecole Polytechnique Federale de Lausanne, Switzerland

Published by

WIT Press
Ashurst Lodge, Ashurst, Southampton, SO40 7AA, UK
Tel: 44 (0) 238 029 3223; Fax: 44 (0) 238 029 2853
E-Mail: witpress@witpress.com
http://www.witpress.com

For USA, Canada and Mexico

Computational Mechanics International Inc
25 Bridge Street, Billerica, MA 01821, USA
Tel: 978 667 5841; Fax: 978 667 7582
E-Mail: infousa@witpress.com
http://www.witpress.com

British Library Cataloguing-in-Publication Data

A Catalogue record for this book is available
from the British Library.

Library of Congress Catalog Card Number: 2016917268

ISBN: 978-1-78466-243-1
eISBN: 978-1-78466-244-8

The text of the papers in this volume were set individually by the authors or under their supervision. Only minor corrections to the text may have been carried out by the publisher.

Printed in Great Britain by Lightning Source, UK.

The material contained herein is reprinted from a special edition of the International Journal of Transport Development and Integration, Volume 1, Number 3, published by WIT Press.

Preface

The papers contained in this Volume are a selection of those presented at the Conference on Railway Engineering, Design and Operation (COMPRAIL) which took place in Madrid, organised by the Wessex Institute of Technology. The Meeting, which started in Frankfurt in 1987, has been successfully reconvened every two years in different locations around the world.

The Conference aims to present and discuss the use of advanced railway systems, promoting general awareness of new developments in the management, design, manufacture and operation of railways and related transport systems. It emphasises the use of computer systems in advanced railway engineering.

Modern railway engineering attracts a variety of disciplines including managers, consultants, designers of train control systems and computer specialists, in addition to railway engineers. This is reflected in the contents of this Volume, which also demonstrates the diversity and the international character of modern railway engineering development.

The Editors
2017

Contents

STOCHASTIC MODEL FOR THE REAL-TIME TRAIN RESCHEDULING

B. DAVYDOV[1], V. CHEBOTAREV[2] & K. KABLUKOVA[2]
[1]Far Eastern State Transport University, Russia.
[2]Computing Center, Far Eastern Branch, Russian Academy of Sciences, Russia.

ABSTRACT
The article explores the problem of train rescheduling based on the actual situation. The proposed stochastic model uses specific distributions of operating times which are dependent on the current traffic conditions. The arrival time distribution is considered as a result of adjusting the train trajectory by speed control. The results of modelled arrival distributions correspond well with the experimental data received at the Russian Railways. The proposed model is used for prevention of sequence-of-trains conflicts and violations of connections. The basis of deviation prediction is two-train model of mesa-level which uses actual features of scattering of the operation times both at sites and at stations. The article also proposed a new measure of arrival delay which considers the share of satisfied passengers.
Keywords: local adjustments, online rescheduling, stochastic mesa-mode, train traffic.

1 INTRODUCTION

Modelling the train traffic on the railway section is necessary for the optimal schedule creation. Ensuring a high level of traffic stability is the aim of constructing a tactical schedule. There is a need to use the online rescheduling when there is assumed a disrupting process. Modelling in both of these cases has significant differences. Tactical graphic takes into account the average process conditions. A stochastic model for the schedule calculation is based on fixed probability densities that characterize the operating times. Such a model very approximately describes the real traffic options because of using a large diversity of situations and averaged densities. It is necessary to include sufficiently big buffer and running time supplements in the schedule to partially eliminate this disadvantage.

Distribution of the train arrivals at a certain station is formed from the scattering of travel and stopping times. There are prevalent models that take into account only random deviations of travel times. The known models of other type suggest that adjustment of traffic is made at the stations only in order to eliminate delays. This activity is to relocate the departure moments or change the order of trains. In each individual case, use rigid rules to resolve conflicts between trains which arrive at the transfer station. Such a regulatory framework is used in the model of Berger et al. [1]. Using the simplified model and hard algorithms of conflict resolution is useful when creating a normative schedule which however is a rough approximation to reality. This assumes the absence of operative dispatching which is a fundamental drawback of the existing methods of solving long-term scheduling problems. Section 3 discusses the main points of creation of the generic stochastic model which combines the two specified approaches.

DOI: 10.2495/TDI-V1-N3-307-317

Operative correction of train traffic can significantly improve the quality of functioning of the railway section. The main feature is to carry out an elementary cycle of modelling for that local area where there is a single disturbance. The deviation arises in some a 'cone of delays' if of moving an intense train flow. It is necessary to operatively change the trajectory of each of the trains to eliminate the spread of delays. This process is realized through the efforts of train manager and drivers. The train dispatcher implements centralized control of the entire system function, while the train drivers solve local problems on delays compensation. This all-inclusive technique is not reflected in the existing models.

The models used in the known papers have two drawbacks. One drawback is the lack of a framework that reflects a purposeful driver's operation (or of on-board unit control) for the perturbation compensation. Another disadvantage is that they ignore the following facts. The distributions that characterize partial travel times are distinct at different times of day and in different situations. Obviously, in the model which is used for online regulation, the parameters of these distributions should be established based on the emerging conditions. Section 4 discusses the impact of the specific situation at mentioned parameters.

The stochastic model which comes really close to our approach is in Büker and Seybold [2]. They basically have the same form of distribution of the train arrivals and respective waiting policies. However, they concentrated on train interactions independent of the specific operating conditions and of activities of the personnel. We will enhance their model to analyse the impact of actual factors on the distribution of arrival times.

Computer experiments show that an output distribution width increases with the expansion of the forecast horizon in case of weak control algorithm. In many cases, this modelling result does not correspond to the actual movement of trains. When using the effective adjustments which are implemented by managers and drivers, the distribution width is stabilized over time. The problem of the effect of the driver operation on the distribution of the train arrival moments is explored in Section 5. Combining techniques of forming delays on sections and stations allows you to create the generalized stochastic model of train traffic. The basis of the formal description of the motion is two-train model that is consistently applied to all pairs of train. A generalized representation of the traffic and the process of conflicts arising is a probability graph.

2 LITERATURE REVIEW

Many scholars study the rescheduling problem which considers the way to recover from a disturbed movement of trains. The researches were devoted to the algorithms that optimally adjust the schedule. The well-known papers [3, 4] provide an extensive survey of recovery models and algorithms for real-time rescheduling. Depending on the focus, the models used to solve the scheduling problem that differs by levels of infrastructure and traffic detail. A line appears as a sequence of segments between two major stations with few intermediate stations, while a network is composed of one or several junctions of lines. Macroscopic models contain least details and have a more aggregated representation of some resources (e.g. a station), while microscopic representation includes a lot more detail such as block sections. Existing models are not focused on direct description of the points of conflicts between trains.

Analyses of previous studies show an increasing interest in stochastic approaches and models that help create adequate forecasts and make optimal decisions for traffic recovery. One of the first papers considering this problem [5] proposed analytical method for the determination of delays by a convolution of the initial (inlet area) and current delays. The analysis was limited because it used the assumption of uniform distribution of secondary delays. In reality, this assumption is not confirmed. Most stochastic models to analyse propagation of delays focused on single-track sections or on simple junctions. Classical stochastic models for the propagation of delays had been studied intensively, most importantly in Refs [6, 7]. The papers proposed the approximation of delay distributions to reduce the computational effort. The paper [7] represents train traffic in a form of occupying an intermediate position between the macro-models and the models with very detailed description of the process, i.e. micro-models. Cumulative distribution of delays is calculated from the sequence of activities, which is defined by a stochastic event graph.

Stochastic analytical model for a traffic on a double-track line was proposed in Ref. [8]. The paper investigates delays of a fast train caught behind slower ones. The section is modelled as an infinite server. The running time distributions for each service are obtained by solving a system of linear differential equations. This model gives a good insight into delay propagation on a simple line but becomes too complicated to handle when dealing with large-scale real-world networks. The approach for predicting waiting times using queuing system with a semi-Markovian kernel was presented in Ref. [9]. Description of the service process is based on an application of the theory of blocking times and minimum headway times.

A stochastic model for delay propagation and forecasts of arrival and departure events was proposed in Ref. [1]. The model includes the general train waiting policies on stations and considers discrete distributions of travel time profiles. The model is formulated with respect to a directed event graph. Another model proposed in Ref. [10] is also designed to assess the knock-on delays. The analytical probabilistic model takes into account the duration of stops, stochastic relationship between the trajectories of trains, speed fluctuations and dynamic delay propagation. A new scheme was proposed for the exact determination of parameters of the distribution model based on the maximum likelihood method.

Stochastic simplified model can be used to find an optimal allocation of the running time supplements of a train on a number of consecutive trips along the same line [11]. The result of this study is the optimal distribution of supplements between the site elements. The approach is based on the periodic event scheduling developed in Ref. [12]. An analytical method such as the max-plus algebra (MPA) is used to evaluate stability of the schedule.

Various known models describe the functions of frequency and length of random traffic deviations. Thus, the exponential distribution is used as a stochastic model of every component [7–9]. Changes of distribution are due to ‘internal’ reasons. Obviously, accumulation of the individual random variables leads to exponential behaviour of delays. This problem is explored in Ref. [13] and requires further study especially for the mixed flow of passenger and freight trains.

Berger et al. [1] solve the problem of train traffic rescheduling on a large network. Scattering of arrival moments is treated as a random variable generated by a spread of the departure times and the running intervals. The basis of the analysis is two-train model. This model allows to define the order of departure of the train from the initial station and distribution of the respective moments of departure. The authors use a discrete representation of distributions.

A number of studies considered a model that comprehends two train runs, each with transferring passengers at some stations [14]. The model is illustrated as a Petri net graph. The approach proposed in Ref. [2] uses a probabilistic operational graph, which considers conflicts within transfer operation. The interlinking conflicts are mapped in the model by manipulating the time distributions of the elementary activities. The paper uses two-train mesoscopic model and the FIFO rule for conflict resolution. Use of both conditional and unconditional convolution is specifically made. Various types of delay are modelled by means of cumulative distribution function. Described model is logically incomplete. The authors consider only the station and crossing as the conflict points. To complete the model, you must use a resolution of the sequence-of-trains conflicts at a section.

A significant number of studies are devoted to the development of criteria for determining the optimal solutions for train traffic management. Top-level rail experts created a set of railway performance indicators that measure efficiency from the perspectives of the government, passenger/client, infrastructure manager and train operating company [15]. A very common measure to quantify the service quality is to calculate the punctuality at different nodes in the system [12]. Using only the punctuality criterion has some drawbacks. It does not take into account the number of passengers affected by delays nor does it cover the case where passengers are missing connecting trains. The new approach overcomes the above drawbacks, as it accounts for passenger delays (personal passenger time is measured in minutes) instead of only delays [10]. However, the total duration of delays weakly reflects satisfaction of the passengers. Obviously, there is a certain threshold mechanism, where the allowable time limits depend on the non-nervous traveller sentiment.

Previous studies have used the distributions of typical operations (running, stopping) in the probabilistic model that are unchanged in time and do not depend on the coordinates of a particular path element. It is required to take into account the circumstances in a particular local problem, individual distributions of which depend on the influencing factors. It is also necessary to model the impact of the online system adjustment at the quality of train traffic. Ways to solve these problems are described in the following sections.

3 MESOSCOPIC STOCHASTIC MODELLING OF TRAIN TRAFFIC

Two types of deterministic models are used in the analysis of intensive train traffic. The micro-model reflects the train running with a readability corresponding to the block sections. Speed within block section is considered as a constant value and it changes when the corresponding signal is received from the next section. The micro-model is losing its effectiveness in the presence of random disturbances.

From the viewpoint of queuing theory all the points of conflict on open tracks and at stations may be interpreted as server systems. This is considered as the mesoscopic approach for the creation of train traffic model. The mesa-model assumes that an elementary checking element significantly exceeds the length of a block section. Checkpoints such as stations, stops and crossings are the element boundaries. The speed profile and a scheduled time interval between checkpoints here are taken as fixed values.

Random components, which are due to peculiar specific conditions of traffic and style of work of the driver, are reflected in a common distribution of the running time. In reality, coordinate control and centralized management of train runs (i.e. introduction of corrections) are made at the moments of train passage at the checkpoints. Decision to change the mode of run is made in the dispatch centre and is based on a mesa-model. Movement of a train along a spatial trajectory is represented as a set of elements between the control points. Complete time of a train passage by the trajectory is an additive value namely a sum of the elementary operation times. The trajectory of the train is modelled by a directed graph in the time domain (an activity graph), some nodes of which gather several arcs (Fig. 1).

The nodes of the graph represent the states of the process such as arrivals and departures on the station and the points of conflicts. Time deviation from the timetable is considered as an indicator of every state. The arcs of the graph reflect transition of the train from one state to another that is an activity operation. Some authors (e.g. Büker and Seybold [2]) include connections for

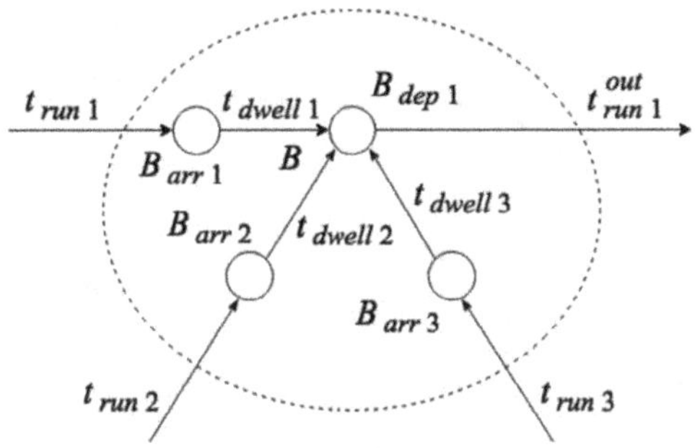

Figure 1: Fragment of the graph which reflects the passenger interlinking.

the passenger transfer and inter-train conflicts in this list. Length of the arcs reflects operating times that characterize the transition between checkpoints. Additional arcs correspond to delays that occur due to the influence of other trains, i.e. conflicts on sites or stations.

The approach considered is the basis for a generic stochastic model of train traffic on the railway section [13]. The process of accumulating the delays is represented as a tree of interferences, which is the graph whose nodes reflect the presence of conflicts between trains at the control points. It is assumed this points can be of two types, namely with primary (i.e. input) disturbances or with secondary delays. The arc length is considered as a random variable that distinguishes it from the most of previously reported frames.

4 STOCHASTIC MODELLING OF CONTROL ACTIVITY

The driver or the on-board system produces the permanent control over the motion parameters of the train. Speed is adjusted so as to perform the schedule. The time trajectory of the train is adjusted occasionally by the infrastructure manager. The purpose of regulation is timely arrival at checkpoints.

The running time is affected by random factors that are characterized by a probability density function. Over time, this function is changing, and its support becomes wider if there is no adjustment. At the same time adjustment by the system or the operator can stabilize the distribution and even to narrow its support.

Now we consider modelling the process of forming the probability distribution of deviation of the arrival time from the scheduled one, when there is a driver's control work. Let t_0 be the time of departure, and $s = s(t)$ and $v = v(t)$ traversed path and speed of the train, respectively. It is assumed that the minimum $v_1 = v_{min}$ and maximum $v_2 = v_{max}$ allowable speeds are known. We denote by v_0 average speed of a train. Random factors lead to the scattering of trajectories (Fig. 2).

Let $\tau(s)$ be the actual arrival time of the train at *P*. If the train moves at a speed of v_j, the moment of its arrival is

$$a_j(s) = t_0 + \frac{s}{v_j},\ j = 0, 1, 2, \quad 0 < s_1 \le s \le s_2. \tag{1}$$

The random variable $Y(s) = \tau(s) - a_0(s)$ is the deviation from the scheduled arrival time $a_0(s)$. Note that $Y(s)$ takes the values of different signs.

Denote by $f(t;s)$ the density function of the random variable $\tau(s)$, and by $g(t;s)$ the density function of the random variable $Y(s)$. Obviously,

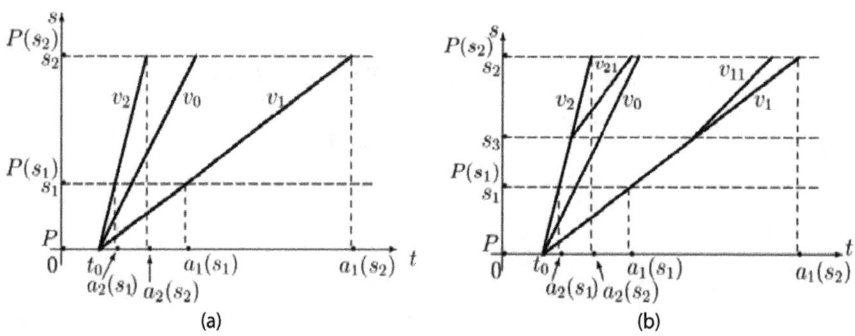

Figure 2: Graphs of functions $s = v_j(t - t_0), j = 0, 1, 2.$

$$g(t;s) = f(t + a_0(s);s) \cdot \tag{2}$$

The interval $[a_2(s), a_1(s)]$ is the support of $f(t;s)$, i.e. $f(t;s) = 0$ when $t \notin [a_2(s), a_1(s)]$, $[a_2(s) - a_0(s), a_1(s) - a_0(s)]$ is the support of $g(t;s)$.

Given the limitations on the speed, we obtain a one-to-one correspondence between the points of these intervals: $t' = A(s_2)t + B(s_2)$, where $t \in [a_2(s_1), a_1(s_1)]$, $t' \in [a_2(s_2), a_1(s_2)]$,

$$A(s) = \frac{a_1(s) - a_2(s)}{a_1(s_1) - a_2(s_1)}, \qquad B(s) = \frac{a_2(s)a_1(s_1) - a_2(s_1)a_1(s)}{a_1(s_1) - a_2(s_1)},$$

$$s_1 \le s \le s_2. \tag{3}$$

Note that the length of the interval $[a_2(s), a_1(s)]$ increases when the boundary condition on the permissible speed is not changed, i.e. if there is no intervention to regulate the movement. In this case, the scattering of the considered probability distribution increases. Speed change in some points, which produces the driver, results in a decrease of scattering.

Assumption. Suppose that for each fixed s $\tau(s) \overset{d}{=} A(s)\tau(s_1) + B(s)$, where $A(s) \neq 0$ and $B(s)$ are defined in eqn (3). This is equivalent to $Y(s) \overset{d}{=} A(s)(Y(s_1) + a_0(s_1)) + B(s)$. Here we have used the notation $W_1 \overset{d}{=} W_2$ as random variables $\tau(s)$ and $[A + B]$ are identically distributed.

Note that if $f(t)$ is the density of the random variable X, $A \neq 0$ and B are constants, then $\frac{1}{|A|} f\left(\frac{t - B}{A}\right)$ is the density of the random variable $AX + B$.

From this observation, assumption 1 and eqn (2), we deduce that

$$f(t;s) = \frac{1}{|A(s)|} f\left(\frac{t - B(s)}{A(s)}; s_1\right), \tag{4}$$

$$g(t;s) = f(t + a_0(s);s) = \frac{1}{|A(s)|} g\left(\frac{t + a_0(s) - B(s)}{A(s)} - a_0(s_1); s_1\right). \tag{5}$$

Thus, we conclude that the density of the random variable $Y(s)$ with increasing s is subjected to a linear transformation. With increasing $A(s)$ supports of densities $f(t;s)$ and $g(t;s)$ extend.

The intervention of the driver consists in change of speed at the point s_3. Speed is reduced, if the train is ahead of schedule, or increases in the opposite case (Fig. 2(b)). The distribution of the time of arrival becomes narrower.

The analysis shows that the limitations imposed, $v_{21} < v_2$, $v_{11} > v_1$, also result in linear transformation of $f(t;s_3)$ density. The form of the distribution $g(t;s_2)$ in the point P is maintained, changing only its parameters. Experimental study on the Russian Railways shows that the real scattering of arrival times is well approximated by a gamma distribution. In the sections of railway, which carried out effective local control, the scattering width of the arrivals differs a little at different sites. The dispersion increases through the train route if the regulation is weak.

5 ACTUALIZATION OF OPERATING TIME DISTRIBUTIONS

In the real world, the actual train arrivals and failures of infrastructure in the coming hours are known quite accurately. This is a prerequisite to develop a more accurate prediction of traffic. The operational model is constructed by taking certain influencing factors into account. Thus, the specific parameters of operating times scattering can be deduced.

You must use the specific probability distribution of operating intervals to create a predictive model for each real situation. Such distributions for the running and stopping times are determined from historical data. By the historical data we mean the situations (scenarios) that occurred in previous periods on considered track section. Parameters of the operating time distributions depend on key factors like constraints on the train passage or weather conditions.

It is necessary to find out how to use historical data for a complete formalization of stochastic model of the train traffic. The idea of the early proposed method for determining the forecasted indicators on the basis of previous experience is known as *situational heuristic predicting* [16]. The method is based on constructing the *array of experience* Z, which is a set of indicators of the line functioning in certain real-life situation. This situation is determined by the values of influencing factors, such as the number of cars or trains at the adjacent stations, etc. The array Z is a multidimensional matrix for various combinations of the factors and their associated cells that contain the realization times τ_i, $1 \le i \le n$, of a certain activity (e.g. travel time) in every historical situation. The array Z is considered in the following general form:

$$Z = \left(\begin{array}{cccc|c} \Phi_1^1 & \Phi_2^1 & \cdots & \Phi_m^1 & \tau_1 \\ \Phi_1^2 & \Phi_2^2 & \cdots & \Phi_m^2 & \tau_2 \\ \vdots & \vdots & \ddots & \vdots & \vdots \\ \Phi_1^n & \Phi_2^n & \cdots & \Phi_m^n & \tau_n \end{array} \right), \tag{6}$$

where Φ_j^i denotes ith particular observation of the influencing factor with number j, $1 \le j \le m$. Thus, sequence of values Φ_1^r, Φ_2^r, … , Φ_m^r represents rth historical situation with $r \in N$.

The paper [17] solves the problem of estimation of the process time τ_{n+1} pointwise. This assessment is carried out by means of the selection of situations, which are close to the $(n + 1)$-th scenario. Note that two situations i_1 and i_2 are considered as close in case when the concrete realizations, $\Phi_j^{i_1}$ and $\Phi_j^{i_2}$, are similar in some predetermined way. The inequality $\left|\Phi_j^i - \Phi_j^{n+1}\right| \le \varepsilon_j$ can be used as a criterion of closeness between the pair of situations, where ε_j represents allowable deviation value of jth factor. Seeking forecasted value τ_{n+1} is defined as the mean between τ_k from the close scenarios, i.e. by using the interpolation.

We propose a slightly different approach to the construction and use of the Z array. The matrix includes process times that have been implemented for a certain set N of previous actual situations. We assume the factors Φ_j are mutually independent random variables, and each of them has a distribution $F_j(x) = \mathrm{P}(\Phi_j < x)$. In contrast to the mentioned paper we do not compute a specific value τ_{n+1} but restore the distribution function of the random variable τ. Method (a) of distribution restoring assumes the random variable τ considering the values τ_1 , ..., τ_n can be represented as a linear combination of all the factors, i.e.

$$\tau = \sum_{j=1}^{m} \lambda_j \Phi_j , \tag{7}$$

where λ_j are coefficients of sensitivity to changes in the factors Φ_j.

We have the following system of equations:

$$\begin{cases} \Phi_1^1\lambda_1 + \Phi_2^1\lambda_2 + \ldots + \Phi_m^1\lambda_m = \tau_1, \\ \Phi_1^2\lambda_1 + \Phi_2^2\lambda_2 + \ldots + \Phi_m^2\lambda_m = \tau_2, \\ \ldots \quad \ldots \quad \ldots \quad \ldots \quad \ldots \quad \ldots \\ \Phi_1^n\lambda_1 + \Phi_2^n\lambda_2 + \ldots + \Phi_m^n\lambda_m = \tau_n. \end{cases} \tag{8}$$

This system of equations is considered as overdetermined, because the number of equations n exceeds the number of unknown variables m. Some 'optimal' vector solution $\lambda = (\lambda_1, \lambda_2, \ldots, \lambda_m)$ can be found by the method of least squares. The following relation is considered as a measure of closeness of the left and right sides of eqn (8):

$$\sum_{i=1}^{n} \left(\tau_i - f_i(\lambda)\right)^2 \to \min_{\lambda} , \tag{9}$$

where $f_i(\lambda) = \sum_{j=1}^{m} \Phi_j^i \lambda_j$. We calculate the required distribution after finding the vector λ by using eqn (7). Determination of the density function of τ is reduced to $(m-1)$-fold convolution of the densities of new random variables $(\lambda_j \Phi_j)$.

The density f_τ can be determined by the other approach (b). The method is based on the fact that an array of experience Z contains a sample of values from the distribution of the random variable τ. For this sample you can restore the law of distribution using the goodness of fit (e.g. Kolmogorov–Smirnov test or Chi-squared test). Thereby our approach is more common and informative than that in Ref. [16] and also allows the best use of the initial data Z.

Experimental studies provide the data that show the density of the arrival time to have a little asymmetric Gaussian. A closer approximation is a Weibull distribution. The main results are the following. The parameters of the arrival times density are almost identical for all the stations in the standard periods. But the dispersion increases an order of magnitude from 0.6 to 6.9 min^2 at night (for the freight flow). At the same time the average delays are small. A similar pattern is in the suburban traffic.

6 THE QUALITY MEASURE OF PASSENGER TRAIN TRAFFIC

Punctuality in the passenger traffic segment is an important component of quality of the rail service. High level of punctuality increases the number of users of transport services and, consequently, income of the carrier and operator companies. Most studies on optimal traffic management have used the total cumulated delay as a measure of punctuality. However, this approach does not fully reflect the situation – how to completely satisfy the demands of the passengers.

The main characteristic of an operator which should be observed by him when he organizes the traffic control is as follows. The passenger begins to perceive the late arrival of the train as a significant delay only for sufficiently large values of delay. This measure has the uniqueness for every station and for different times of day. For example, a person perceives the delay more acutely during peak hours. Therefore, for the *i*th situation in this period the threshold of perception $T_{d_{ij}}^{lim}$ is smaller. The measure of quality of the service should not contain all the delays, but that lateness exceeds a specified threshold.

Another circumstance that must be considered is the damage caused by delays for different groups of passengers. Obviously, the approach is most appropriate when the priority service is given to numerous groups. One example is the traffic management on the suburban line. During rush hours, a manager devotes most attention to stations where there is mass boarding and alighting of passengers. He instinctively tries to make such adjustments to reduce the delays. Therefore, if the delay exceeds the threshold value, it is advisable to assess the damage by the number of passengers $M_{d_{ij}}$ who fell under the delay.

We formulate the following criterion for the quality of traffic in the heavy passenger train flow, using the approach just described. *The scenario is considered the best, if its implementation gives delayed arrival of the minimum number of passengers at all the stations*. In this case, the delays that exceed certain defined thresholds are taken into account:

$$\sum M_{d_{ij}} \rightarrow min\,, \quad \tau_{ij} \geq T_{d_{ij}}^{lim}\,, \; i = 1, 2, \ldots, m, \; j = 1, 2, \ldots, n, \tag{10}$$

where the index (ij) denotes the value of describing events occurring in the *i*th interval at the time of day at the *j*th station. Obviously the specified threshold must be peculiar for different types of trains such as high speed, intercity or commuter.

For example, the threshold value of delayed commuter trains during rush hours can be set from 1 min to 2 min in other periods. For those of local trains that have to deliver passengers on time at the connection station, this interval should be shortened to 0.5 min. Using this rule it enables the separation of any delays into the groups, each of which obeys the following relation:

$$\tau_{\mathrm{d}} \geq \tau_{all\,\mathrm{k}}\,. \tag{11}$$

In this case the objective function takes the following form:

$$J(\Theta) = \sum_{k} n_{del\,\mathrm{k}}\,, \quad k = 1, \ldots, K, \tag{12}$$

where K is the number of train groups with different boundary values $\tau_{ac\,\mathrm{k}}$.

7 CONCLUSIONS

In this article, we present an approach that allows considering the effect of adjustments on the traffic characteristics. The model developed takes into account the actual situation and its impact on operating times. The probabilistic forecast uses historical data to determine the improved time distributions. The calculated results agree well with the real data on the train traffic. The proposed new criterion of punctuality provides more correct adjusting solutions.

The next step of this study is to use the model developed in solving a specific real-time problem. In particular, we will carry out a detailed study on a linear transformation of the arrival time distribution when the local adjustment is made.

REFERENCES

[1] Berger, A., Gebhardt, A., Müller-Hannemann, M. & Ostrowski, M., Stochastic Delay Prediction in Large Train Networks. *Proceedings of 11th Workshop on Algorithmic Approaches for Transportation Modelling, Optimization, and Systems* (ATMOS'11), eds. A. Caprara & S. Kontogiannis, pp. 100–111, 2011.

[2] Büker, T. & Seybold, B., Stochastic modelling of delay propagation in large networks. *Journal of Rail Transport Planning & Management*, **2**, pp. 34–50, 2012. DOI: 10.1016/j.jrtpm.2012.10.001.

[3] Cacchiani, V., Huisman, D., Kidd, M., Kroon, L., Toth, P., Veelenturf, L. & Wagenaar, J., Overview of recovery models and algorithms for real-time railway rescheduling. *Transportation Research, Part B*, **63**, pp. 15–37, 2014. DOI: 10.1016/j.trb.2014.01.009.

[4] Törnquist, J., Computer-based decision support for railway traffic scheduling and dispatching: A review of models and algorithms. *Proceedings of the 5th Workshop on Algorithmic Methods and Models for Optimization of Railways*, (ATMOS'05), eds. L. Kroon & H. Rolf, 23 p., 2005.

[5] Muhlhans, E., Berechnung der verspatungsentwicklung bei zugfahrten. *Eisenbahntechn. Rundschau ETR*, **39(7/8)**, pp. 465–468, 1990.

[6] Carey, M. & Kwieciński, A., Stochastic approximation to the effects of headways on knock-on delays of trains. *Transportation Research, Part B*, **28**, pp. 251–267, 1994. DOI: 10.1016/0191-2615(94)90001-9.

[7] Meester, L.E. & Muns, S., Stochastic delay propagation in railway networks and phase-type distributions. *Transportation Research, Part B*, **41**, pp. 218–230, 2007. DOI: 10.1016/j.trb.2006.02.007.

[8] Huisman, T. & Boucherie, R.J., Running times on railway sections with heterogeneous train traffic. *Transportation Research, Part B*, **35**, pp. 271–292, 2001. DOI: 10.1016/S0191-2615(99)00051-X.

[9] Wendler, E., The scheduled waiting time on railway lines. *Transportation Research, Part B*, **41(2)**, pp. 148–158, 2007. DOI: 10.1016/j.trb.2006.02.009.

[10] Yuan, J., Stochastic Modelling of Train Delays and Delay Propagation in Stations, PhD Thesis, TRAIL Thesis Series no. T2006/6, the Netherlands, 142p., 2006.

[11] Kroon, L.G., Dekker, R. & Vromans, M., Cyclic Railway Timetabling: A Stochastic Optimization Approach. *Proceedings of the 4th Workshop on Algorithmic Methods and Models for Optimization of Railways* (ATMOS'04), eds. F. Geraets, L. Kroon, A. Schoebel, D. Wagner & C. Zaroliagiis, pp. 41–66, 2004.

[12] Vromans, M., Reliability of Railway Systems, PhD Thesis. TRAIL Thesis series no. T2005/7, The Netherlands TRAIL Research School, 244p., 2005.

[13] Chebotarev, V., Davydov, B. & Godyaev, A., Stochastic Traffic Models for the Adaptive Train Dispatching. *Proceedings of the 1st International Scientific Conference on "Intelligent Information Technologies for Industry"* (IITI'16), Rostov-on-Don, Russia, May 16–21, 2016.

[14] Goverde, R.P.M., A delay propagation algorithm for large-scale railway traffic networks. *Transportation Research, Part C*, **18(3)**, pp. 269–287, 2010. DOI: 10.1016/j.trc.2010.01.002

[15] Beck, A., Bente, H. & Schilling, M., Railway Efficiency: An Overview and a Look at Opportunities for Improvement. The International Transport Forum, Discussion paper no. 2013–12, 44p., 2013.

[16] Shapkin, I.N., Yusipov, R.A. & Kozhanov, E.M., Modelling train traffic based on multifactor setting technological operations. *Vestnik VNIIZhT*, **4**, pp. 30–36, 2006.

[17] Törnquist Krasemann, J., Design of an effective algorithm for fast response to the rescheduling of railway traffic during disturbances. *Transportation Research, Part C*, **20(6)**, pp. 62–78, 2012. DOI: 10.1016/j.trc.2010.12.004.

REDUCING MAINTENANCE COSTS FOR BALLASTED TRACK: EVIDENCE FROM BRITAIN

ALEJANDRO ORTEGA, SIMON BLAINEY & JOHN PRESTON
Transportation Research Group, University of Southampton, United Kingdom.

ABSTRACT

Over the last twenty years, the railways in Britain have seen rapid growth, with patronage doubling and the level of train service increasing by 50%. However, these successes have also led to challenges. In particular real unit costs are estimated to have increased by 50%, with particularly marked increases in infrastructure renewal and enhancement costs. Against this background, the University of Southampton have been leading the Track 21 and Track to the Future projects, funded by the Engineering and Physical Sciences Research Council. These projects are assessing how to make ballasted track systems more durable given higher traffic levels and how to reduce the costs of maintenance and renewal.

Track 21 considered a number of engineering interventions of which this paper focusses on one, under sleeper pads (USPs). The cost implications of this intervention are assessed for the South West Main Line using an industry-specific model, VTISM (vehicle track interaction strategic model), adapted in two ways. First, the results of laboratory experiments from specialist test rigs are incorporated into the model. Secondly, the wider effects of increased reliability, improved ride quality and reduced vibration are also considered within a cost-benefit analysis framework. It is found that USPs can lead to substantial financial savings and, depending on assumptions concerning noise impacts, are likely to have substantial wider social benefits.

Keywords: CBA, maintenance costs, under sleeper pads

1 INTRODUCTION

Track 21 and its successor Track to the Future are projects funded by the UK Engineering and Physical Sciences Research Council (EPSRC) which explore several innovative interventions which are intended to deliver better engineering, economic and environmental performance of railway track (see more information on http://www.track21.org.uk/). As part of a good transport planning process, every transport infrastructure investment should be subject to some kind of economic appraisal. Cost benefit analysis (CBA) is the most commonly used tool for this purpose. When applying these tools, cost modelling is a crucial issue because it helps to determine the value of cost to be used. This financial valuation might have an important effect on the outputs of the analysis (see, for instance, Ortega et al. [1]). In the case of railways, potential advances in track and sub-base construction and maintenance can lead to improvements in reliability or reductions in perceived noise, for example. The values adopted for these parameters could influence the result and profitability of track upgrades, with errors having the consequence of misperceptions of the return gained from investment. In order to conduct such an economic appraisal, a review of the valuation of some of the most relevant parameters was carried out as the first stage of this research, before using the findings to inform modelling of track system costs.

The remainder of this paper is divided into five sections. Firstly, a review of values of noise is provided. Secondly, values of vibration and ride quality are given. Thirdly, a review

DOI: 10.2495/TDI-V1-N3-318-328

of reliability and journey time is undertaken. Fourthly an illustrative application of the cost modelling methodology for the South West Main Line in the UK is provided. Finally, derived from this review, we are able to offer a set of conclusions for policy makers.

2 VALUES OF NOISE

Noise cost estimates can be difficult to generalize due to their local nature and dependence on background noise levels [2]. Furthermore, the impact of a given noise level will depend on people's activities and their attitudes towards the railway, and as people become accustomed to a given level of noise exposure their annoyance level may decrease [3]. Four different methods were used in the studies reviewed to evaluate the cost of noise, including hedonic pricing (HP), i.e. the real market is used to obtain the economic value of one particular feature, willingness to pay (WTP) and contingent valuation (both generally carried out through stated preference (SP) surveys), and finally estimation of abatement costs. Willingness to pay values may be derived from HP as well as from SP surveys, although as Strand and Vagnes [4] point out there are a number of problems associated with such a derivation. In general, noise nuisance valuations obtained through HP methods tend to be larger than those from SP methods such as Contingent Valuation Method (CVM) or Stated Choice (SC) experiments. In fact, there are also differences between CVM and SC [5]. In addition, Janic [6] considered that the cost of noise can be calculated as the marginal cost of setting up noise protection barriers, with evidence suggesting that each barrier could reduce noise over a range of 8–10 dB [7]. Finally, it is worth noting that there was also little consistency between the formats in which the results were presented. These groups are cost of annoyance per person, property value depreciation, marginal cost of additional vehicles and other valuations.

3 VALUES OF VIBRATION, RIDE QUALITY AND COMFORT

Despite an extensive search, relatively few studies which assigned financial values to vibration or ride quality could be found. This might be explained by the fact that passenger comfort and perception of service quality are affected by several factors [8], such as time, country, culture or the physical condition of passengers, and when the subjective contribution from all aspects is approximately the same then the optimum level of comfort will be achieved [9], regardless of the relative contribution of different factors. Schwanitz et al. [10] identified the four most important attributes of comfort as being odour, air quality, air ventilation and temperature. Moreover, users have a higher sensitivity to vertical acceleration than to longitudinal or transversal acceleration [11].

The first research found giving financial valuations of ride quality was based on both RP and SP methods and focussed on alternative types of rolling stock [12]. The research analysed seating comfort, ambience, ride quality, seating layout, ventilation and noise. The former three aspects were found to be the most important aspects with a valuation ranging from 3.41% to 5.22% of the single fare for seating comfort, 1.88% to 3.05% for ambience and 1.89% to 3.06% for ride quality. However, in a literature review conducted by Johnson et al. [13], higher valuations were given for the case of trams. Although a packaging effect has been demonstrated and could therefore mean that the valuation of individual attributes is too high, this effect does not impact on the rank of each attribute in relation to the others and the classification of [13] was also obtained by the same authors [14]. Paulley et al. [15] found that refurbishment which changed train characteristics from those associated with old 'slam door' stock to new air-conditioned stock in SE England were worth around 2.5% of the fare paid, with most train refurbishments worth somewhat less than this (~1.5%). Finally, the Passenger Demand Forecasting Handbook (PDFH) [16] measured improvements in ride quality in terms of a decrease in in-vehicle time (IVT). Going from an extremely bumpy ride to a very smooth ride would lead to a benefit equivalent to 3.3% of IVT (commuting) to 3.7% (business/leisure), whereas going from an extremely bumpy ride to a train with a lot of movement would have benefits of 2.4% and 2.6% of IVT.

4 VALUES OF RELIABILITY

Frequency of travel may affect perceptions of reliability, with regular passengers likely to have a clear idea of reliability levels on their usual services, while infrequent travellers are more likely to be influenced by their most recent journey (however atypical) or by press reports and hearsay [16]. Bates et al. [17] suggested that a multiplication factor of 2.5 for delay minutes was a reasonable measure of the direct disutility of delays but also pointed out that this was an average value based on passengers having full knowledge of random delay distributions and concealing a range between 1 and 5 depending on journey length and purpose. Chang [18] recommended including in the CBA a value of journey time reliability which depends on the type of user, i.e. interregional or urban and working or non-working.

Most studies of the demand impacts of reliability have been based on SP methods. After a comparison between implied elasticities by SP valuations of late time and directly estimated lateness elasticities, Wardman and Batley [19] concluded that implied lateness elasticities are considerably higher than those which are directly estimated. Apart from overvaluation of lateness, there are other reasons which can help to explain these results such as the possibility that despite poor service performance rail travellers are unwilling or unable to reduce their rail travel. Preston et al. [20] hypothesized that users may have got used to a certain level of delay or be more tolerant to delays due to a high degree of delay associated with road travel. One of the few investigations based on RP data, conducted by Rail OR [21], looked at three significant runs of poor performance on the London–Northampton line using TRUST and CAPRI data, deriving implied multiplication factors which were around the generally accepted figure of 2.5.

5 CASE STUDY: THE SOUTH WEST MAIN LINE

The South West Main Line is a railway route between London Waterloo and Weymouth on the south coast of England. It was built in the 19th century and is one of the most used lines in the UK. London is the origin or destination of approximately 90% of all journeys, trains operate at medium to high speed, and there are important proportions of freight and commuter traffic. In this paper we are going to analyse the economic effect of installing under sleeper pads (USPs) in the stretch between Portsmouth and London Waterloo. For these calculations we only consider passengers from the main stations in the corridor which are at least 30 min journey time to/from Waterloo (i.e. Portsmouth Harbour, Portsmouth & Southsea, Fratton, Havant, Petersfield, Haslemere, Guildford and Woking). The length of the stretch considered is approximately 74 miles, the project life for the analysis is 60 years, the initial year 2009 and the discount rate 3.5%. Table 1 offers the values we suggest should be used for this route:

Table 1: Suggested valuations for the SWML.

Parameter	Measure	Value
Noise	£ per dB of household affected per annum	£20 (Year 2009)
Comfort/ride quality	Value per passenger	1% of the ticket price
Reliability	Value of late time (value in minutes per minute of average lateness)	2.5 times value of travel time on average

The effect stemming from the installation of USPs can be split into installation cost of USPs, track maintenance, rolling stock, noise, comfort/ride quality and reliability. The methodology carried out to calculate the cost of each impact is explained within the next paragraphs:

- Investment. This is the cost of installing USP. We assume a hypothetical renewal of the whole stretch in the initial modelled year, 2009. By multiplying the number of sleepers by the unit cost of each sleeper the investment is obtained. There are 91 sleepers per 55.21 m of single track [22]. Taking also into account switches, and crossings and stations we assume that 400,000 USPs will be installed. We know from the Track 21 project that the unit cost per USP is £15 which leads to an investment of £5.9 million.
- Track maintenance and renewal. The cost of track maintenance and successive renewals, also known as life cycle cost, is given by the vehicle track interaction strategic model (VTISM) [23]. VTISM calculates the cost for the stretch London–Woking–Portsmouth if 40% of that route is renewed at the base year. By recalculating the average cost of the whole route (i.e. for illustrative calculations we multiply the result by 2.5) and applying the updated average cost the impact of track maintenance is determined. According to the lab tests conducted by Le Pen [24] the local track section modifier (LTSF) would be reduced by 25% after the installation of stiff USPs of 4 mm thickness and the main benefit would arise from an increased service life of the track and a reduced maintenance and renewal volume [25]. After the hypothetical investment not many renewals would be needed on the route, but VTISM considers additional renewals in some stretches along the infrastructure life. Therefore, in order to be conservative in our analysis we introduced the cost of installing USPs in each successive renewal calculated by VTISM, which is roughly an increase of 3% for each renewal. We estimate a benefit of around £45 million in terms of reduced maintenance and renewal costs.
- Rolling stock. Based on the results from VTISM under the USPs scenario around 10% more rail renewals would be needed. However, we have not been able to find any evidence that more rail renewal could lead to a deterioration of rolling stock. Therefore, we have decided to leave it out of the analysis.
- Noise. We estimated with ArcGIS that in 2009 there were slightly above 108,000 buildings within 300 m of the track and 30,500 within 80 m of the track. These buildings were therefore assumed to be affected by air-borne noise and ground-borne noise, respectively [26]. Moreover, the use of USPs could lead to a reduction of ground noise of up to 15 dB [27] but would slightly increase air-borne noise [28]. For those buildings within 80 m the dominant effect may be reduction in ground-borne noise whereas for those houses between 80 and 300 m the increase in air-borne noise would be the dominant impact. Noise and vibration due to ground-borne noise are more annoying than just air-borne noise, albeit taking into account there are no monetary values for ground-borne noise, we have to assume this has the same valuation as air-borne noise. In order to be conservatives we have adopted a reduction of 5 dB in ground-borne noise and an increase of 1 dB in air-borne noise. The annual increase in the number of houses and in the values for noise has been chosen from [29]. We selected these values because are they lower than the WebTAG (transport analysis guidance in the UK) parameters. By multiplying the reduction of noise per house, the number of houses and the value of noise per house in each year we obtain the balance for this externality.
- Comfort/ride quality. USPs would also reduce noise inside the train and improve ride quality due to the reduction of vertical acceleration [30] but would have no impact on

other attributes of comfort. The value of changing from a very uncomfortable to a very comfortable ride could be up to 10% of the ticket price. Thus, and according to the literature review, the value chosen for an increase in comfort as a result of installing USPs is 1% of the ticket price. We have differentiated between anytime or full, season, and off-peak or cheapest tickets. Commuters pay season ticket in a proportion of 85%/15% standard/first class which is the same proportion as of seats offered in the train and use the season ticket 480 times each year. The same proportion of standard/first is applied to anytime users. However, we assume that holders of reduced tickets do not travel in first class and that 25% of travellers within this category have a card with an average discount of 33%. The proportion of each type of ticket sold (full, season or reduced) is given in the Office of Rail and Road station usage database. Finally, by multiplying the number of users in each category by the percentage of their ticket price attributed to the improvement, the value of comfort/ride quality is known.

- Reliability. 30% of delays are due to track defects and networks management [31] and according to VTISM these disruptions would be reduced by around 20% after the installation of USPs, which would lead to an overall improvement in reliability of 6%. Nevertheless some items of the railway infrastructure would be relatively unaffected by USPs and therefore we have decided to reduce the figure by 50%, i.e. giving a final improvement of 3%. We have considered that 92% of trains would arrive on time, that is within 5 min of expected time, 5.6% would be delayed up to 30 min and 2.4% would be cancelled or with a significant lateness [31]. The values of time have been chosen from WebTAG and depend on the type of user: working, non-working commuting and non-working others. To conduct the appraisal, the kind of ticket was transformed into the type of user using the figures given by the Passenger Demand Forecasting Handbook (PDFH). So we linked commuting users with non-working commuting users, business users with working travellers, and finally leisure travellers with non-working other users. Taking into account all these figures the improvement of 3% in reliability generates a benefit of £35 million.

Finally, other general economic parameters such as demand, evolution of demand and Retail Price Index (RPI) have been chosen from WebTAG and the National Rail Trends (NRT) web sites. Demand growth and RPI are capped after 20 years, and their annual average values for the first 20 years are settled around 2.05% and 3.5% respectively. Following the WebTAG recommendations, RPI has been used to calculate tickets price and rolling stock investment in nominal terms, but converted afterwards into real terms through the GDP deflator.

To conduct this economic analysis some hypotheses have been assumed, and it is often remarked that risks and uncertainty are not easily taken into account in CBA assessment [32]. For instance, the level of renewals could be slightly different from that considered by VTISM because the program assumes the renewal of some stretches in the initial year instead of the renewal of the whole route, or the value of the ground-borne noise reduction or reliability improvement could also be different. So, in order to overcome the intrinsic weakness of the CBA methodology, we use a stochastic approach rather than the deterministic one that is commonly used. This is done with a Monte Carlo Simulation underpinned by a sensitivity analysis aimed at determining the level of certainty and identifying the most important parameters for policy lessons. The goal of running a Monte Carlo Simulation is to assess risk more accurately, choosing randomly modifiable values in each iteration, whereas the goal of sensitivity analysis is to know the most important parameters influencing the result.

In order to display the range of possible outcomes, the results of the CBA are sorted and plotted as histograms and cumulative probability distribution curves (also called value at risk and gain curve (VARG)). We have run two models with different assumptions with respect to noise. In assumption (a) we only consider an increase of air-borne noise for all houses within 300 m of the track, whilst in assumption (b) we consider the reduction of ground-borne noise for houses within 80 m of the track and the slightly increment of air noise for the remaining houses between 80 m and 300 m from the track. Figure 1 shows how the cumulative probability distribution curves have the shape of a cumulative normal distribution, since we selected each variable to fluctuate between minimum and maximum values according to a normal distribution. In the most likely option, the social benefits of installing USPs in the selected route would be around £89 million. The final outcome in terms of social welfare is almost always positive and fairly stable over the MonteCarlo analysis. Only the worst-case scenario under assumption (a) brings some welfare loss whilst the worst-case scenario under assumption (b) still produces a global benefit greater than £135 million. These results demonstrate that the outcome of the analysis is quite robust.

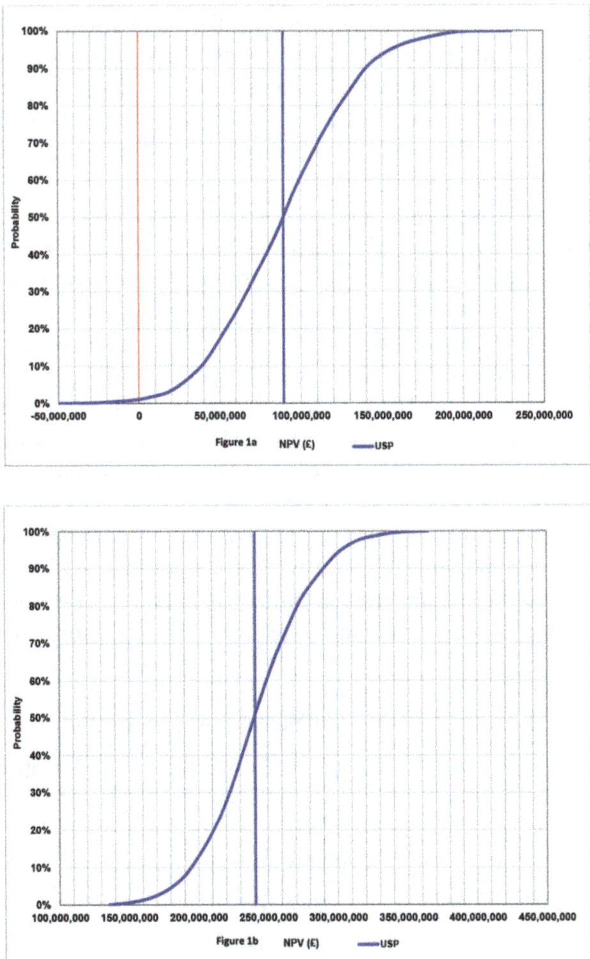

Figure 1: Net present value (NPV) cumulative distribution function.

Two conclusions arise from the last figure. Firstly, under the majority of scenarios the expected net present value (NPV) will be always positive. Only taking into account just air-borne noise (Fig. 1a) the NPV might be negative with a probability of less than 1%. In fact, under this assumption the benefit can be set between £38 million and £140 million with a probability of 80% and the mean of the sample is in the middle of that range, £89 million. Under the more realistic assumption of considering that the reduction of ground-borne noise dominates for buildings within 80 m of the track whereas the increase of air-borne noise affects the remaining houses up to 300 m, the benefit is even greater. The difference between the mean of both assumptions is £152 million. Secondly, there is a high variability between the minimum and the maximum value. For instance, under scenario b the maximum value is almost three times higher than the minimum value. Table 2 shows the results of the two CBA carried out.

Figure 2 shows the outcome of a test carried out to analyse the sensitivity of the results to the changes of the variables studied before. The analysis was conducted through a tornado diagram to show the sensitivity of the NPV to the variables used in the methodology while keeping the remaining variables constant (Fig. 2). On the left side of this figure, we find all the variables determining the final outcome in terms of net present social value. Next to the definition of each variable, in brackets, we display the variation range tested for each variable. The variables with the largest influence on the NPV have been plotted in red, whilst the variables with little impact on the final outcome have been depicted in blue. The economic variables have been represented at the top of the figure whilst the variables that could be managed by train operating companies (TOCs) and the infrastructure manager are drawn in the middle and at the bottom of the graph.

It is noteworthy that all stakeholders benefit: train operating company, infrastructure manager, travellers and non-users. The financial benefit to Network Rail is around £40 million and might mean a risk of under-spending on maintenance given the fact that the social NPV is at least £100 million. As expected VTISM indicates that there are less interventions in the track than in the base case. The improvement in comfort and ride quality could be easily internalized through a higher ticket price and the housing market can internalize the effect of noise. The upgrades studied in Track 21 project (e.g. installation of USPs or random fibre reinforcement to name but two) are realistic and could be beneficial for the UK.

Three variables dominate the outcome: RPI, air noise increase and finally improvements in comfort/ride quality. The remaining variables have less impact on the NPV. The large influence of RPI is explained by inconsistencies with the GDP deflator and for this reason RPI is usually removed from CBA calculations. Each one dB increase in air-borne noise alters the NPV by almost £90 million under assumption (a) or £66 million under assumption (b). Ground-borne noise is also very important since each 1 dB reduction brings a benefit of £25

Table 2: NPV from the CBA parameter values.

Probability	NPV option (a) (£ million)	NPV option (b) (£ million)
Minimum value	–48.676	135.683
VAR 10%	38.079	194.374
Expected value (mean)	89.331	241.279
VAG 90%	140.257	290.150
Maximum value	230.139	364.720

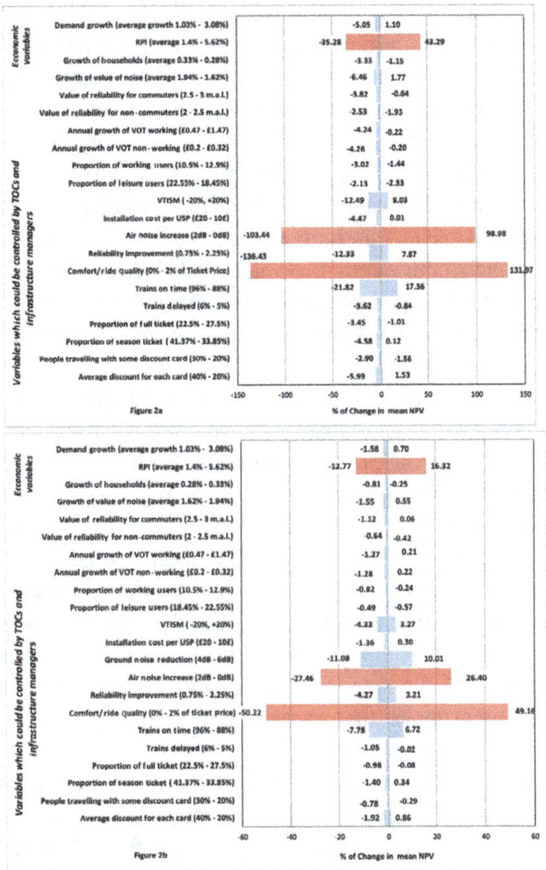

Figure 2: Sensitivity analysis.

million, mainly for the buildings within 80 m of the track. Any track intervention improving comfort/ride quality just by 1% of the ticket price leads to a social gain of £120 million. A reliability improvement of 1% would lead to a smaller benefit of nearly £10 million. This is also the case for the proportion of trains on time; each improvement of 1% means a NPV decrease of around £4 million. Logically, the better the punctuality of the system, the less the effect of USP, and this is what the figure is telling us. This result is based on the fact that this proportion is assumed to be constant along the infrastructure life, but different for USPs and base case. Fortunately these variables can be managed by TOCs and the infrastructure owner (i.e. Network Rail) and therefore there is room for realistic upgrades of the track. The UK railways are among the oldest in Europe and it is important that in order to avoid welfare losses from users with a high valuation of their trip or from their externalities the railway system should be upgraded to the 21st century.

6 CONCLUSIONS

In order to carry out effective cost modelling of potential advances in track and sub-base construction and maintenance, and to reliably optimize the timing of maintenance interventions, it is necessary to place values on the various factors that are affected by such interventions. These include the effects of noise and vibration on both passengers and non-users, and of reliability, journey time and ride quality on both passengers and rail industry stakeholders. A review of the evidence on valuations of these factors has therefore been carried out in this paper, as the basis for undertaking a meta-analysis of the valuations. It should be noted that the extent of the evidence-base was found to be highly variable, with little research having been undertaken on valuing the effect of some of the factors.

The high degree of variability can be explained by several reasons. Firstly, differences between regions, types of travellers and affected non-travellers have been found. Secondly, depending on the approach considered the value to be placed in the analysis has to be different; such is the case of ride quality which can be valued through IVT or as a percentage of the ticket price. Finally, HP methods and SP surveys usually lead to different values, whilst bias in some results can also be found.

It is also noteworthy that noise can be valued by two approaches. On the one hand, noise should be taken into account as an externality that affects non-rail users such as neighbourhoods close to a rail line. On the other hand, noise experienced by rail users must also be considered as part of passenger comfort. An example which clarifies this can be found in noise protection barriers. If noise protection barriers are installed along the track, the external effect of noise would be greatly reduced, but travellers would continue to be negatively affected by noise while also losing potential benefits gained from appreciation of the view from the train. Regarding valuation of ride quality and passenger comfort, despite the improvement on the topic, there is still a lack of consistent values and therefore further research is needed. With respect to reliability, it is well acknowledged that rail users are adversely affected by lateness. Nevertheless, some findings suggest that travellers are becoming less sensitive to delays and consequently its value should be lower.

So, in order to carry out economic appraisal and avoid misinterpretations of the results, we suggest adapting the values to the particular case study reflecting a realistic situation and also conducting a sensitivity analysis which can help identify the key parameters. Finally, in developed countries with old rail infrastructures and high demand, such as is the case in the UK, it appears that minor interventions which improve reliability and comfort/ride quality and at the same time reduce externalities could potentially bring important social benefits.

REFERENCES

[1] Ortega, A., Vassallo, J., Guzmán, A. & Pérez-Martínez, P., Are Longer and Heavier Vehicles (LHVs) beneficial for society? A cost benefit analysis to evaluate their potential implementation in Spain. *Transport Reviews: A Transnational Transdisciplinary Journal*, **34(2)**, pp. 150–168, 2014. DOI: 10.1080/01441647.2014.891161.

[2] Nash, C., Unification of accounts and marginal costs for transport efficiency: Final report, s.l.: ITS, University of Leeds, 2003.

[3] Brons, M., Nijkamp, P., Pels, E. & Rietveld, P., Railroad noise: economic valuation and policy. *Transportation Research D*, **8**, pp. 169–184, 2003. DOI: 10.1016/S1361-9209(02)00048-2.

[4] Strand, J. & Vagnes, M., The relationship between property values and railroad proximity: a study based on hedonic prices and real estate brokers' appraisals. *Transportation*, **28**, pp. 137–156, 2001.

[5] Bristow, A., Wardman, M. & Chintakayala, V., International meta-analysis of stated preference studies of transportation noise nuisance. *Transportation*, **42**, pp. 71–100, 2015. DOI: 10.1007/s11116-014-9527-4.

[6] Janic, M., An assessment of the performance of the European long intermodal freight trains (LIFT). *Transportation Research Part A*, **42**, pp. 1326–1339, 2008. DOI: 10.1016/j.tra.2008.06.008.

[7] Clausen, U., Doll, C., Franklin, F.J., Franklin, G.V., Heinrichmeyer, H., Kochsiek, J., Rothengatter, W. & Sieber, N., Reducing railway noise pollution, Brussels: European Parliament, Directorate-General for Internal Policies, 2012.

[8] Grujičić, D., Ivanović, I., Jović, J. & Đorić, V., Customer perception of service quality in public. *Transport*, **29(3)**, pp. 285–295, 2014. DOI: 10.3846/16484142.2014.951685.

[9] Hardy, A., Measurement and assessment of noise within passengers trains. *Journal of Sound and Vibration*, **31(3)**, pp. 819–829, 2000. DOI: 10.1006/jsvi.1999.2565.

[10] Schwanitz, S., Wittkowski, M., Rolny, V. & Basner, M., Pressure variations on a train – Where is the threshold to railway passenger discomfort? *Applied Ergonomics*, **44(2)**, pp. 200–209, 2013. DOI: 10.1016/j.apergo.2012.07.003

[11] Castellanos, J. C. & Fruett, F., Embedded system to evaluate the passenger comfort in public transportation based on dynamical vehicle behavior with user's feedback. *Measurement*, **47**, pp. 442–451, 2014. DOI: 10.1016/j.measurement.2013.08.068.

[12] Wardman, M. & Whelan, G., Rolling stock quality – Improvements and willingness of users to pay, s.l.: Working paper. University of Leeds, 1998.

[13] Johnson, D., Abrantes, P. & Wardman, M., *Analysis of Quantitative Research on Quality Attributes for Trams*, s.l.: University of Leeds, 2009.

[14] Wardman, M. & Whelan, G., Valuation of improved railway rolling stock: A review of the literature and new evidence. *Transport Reviews: A Transnational Transdisciplinary Journal*, **21(4)**, pp. 415–447, 2001. DOI: 10.1080/01441640010020115.

[15] Paulley, N., Balcombe, R., Mackett, R., Titheridge, H., Preston, J., Wardman, M., Shires, J. & White, P., The demand for public transport: The effects of fares, quality of service, income and car ownership. *Transport Policy*, **13**, pp. 295–306, 2006. DOI: 10.1016/j.tranpol.2005.12.004.

[16] Association of Train Operating Companies. Passenger Demand Forecasting Handbook v.5. London: ATOC, 2009.

[17] Bates, J., Jones, P., Polak, J. & Cook, A., *The Investigation of Punctuality and Reliability*, s.l.: s.n, 2000.

[18] Chang, J.S., Assessing travel time reliability in transport appraisal. *Journal of Transport Geography*, **18**(3), pp. 419–425, 2010. DOI: 10.1016/j.jtrangeo.2009.06.012.

[19] Wardman, M. & Batley, R., Travel time reliability: A review of late time valuations. *Transportation*, **41**, pp. 1041–1069, 2014. DOI: 10.1007/s11116-014-9526-5.

[20] Preston, J., Wall, G., Batley, R., Ibáñez, J. & Shires, J., Impact of delays on passenger train services: evidence from Great Britain. *Transportation Research Record: Journal of the Transportation Research Board*, **2117**, pp. 14–23, 2009. DOI: 10.3141/2117-03.

[21] Rail, O.R., *Performance, Perceptions and Revenue*, s.l.: s.n, 1996.

[22] Pieringer, A., Innovative sleeper design with under sleeper pads as an efficient method to reduce railway induced vibration propagation. Research results of RIVAS. *Urbana-Champaign, Illinois, UIUC: Crosstie & Fastening Symposium*, 2014.

[23] SERCO., *VTISM Version 2.6.6 User Guide*, s.l.: SERCO, 2014.

[24] Le Pen, L., Proposed LTSF modification procedures. Southampton: University of Southampton, Working Paper, 2015.

[25] Marschnig, S., Lowering track lifecycle costs with sleeper pads. EURAILmag Business & Technology, The magazine for European Rail decision makers, 23, pp. 146–147, 2011.

[26] Zapfe, J., Saurenman, H. & Fidell, S., *Ground-Borne Noise and Vibration in Buildings Caused by Rail Transit*, s.l.: TCRP, 2009.
[27] Deutsche Bahn, A. & Hans-Joerg, T., State of the art review of mitigation measures on track. Deliverable D 3.1, s.l.: Railway Induced Vibration Abatement Solutions Collaborative project, 2011.
[28] Triepaischajonsak, N., *The Effect of Sleeper Soffit Pads on Railway Noise*, University of Southampton: Southampton, 2008.
[29] Nellthorp, J., Bristow, A.L. & Day, B., Introducing willingness-to-pay for noise changes into transport appraisal: An application of benefit transfer. *Transport Reviews: A Transnational Transdisciplinary Journal*, **27(3)**, pp. 327–353, 2007. DOI: 10.1080/01441640601062621.
[30] Lakuši, S., Ahac, M. & Haladin, I., Experimental investigation of railway track with under sleeper pad. Portoroz, Slovenian congress on the roads and traffic, 2010.
[31] Network Rail. Annual Return 2010. Moving ahead. Investing in the future, Milton Keynes: Network Rail, 2010.
[32] De Palma, A., Picard, N. & Andrieu, L., Risk in transport investments. *Networks and Spatial Economics*, **12(2)**, pp. 187–204, 2012. DOI: 10.1007/s11067-009-9109-8.

A LONG-TERM ANALYSIS OF PASSENGER FLOWS ON A REGIONAL RAIL LINE

C. CAROPRESO, C. DI SALVO, M. BOTTE & L. D'ACIERNO
Department of Civil, Architectural and Environmental Engineering, Federico II University of Naples, Italy.

ABSTRACT
Promoting rail systems can represent a useful policy for rebalancing modal choices and reducing private car use, especially in high-density contexts. Obviously, an increase in passenger numbers is only possible if generalized costs (i.e. a weighted sum of times and monetary costs) associated to public transport are abated. According to the recent literature and current professional practice, most strategies for achieving this objective are based only on infrastructural interventions which may be unfeasible or inadequate in densely populated contexts. Likewise, the adoption of policies based on replacing existing fleets or reducing fare levels entails increases in national or regional subsidies, which would be difficult to achieve in the current economic climate.

Hence, our proposal is based on investigating effects on travel demand arising from the replacement or upgrading of existing signalling systems (both in terms of trackside and on-board equipment). Indeed, the recent European Union policy to create a single transnational interoperable rail network imposes the development of innovative signalling systems. In this context, since cost–benefit analysis has to be implemented to verify the economic and environmental feasibility of the proposed intervention strategy, an appropriate method should be developed to estimate passenger flows according to future configurations. In this article, we propose a method to determine travel demand in current and future contexts by appropriately processing data from Italy's national census on mobility, population growth forecasts and turnstile counts. The proposed approach is applied to the regional Naples–Sorrento rail line serving the metropolitan area of Naples in southern Italy in order to show its feasibility.
Keywords: environmental impacts, microscopic rail system simulation, public transport management, signalling system, travel demand estimation.

1 INTRODUCTION

Public transport, especially in the case of a rail-based system, may be considered a useful resource for managing travel demand and reducing the use of road (car and truck) systems. Obviously, since according to the assumptions of rational decision-maker (see, for instance, [1–5]) each user tends to choose the alternative of maximum utility (i.e. minimum disutility), road system costs need to be significantly increased and/or public transport costs considerably reduced.

Former strategies were based on the application of real pricing (such as road and parking pricing) or induced pricing (such as driving restrictions, restricted traffic zones and pedestrian areas). Details on the applications of such strategies and related implications can be found in Refs [6–11]. Subsequent policies were based on interventions to reduce travel times and ticket costs and improve service quality both in ordinary conditions and in disruption contexts (see Refs [12–23] for details). However, although the purpose of this article is to deal with the latter class of measures (i.e. improving public transport performance), it should

DOI: 10.2495/TDI-V1-N3-329-338

be pointed out that the literature also contains a third class of strategies based on reductions in social and environmental impacts of road systems, as reported by Refs [24–27].

The increase in public transport attractiveness can be obtained by reducing user generalized costs whose terms are the access and egress time (depending on station and stop location), the waiting time (depending on the headway between two successive rail convoys), the on-board travel time (depending on infrastructure features and rolling stock performance), the transfer time (depending on the layout of rolling stock, stations and platforms) and the ticket cost (depending on the adopted pricing scheme and subsidy policy). This means that improving interventions on a rail transport system are based on infrastructural measures (new lines or modification of existing lines), fleet improvement (partial or complete replacement of rolling stock) or signalling system modification (replacement or upgrade of trackside and on-board equipment). Although, especially in densely populated environments, the optimal intervention class consists in signalling system improvements, in some cases the infrastructural component may be essential in order to make any kind of further intervention effective. Obviously, in order to verify the benefits of each proposal it is necessary to simulate effects on travel demand explicitly (see, for instance, Refs [28–33]) for long-time horizons.

In this article we propose a methodology based on the use of different data sources (censuses, historical data, forecasts, counts, etc.) to estimate travel demand in a wide time period (several decades), taking into account variations due to infrastructure and signalling system interventions. The proposed approach was applied to a real regional line in southern Italy so as to show its feasibility. Hence, our proposal is organized as follows: in Section 2, we describe the general framework of the methodology for estimating travel demand; in Section 3, the application in the case of a regional rail line is proposed; finally, conclusions and research perspectives are summarized in Section 4.

2 TRAVEL DEMAND DEFINITION

Defining travel demand may be considered to be of prime importance in order to evaluate effects of any intervention on transportation systems. However, whatever the methodology adopted, the following requirements have to be met:

- accurate reproduction of the current situation;
- prediction of future conditions arising from demographic changes and/or different performance of transportation systems;
- travel demand must be considered a random variable and hence not only average values but also their distribution must be analysed.

This means that the model has to be elastic at least at the level of modal choice (in the case of transportation system variations) and trip generation (in the case of demographic changes).

Hence, in order to meet these needs, we propose the following methodology based on the use of different data sources. However, it should be pointed out that our proposal is based on Italian data sources even if generalization to a different context may easily be obtained.

The first phase consists in using data from the national census [34]. These data provide revealed information (i.e. related to behaviour actually occurring in the days prior to the survey) concerning mobility choices in terms of origin, destination, daily time period and transport mode. Census data concern systematic trips (i.e. for work or school purposes) during the average working day; origins and destinations are expressed in terms of municipalities; daily times are indicated as the morning peak hour (i.e. from 07:30 to 09:29) and the rest of the day. Obviously since trips are generally bidirectional (i.e. from home to the workplace and return), data provide only outward trips. In order to increase our dataset and satisfy the third requirement (i.e. a wide distribution of considered values), we propose to analyse data from at least two decades (i.e. data from the 2001 and 2011 Italian censuses). Finally, data related to our area of interest have to be selected by extracting from the whole national database only data related to internal trips (i.e. with origin and destination both in the study area) and exchange trips (i.e. with the origin or the destination in the study area).

The second phase consists in using data from mobility observatories (such as Ref. [35]) which provide further information such as total daily regional trips, rates of trips during morning peak hours, rates of trip chains (i.e. trips with intermediate destinations) and regional modal split. By combining these data it is possible to generate non-systematic trips during the average working day classified by origin and destination municipality, time period (i.e. peak hour or rest of the day), transport mode used and reference year (i.e. 2001 or 2011).

In the third phase, by using historical data from the resident population [36], data for systematic and non-systematic trips may be extended from the census period to a successive period by considering the trip generation model as elastic and adopting an increase or decrease rate equal to population variation (i.e. a variation in $\alpha\%$ of population in municipality *A* provides a variation in $\alpha\%$ of all trips with origin in *A*).

The following phase consists in generating travel demand matrices related to all-day trips where the origin and destination are the stations of the rail line in question. This means that it is possible to identify two sub-phases: the first for obtaining round trips from outward trips in the case of all-day trips; the second for transforming trips expressed in terms of origin and destination municipalities into origin and destination stations.

The first sub-phase can be implemented by adopting the following formula:

$$OD_{rt}^{i,m} = OD_{ot}^{i,m} + \left[OD_{ot}^{i,m}\right]^{\mathrm{T}}, \tag{1}$$

where $OD_{rt}^{i,m}$ is the origin-destination matrix related to round trips (*rt*) throughout the day associated to purpose *i* (i.e. systematic or non-systematic) and mode *m*; $OD_{ot}^{i,m}$ is the origin-destination matrix related to outward trips (*ot*) all day associated to purpose *i* and mode *m*; $\left[OD_{ot}^{i,m}\right]^{\mathrm{T}}$ is the transposed matrix of $OD_{ot}^{i,m}$.

The second sub-phase consists in defining a regional network model in order to implement a minimum path approach for associating each municipality to each station. Obviously, if there are no stations in a municipality, we associate the nearest station; if there is only one station in a municipality, the problem is trivial (i.e. we associate the single station); finally, if there are two or more stations in a municipality, we hypothesize some distribution coefficients (for instance, related to turnstile counts).

The fifth phase consists in correcting origin-destination matrices associated to rail mode (*r*) by using turnstile counts. Obviously since these counts are generally aggregated in a daily scale, we propose to correct the all-day matrices and modify initial matrices by adopting the same variation rates with the assumption that differences are related only to a different modal split (i.e. total travel demand is assumed as constant), that is,

$$OD_{rt}^{r} = \sum_{i} OD_{rt}^{i,r}, \tag{2}$$

$$\overline{OD_{rt}^{r}} = \arg\min_{x \geq 0} Z\left(d_1\left(x\ ;\ OD_{rt}^{r}\right)\ ;\ d_2\left(\Lambda(x)\ ;\ f^{r}\right)\right), \tag{3}$$

$$\delta_j = \overline{d^r_{j,rt}} \Big/ d^r_{j,rt} \;\; \text{with} \;\; \overline{d^r_{j,rt}} \in \overline{OD^r_{rt}} \;\; \text{and} \;\; d^r_{j,rt} \in OD^r_{rt}, \tag{4}$$

$$\overline{OD^{i,r}_h} = \left\{ \overline{d^{i,r}_{j,h}} \; : \; \overline{d^{i,r}_{j,h}} = d^{i,r}_{j,h} \cdot \delta_j \right\} \;\; \forall h \in [ph \; ; \; ad] \;\; \forall i \in [s \; ; \; ns], \tag{5}$$

$$d^i_{j,h} = \sum_m d^{i,m}_{j,h}, \tag{6}$$

$$\overline{OD^{i,m}_h} = \left\{ \overline{d^{i,m}_{j,h}} \; : \; \overline{d^{i,m}_{j,h}} = d^{i,m}_{j,h} \cdot \left(d^i_{j,h} - \overline{d^{i,r}_{j,h}} \right) \Big/ \sum_{m \neq r} d^{i,m}_{j,h} \right\}$$

$$\forall m \neq r \;\; \forall h \in [ph \; ; \; ad] \;\; \forall i \in [s \; ; \; ns], \tag{7}$$

where OD^r_{rt} is the origin-destination matrix related to all-day round trips (*rt*) associated to mode *r* (i.e. rail mode), $\overline{OD^r_{rt}}$ is the correction of matrix OD^r_{rt}; *x* is the variable expressing in the optimization problem (3) the generic value of matrix OD^r_{rt}; $Z(\cdot)$ is the objective function to be minimized; d_1 is a function which expresses the distance between matrix *x* and the a-priori estimation of matrix OD^r_{rt}; $\Lambda(\cdot)$ is the assignment function which provides passenger flows associated to origin-destination matrix *x*; f^r is the vector of turnstile counts; d_1 is a function which expresses the distance between flows obtained by assigning matrix *x* and flows provided by turnstile counts (i.e. f^r); δ_j is the variation rate of travel demand associated to origin-destination *j*; $\overline{d^r_{j,rt}}$ is the generic element of matrix $\overline{OD^r_{rt}}$; $d^r_{j,rt}$ is the generic element of matrix OD^r_{rt}; $\overline{OD^{i,m}_h}$ is the corrected origin-destination matrix in the time period *h*, for purpose *i* by using mode *m*; $\overline{d^{i,m}_{j,h}}$ is the generic element of matrix $\overline{OD^{i,m}_h}$ associated to origin-destination *j*; $d^{i,m}_{j,h}$ is the a-priori estimation of trips in the case of origin-destination *j*, in the time period *h*, for purpose *i* by using mode *m*; $d^i_{j,h}$ is the a-priori estimation of trips in the case of origin-destination *j*, in the time period *h*, for purpose *i* by using all transportation mode.

It is worth noting that since variable *i*, expressing the purpose of the trip, may assume *s* for systematic and *ns* for non-systematic and variable *h*, expressing time period, may assume *ph* for the morning peak hour and *ad* for all day, we obtain that, in the case of corrected matrices, eqn (1) can be expressed as follows:

$$\overline{OD^{i,m}_{rt}} = \overline{OD^{i,m}_{ad}} + \left[\overline{OD^{i,m}_{ad}} \right]^{\mathrm{T}}, \tag{8}$$

where $\overline{OD^{i,m}_{rt}}$ is the corrected origin-destination matrix related to all-day round trips (*rt*) associated to purpose *i* and mode *m*.

The sixth phase consists in the temporal extension to one or more analysis periods of corrected matrices. In particular, the new matrices may be obtained by considering (real or estimated) demographic variations as in the case of the third phase.

In order to make demand elastic at least at modal choice level, it is possible to specify, calibrate and validate a suitable choice model by adopting traditional methodology proposed in the literature (see, for instance, Ref. [4]). In particular, it is necessary to:

- specify a utility formulation and a probability choice model such as:

$$V_{j,h}^{i,m} = V_{j,h}^{i,m}\left(\beta_k^m\right) = \sum_k \beta_k^m \cdot X_{k,j,h}^{i,m}, \tag{9}$$

$$p_{j,h}^{i}[m] = p_{j,h}^{i}[m]\left(V_{j,h}^{i,m}\left(\beta_k^m\right)\right) = p_{j,h}^{i}[m]\left(\beta_k^m\right), \tag{10}$$

where $V_{j,h}^{i,m}$ is the utility associated to mode *m* in the case of purpose *i* during the time period *h* for travelling between the origin-destination *j*; β_k^m is the parameter associated to *k*th attribute of the mode *m* ; $X_{j,h}^{i,m}$ is the *k*th attribute associated to mode *m* in the case of purpose *i* during the time period *h* for travelling between the origin-destination *j*; $p_{j,h}^{i}[m]$ is the probability of choosing mode *m* for travelling between the origin-destination *j* in the case of purpose *i* during the time period *h*;

- calibrate the values of parameters β_k^m by solving the following optimization problem:

$$\hat{\beta}_{k,}^{m} = \arg\max_{\beta_k^m} \left(\ln L\left(\beta_k^m\right)\right) = \sum \ln\left(p_{j,h}^{i}[m]\left(\beta_k^m\right)\right), \tag{11}$$

where $\hat{\beta}_k^m$ is a calibrated value of parameter β_k^m ; $L(\cdot)$ is a likelihood function to be maximized;
- validate the results by means of suitable statistical tests.

The following phase consists in determining hourly matrices being consistent with the corrected matrices, that is:

- in the time period 07:30–09:29, the hourly travel demand can be obtained by dividing by 2 the peak hour origin-destination matrix $\overline{OD_{ph}^{i,r}}$;
- in the morning period (for instance, until 13:30), except in the peak hour period already analysed, the hourly travel demand can be obtained by dividing the outward matrix minus the peak hour matrix (i.e. $\overline{OD_{ad}^{i,r}} - \overline{OD_{ph}^{i,r}}$) by adopting suitable coefficients (for instance, derived from previous flow studies);
- in the afternoon and evening period (for instance, from 13:30 onwards), hourly demand may be obtained by dividing the transposed of the outward matrix $\overline{OD_{ad}^{i,r}}$ by adopting suitable coefficients.

In order to consider interaction between travel demand and the rail service, we suggest adopting the approach proposed by D'Acierno et al. [17, 29] which consists in considering explicitly the behaviour of passengers waiting on a platform when a train arrives. The methodology allows boarding, alighting and on-board passenger flows to be determined for each train, average passenger travel and waiting times for each time period and hence the total generalized cost for passengers on the rail system.

The application of the interaction model for each analysed scenario allows all terms of the objective function to be calculated, that is,

$$\Psi = FC + UC + EC, \tag{12}$$

with

$$FC = OC - TR, \tag{13}$$

$$UC = RPC + MTPC + RC \,, \tag{14}$$

$$RPC = TTC + WTC + TC \,, \tag{15}$$

where Ψ is the objective function to minimize; *FC* are the total costs of the firm operating the rail system; *UC* are the user generalized costs; *EC* are the external costs; *OC* are the operative costs of the analysed rail system; *TR* are the traffic revenues due to the sale of rail tickets; *RPC* are the passenger costs on the analysed rail system; *MTPC* are the passenger costs on mass-transit systems except the analysed rail system; *RC* are the user costs on the road system; *TTC* are the passenger costs associated to travel times; *WTC* are the passenger costs associated to waiting times; *TC* are passenger monetary costs. Details on the calculation of these terms can be found in Refs [4, 17, 19, 37].

Obviously, in order to calculate cost terms in the case of other mass-transit and road systems it is necessary to adopt a multimodal approach which consists in generating a complete public transport and private car network in order to calculate flows and related time and monetary costs (see, for instance, Refs [4, 7, 19, 37–39]).

3 REGIONAL LINE APPLICATION

In order to show the feasibility of the proposed methodology, we applied it in the case of a real regional rail line, namely the Naples–Sorrento line which connects the regional capital (i.e. Naples) with the Sorrento peninsula (the city of Sorrento represents the line terminus) in the south of Italy.

As shown in Fig.1, the line, which is 41.5 km long, can be decomposed into four sections: the Naples–Barra double-track section, 4.5 km long, at the end of which there is a branch to Sarno; the Barra–Torre Annunziata double-track section, 15.6 km long, at the end of which there is a branch to Poggiomarino; the Torre Annunziata–Moregine double-track section, 4.4 km long, after which there is a single-track section; and the Moregine–Sorrento single-track section, 17.0 km long, which finishes with the line terminus. Therefore, in the first section there is the overlap of three lines (Naples–Sorrento, Naples–Poggiomarino and Naples–Sarno), in the second section the interference is limited to only two lines and in the third section there are bottlenecks related to the Moregine–Sorrento single-track section.

The application consists in verifying the effects of a different signalling system in terms of variation in the objective function. In the case of a cost–benefit analysis, intervention costs have to be compared with global benefits. Hence, since we want to identify the improving signalling system in terms of performance and not in terms of existing commercial products, it is necessary to calculate long-term benefits so as to define cost thresholds for implementing new technologies.

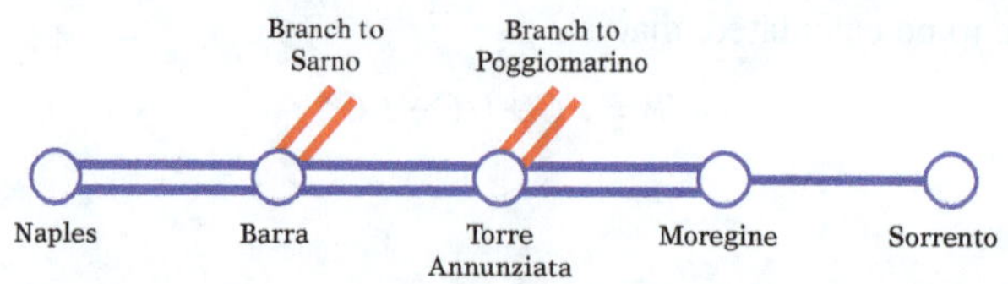

Figure 1: Functional scheme of the Naples–Sorrento line.

However, since the last line section is single track, in order to improve line performance and maximize the effects of a different signalling system, we also explored the possibility of an infrastructural intervention consisting in doubling the line (i.e. construct another track in order to make the line double track). Therefore, in the application, we considered the following seven scenarios:

- Scenario 1 consists in considering the current framework of the line in terms of infrastructure, signalling system and timetable;
- Scenario 2 consists in considering the current framework of the line in terms of infrastructure and signalling system while, in terms of timetable, we maximize the number of runs on the Naples–Sorrento line by considering the timetable of other overlapping lines as invariant;
- Scenario 3 consists in considering the same assumption as Scenario 2 but considering the Naples–Sorrento line as having priority over the other overlapping lines (i.e. a run of the other lines can be done only if a suitable time-window is available);
- Scenario 4 consists in doubling the 'Moregine–Sorrento' section so as to make the whole Naples–Sorrento line double track. In this case, we consider the existing signalling system as invariant and the timetable optimized by maximising the number of runs and considering the timetable of other overlapping lines as invariant;

Table 1: Objective function values.

Scenario	Analysed year				
	2016	2026	2036	2046	2056
1	26,001,375	25,470,409	24,640,361	23,484,308	21,984,231
2	25,966,660	25,438,469	24,612,758	23,462,747	21,970,508
3	25,962,951	25,434,976	24,609,605	23,460,065	21,968,439
4	25,822,162	25,301,345	24,487,163	23,353,207	21,881,802
5	25,694,677	25,178,209	24,370,825	23,246,337	21,787,218
6	25,030,696	24,533,087	23,755,185	22,671,758	21,265,919
7	22,935,558	22,488,373	21,789,297	20,815,657	19,552,276

Table 2: Variations in the objective function.

Scenario	Analysed year				
	2016	2026	2036	2046	2056
1	–	–	–	–	–
2	–0.14%	–0.13%	–0.12%	–0.10%	–0.06%
3	–0.16%	–0.15%	–0.13%	–0.11%	–0.07%
4	–0.74%	–0.71%	–0.67%	–0.60%	–0.49%
5	–1.26%	–1.22%	–1.16%	–1.08%	–0.95%
6	–3.93%	–3.87%	–3.78%	–3.63%	–3.42%
7	–12.20%	–12.11%	–11.96%	–11.73%	–11.40%

- Scenario 5 consists in considering the same assumption as Scenario 4 but considering the Naples–Sorrento line as a priority over the other overlapping lines (i.e. a run on other lines can be undertaken only if a suitable time-window is available);
- Scenario 6 is similar to Scenario 5 but considers a new signalling system which allows a headway of 3 minutes between two successive rail convoys;
- Scenario 7 is similar to Scenario 6 but considers a headway of 2 minutes between two successive rail convoys.

The application of the proposed methodology in the case of the above seven scenarios is summarized in Tables 1 and 2.

Numerical results show that only the timetable optimization (i.e. Scenarios 2 and 3) provides limited improvements, which are lower than 0.2%. In this context, complete replacement of the signalling system is not sufficient to improve line performance since the real issue is the single-track section. For these reasons, although simple doubling of the line (Scenarios 4 and 5) provides even limited improvements (<1.5%), it allows us to benefit fully from creating a different signalling system which allows the current minimum headway of 4.62 minutes to be modified.

4 CONCLUSIONS AND RESEARCH PROSPECTS

The analysis of any transportation system requires implementation of a cost–benefit approach where long-term estimation of travel demand is a major requirement. In this context we proposed a method based on the use of different source data and applied it in the case of a regional rail line.

Analyses showed that, in the context considered, the existence of a single-track section represents the main limitation in line improvements. However, the only infrastructural intervention, whose costs are about €300–800 million, does not provide significant improvements unless coupled with the replacement of the signalling system.

In terms of future research, we suggest applying the proposed methodology, based on data from Italian sources, in other contexts both vis-à-vis other Italian railways (in order to verify the reliability of the adopted datasets) and other non-Italian railways (in order to test the methodology in the case of different source data).

ACKNOWLEDGEMENTS

This article is partially supported under research project FerSat grant no. PON03PE_00159_4.

REFERENCES

[1] Domencich, T.A. & McFadden, D., *Urban Travel Demand: A Behavioural Analysis,* American Elsevier: New York, 1975.

[2] Ben-Akiva, M. & Lerman, S.R., *Discrete Choice Analysis: Theory and Application to Travel Demand.* The MIT Press: Cambridge, MA, 1985.

[3] Cascetta, E., Pagliara, F. & Papola, A., Alternative approaches to trip distribution modelling: A retrospective review and suggestions for combining different approaches. *Papers in Regional Science*, **86(4)**, pp. 597–620, 2007.

[4] Cascetta, E., *Transportation Systems Analysis: Models and Applications*. Springer: New York, 2009.

[5] Pagliara, F. & Timmermans H.J.P., Choice set generation in spatial contexts: A review, *Transportation Letters: The International Journal of Transportation Research*, **1(3)**, pp. 181–196, 2009.

[6] Ho, H.W., Wong, S.C., Yang, H. & Loo, B.P.Y., Cordon-based congestion pricing in a continuum traffic equilibrium system. *Transportation Research Part A*, **39(7–9)**, pp. 813–834, 2005.

[7] D'Acierno, L., Ciccarelli, R., Montella, B. & Gallo, M., A multimodal multiuser approach for analysing pricing policies in urban contexts. *Journal of Applied Sciences*, **11(4)**, pp. 599–609, 2011.

[8] Ibeas, A., Cordera, R., dell'Olio, L. & Moura, J.L., Modelling demand in restricted parking zones. *Transportation Research Part A*, **45(6)**, pp. 485–498, 2011.

[9] Biggiero, L., The impact of transport management on the local activities system: The role of limited traffic zones. *WIT Transactions on the Built Environment*, **138**, pp. 669–678, 2014.

[10] Noordegraaf, D.V., Annema, J.A. & van Wee, B., Policy implementation lessons from six road pricing cases. *Transportation Research Part A*, **59**, pp. 172–191, 2014.

[11] Wang, L., Xu, J. & Qin, P., Will a driving restriction policy reduce car trips? The case study of Beijing, China. *Transportation Research Part A*, **67**, pp. 279–290, 2014.

[12] Brännlund, U., Lindberg, P.O., Nõu, A. & Nilsson, J.E., Railway timetabling using Lagrangian relaxation. *Transportation Science*, **32(4)**, pp. 358–369, 1998.

[13] Goverde, R.M.P., Punctuality of railway operations and timetable stability analysis. *PhD Thesis*, TU Delft, The Netherlands, 2005.

[14] D'Ariano, A., Corman, F., Pacciarelli, D. & Pranzo, M., Reordering and local rerouting strategies to manage train traffic in real time. *Transportation Science*, **42(4)**, pp. 405–419, 2008.

[15] D'Ariano, A., Pacciarelli, D. & Pranzo, M., Assessment of flexible timetables in real-time traffic management of a railway bottleneck. *Transportation Research Part C*, **16(2)**, pp. 232–245, 2008.

[16] Canca, D., Zarzo, A., Algaba, E. & Barrena, E., Confrontation of different objectives in the determination of train scheduling. *Procedia: Social and Behavioral Sciences*, **20**, pp. 302–312, 2011.

[17] D'Acierno, L., Gallo, M., Montella, B. & Placido, A., The definition of a model framework for managing rail systems in the case of breakdowns. *Proceedings of the 16th IEEE Conference on Intelligent Transportation Systems (ITSC)*, The Hague, The Netherlands, pp. 1059–1064, 2013. DOI: 10.1109/ITSC.2013.6728372.

[18] Quaglietta, E. Corman, F. & Goverde, R.M.P., Impact of a stochastic and dynamic setting on the stability of railway dispatching solutions. *Proceedings of the 16th IEEE Conference on Intelligent Transportation Systems (ITSC)*, The Hague, The Netherlands, pp. 1035–1040, 2013. DOI: 10.1109/ITSC.2013.6728368.

[19] Montella, B., D'Acierno, L. & Gallo, M., A multimodal approach for determining optimal public transport fares. *Journal of Applied Sciences*, **14(21)**, pp. 2767–2781, 2014.

[20] Gallo, M., Simonelli, F., De Luca, G. & De Martinis, V., Estimating the effects of energy-efficient driving profiles on railway consumption. *Proceedings of the 15th IEEE International Conference on Environment and Electrical Engineering (EEEIC)*, Rome, Italy, pp. 1059–1064, 2015. DOI: 10.1109/EEEIC.2015.7165269.

[21] Biggiero, L., Pagliara, F., Patrone, A. & Peruggini, F., Spatial equity and high-speed rail systems. *WIT Transactions on the Built Environment*, **164**, 2016.

[22] Botte, M., Di Salvo, C., Placido, A., Montella, B. & D'Acierno, L., A Neighbourhood Search Algorithm (NSA) for determining optimal intervention strategies in the case of metro system failures. *International Journal of Transport Development and Integration*, **1 (1)**, pp. 63–73, 2017.

[23] D'Acierno, L., Placido, A., Botte, M. & Montella, B., A methodological approach for managing rail disruptions with different perspectives. *International Journal of Mathematical Models and Methods in Applied Sciences*, **10**, pp. 80–86, 2016.

[24] Dell'Acqua, G., Russo, F. & Biancardo S.A., Risk-type density diagrams by crash type on two-lane rural roads. *Journal of Risk Research*, **16(10)**, pp. 1297–1314, 2013.

[25] Dell'Acqua, G., Modeling driver behavior by using the speed environment for two-lane rural roads. *Transportation Research Record*, **2472**, pp. 83–90, 2015.

[26] Bifulco, G.N., Cantarella, G.E., Simonelli, F. & Velonà, P., Advanced traveller information systems under recurrent traffic conditions: Network equilibrium and stability. *Transportation Research Part B*, 2016. DOI: 10.1016/j.trb.2015.12.008.

[27] Pariota, L., Bifulco, G.N. & Brackstone, M., A linear dynamic model for driving behavior in car following. *Transportation Science*, 2016. DOI: 10.1287/trsc.2015.0622.

[28] Lo, H.-P. & Chan, C.-P., Simultaneous estimation of an origin–destination matrix and link choice proportions using traffic counts. *Transportation Research Part A*, **37(9)**, pp. 771–788, 2003.

[29] D'Acierno, L., Gallo, M., Montella, B. & Placido, A., Analysis of the interaction between travel demand and rail capacity constraints. *WIT Transactions on the Built Environment*, **128**, pp. 197–207, 2012.

[30] Cascetta, E., Papola, A., Marzano, V., Simonelli, F. & Vitiello, I., Quasi-dynamic estimation of o–d flows from traffic counts: Formulation, statistical validation and performance analysis on real data. *Transportation Research Part B*, **55**, pp. 171–187, 2013.

[31] Ercolani, M., Placido, A., D'Acierno, L. & Montella, B., The use of microsimulation models for the planning and management of metro systems. *WIT Transactions on the Built Environment*, **135**, pp. 509–521, 2014.

[32] Cantarella, G.E., de Luca, S. & Cartenì, A., Stochastic equilibrium assignment with variable demand: theoretical and implementation issues. *European Journal of Operational Research*, **241(2)**, pp. 330–347, 2015.

[33] Di Mauro, R., Botte, M. & D'Acierno, L., An analytical methodology for extending passenger counts in a metro system. *International Journal of Transport Development and Integration*, **1**, 2016.

[34] Istituto Nazionale di Statistica – ISTAT (Italian National Institute of Statistics). *Population and Housing Census*, http://www.istat.it/it/censimento-popolazione (last accessed: March 2016).

[35] Osservatorio sui comportamenti di mobilità degli italiani – AudiMob (Observatory on the Italian mobility behaviour). *Regional Mobility Statistics*, http://www.isfort.it/sito/statistiche/Audimob.htm (last accessed: March 2016).

[36] Istituto Nazionale di Statistica – ISTAT (Italian National Institute of Statistics). *Resident Population*, http://www.istat.it/it/popolazione (last accessed: March 2016).

[37] Gallo, M., Montella, B. & D'Acierno, L., The transit network design problem with elastic demand and internalisation of external costs: An application to rail frequency optimisation. *Transportation Research Part C*, **19(6)**, pp. 1276–1305, 2011.

[38] D'Acierno, L., Montella, B. & Gallo, M. A fixed-point model and solution algorithms for simulating urban freight distribution in a multimodal context. *Journal of Applied Sciences*, **11(4)**, pp. 647–654, 2011.

[39] D'Acierno, L., Gallo, M. & Montella, B., A theoretical multimodal assignment model for simulating intermodal trips. *WIT Transactions on the Built Environment*, **116**, pp. 17–29, 2011.

THE EFFECT OF RAILWAY ACCESSIBILITY ON THE CHOICE OF UNIVERSITY STUDIES

I. HENKE
Department of Civil, Construction and Environmental Engineering, University of Naples, Federico II.

ABSTRACT

The choice of university studies is one of the main socio-economic categories that is affected by the quality of public transport services and in particular rail transport. This category is characterized to cover medium to long distances to reach, every day, the place of study. For this reason, among the urban sustainable policies, the mobility choices of the university students have long been analysed in literature. The aim of this paper was to estimate the effect of railway accessibility on the choice of university studies.

The University of Naples Federico II (Italy) is taken as the case study. In the spring of 2014, using the database obtained from the registry office of the university the following were investigated:

- the origin of the daily trips (residential address);
- the trip frequency and the departure/arriving time;
- the modes or sequence of modes used;
- the perceived public transport quality (e.g. travel time, comfort).

To evaluate the catchment area of the university and for analysing what attributes influence the choice of university studies (in particular the 'weight' of a specific railway accessibility), a regression model was used. Among the model attributes educational course variables, home–university distance and a constant-specific attribute were also considered.

The result of the study shows the surprising role of railway accessibility in the choice of university studies; to travel from a station within 900 m from home is equivalent (in terms of perceived utility) to travel every day a distance greater than 28 km to reach the university.

Keywords: geographical distribution of students' residences, logistic function, railways stations, transport accessibility.

1 INTRODUCTION

Among the urban sustainable policies (e.g. [1–3]), the mobility choices of the university students have long been analysed in literature. This category is characterized to cover medium to long distances every day to reach the university. In many cases the distance between the students' residence and the university, although long, is not encouraging the students to rent a room in a dorm near the university for economic, logistic or cultural reasons. In this case the students choose to travel several kilometres every day to reach the university.

Although much attention has been given by the scientific literature to the accessibility effects on the mobility choices ([4–8]), little attention has been given to the choices of university studies.

In the Italian context, in the recent years, only a few universities (e.g. University of Milan [9, 10]) have carried out a mobility survey to investigate the student transport choice. The results of this study show that the public transport accessibility (in particular railway services) influences the attendance of the students for university activities. For example, from the survey carried out

DOI: 10.2495/TDI-V1-N3-339-347

in Milan [9], it appears that the number of students who come from the areas that are served directly by the railway lines is higher. However, in this study, it is not considered if and how much the railway accessibility influences the choice of the students in deciding which university to choose for their education. The choice of the students in deciding which university to enrol is only analysed with respect to the attributes related to the characteristics of the university (e.g. the institute's scientific reputation, curriculum of teachers, infrastructural qualities, etc.) (e.g. [11]) or with respect to possible job opportunities that the university can offer (e.g. [12]). In this study, however, the variables related to the quality and the accessibility of the public transport are not considered while how this variable influences the users' choices is demonstrated (e.g. [13–16]).

Based on these considerations, in this work, the effects of railway accessibility on the choice of university studies are estimated through the following:

a. Quantify the geographical distribution of the students' residences from the technical and scientific departments of the University of Naples (Campania, Italy).
b. Estimate (and so quantify), through a regression method, the weight of a rail service attribute (railway accessibility) in the choice of university studies.

This paper is divided into three parts: in the first section, for the specific case study, the geographic distribution of students' residences is analysed; in the second part, through a logistic regression model, the weight of railway accessibility is quantified. In the last part conclusion and research perspectives are summarized.

2 CASE STUDY

The University of Naples Federico II is taken for the case study. This university with more than 150 years of history is one of the oldest and most prestigious universities in Italy. It is placed in the city of Naples (Campania region in southern Italy) (Fig. 1).

In this part of the research, the analysis focused only on the technical and scientific departments of the University of Naples: Architecture, Biology and Biotechnology, Chemistry, Physics, Engineering, Mathematics, Computer Science, Natural Sciences, and Earth Science. All these departments are located in the west of Naples. These departments are served by two metro lines: the Line 2 of Naples that reaches the Garibaldi station, the most important interchange point of the city, and Cumana Line that connects the historical centre of Naples and the northern municipalities. To analyse the geographic distribution of the students' residences, the database available at the registry office of the university was consulted. This database gives information about the residences (address), age and gender of all students. The results show that 71% of the students reside in the centre of Naples city and in its province (Fig. 2). It is interesting to note that the percentage of students who are not residents in Campania region is negligible. This result is in line with the other Italian universities (e.g. [17]).

Through a geographic information system software, the geographic distribution of students' residence is presented in a territorial map. For clarity of presentation, in Fig. 3, only the main portion of the region (for which 80% of the university students showed interest) is reported. This area is divided into traffic zones. In Fig. 3, the rate of students (university students for the academic year 2013/2014 among 1,000 inhabitants) for each single traffic zone is shown. It is evident that the distribution of the rate of students is not homogenous; the higher value is in the western part of the territory where the departments analysed are placed. This result would seem to indicate that one of the factors that influences the choice of the university is 'proximity effect'. In other words the distance between the students' residences and the university influences the choice of the university.

Figure 1: Case study.

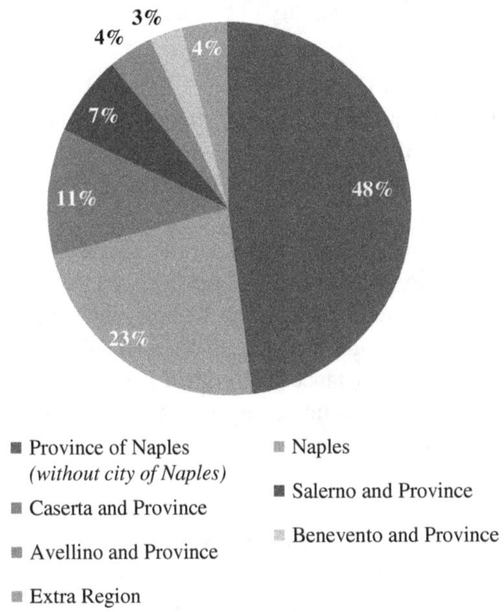

Figure 2: The residence of students of the technical and scientific departments of the University of Naples Federico II.

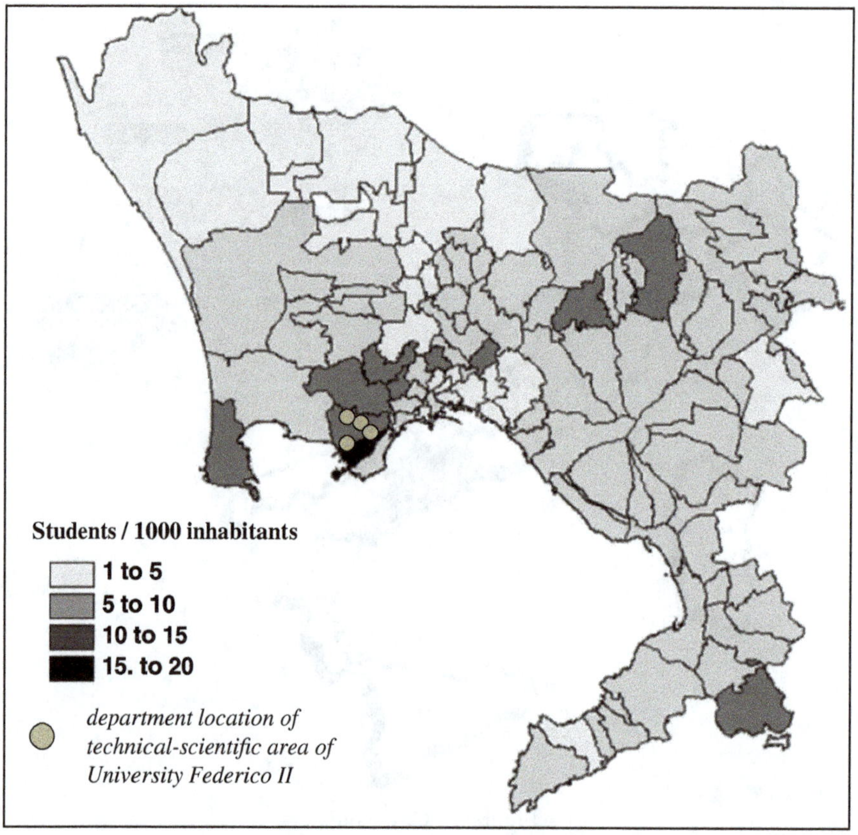

Figure 3: Rate of students (university students/1,000 inhabitants) of the technical and scientific departments of the University of Naples Federico II.

Considering the data about zoning used in Fig. 3, a cumulative distribution of students/inhabitants based on the distances (km) is estimated (Fig. 4). The comparison of two diagrams confirms the proximity effect as introduced earlier. For example, 87% of students live within a distance of 50 km from the departments analysed against the 68% of inhabitants.

The geographic distribution of students' residences is also analysed based on the distance from the railway stations (railway accessibility). Figure 5 shows the railway lines that have a station near the university and the other metropolitan of Naples the Line 1. From Fig. 5 it could be observed that the rate of students is higher in areas that are served by rail service. Mostly 50% of the university students live within a distance shorter than 1.5 km from the railway station.

This result shows that the choice of university is influenced not only by proximity effect, but also by the distance between the students' residences and the railway station (railway accessibility).

3 MODEL ESTIMATION

From the geographical distribution of residences of the students who belong to the technical and scientific departments of the University of Naples (Campania, Italy) (described in the above paragraph "The Case Study"), it is determined that the choice of university is

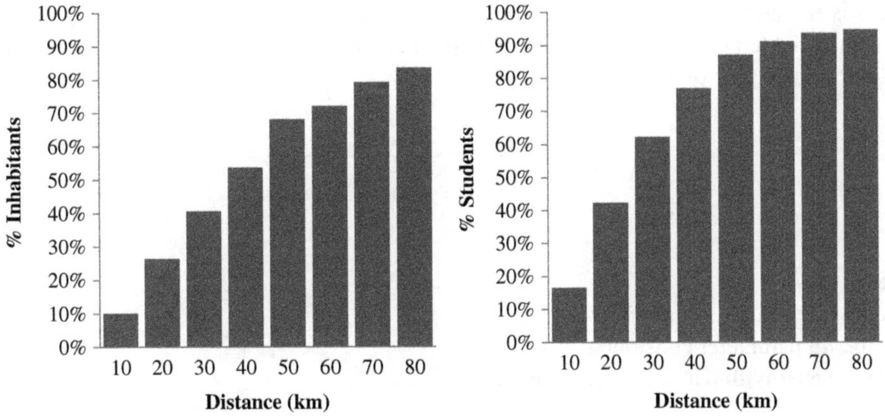

Figure 4: The cumulative distribution of distances (km) between the technical and scientific departments of the university of Naples Federico II and the residence of inhabitants of Campania and enrolled students.

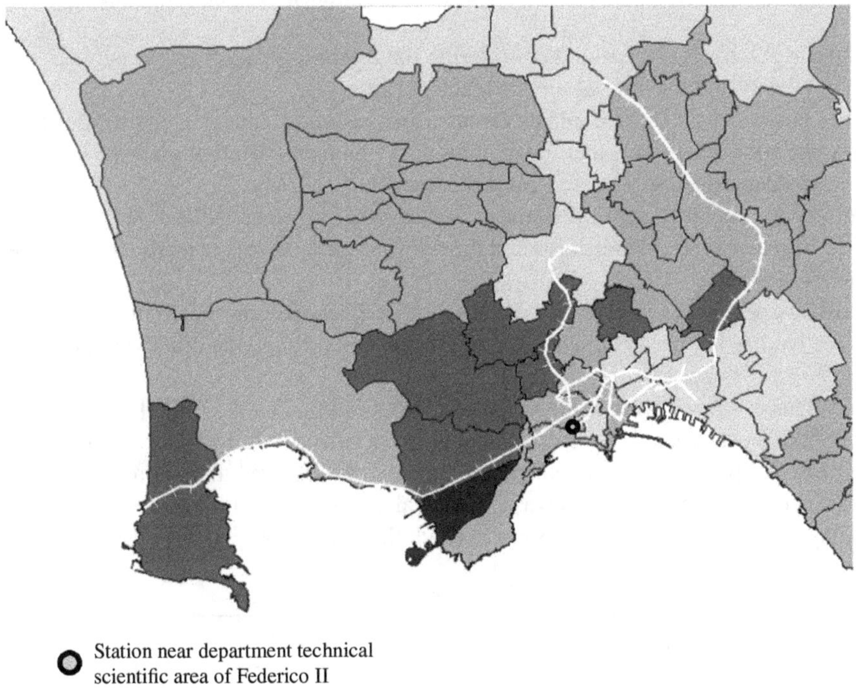

Figure 5: The rate of students (university students/1000 inhabitants) of the technical and scientific departments of the University of Naples Federico II.

influenced by the distance between the students' residences and the university and by the distance between the students' residences and the railway station (transport accessibility). To estimate the effect of railway accessibility on the choice of university studies, a logistic regression model was used [18, 19].

Logistic regression measures the relationship between a dependent variable (y) and one or more independent variables (x_i) [18]. To estimate the probabilities, a logistic function, which is the cumulative logistic distribution, was used. The aim was to estimate the independent variable (x_i) that influences the rate of students in different areas. The formula used to describe this phenomenon is as follows:

$$y_i = \frac{1}{\left(1 + \exp\left(\sum_j - x_j * \beta_j\right)\right)}, \tag{1}$$

where

y_i is the rate students in the *i*th zone.

x_j is the *j*th parameter considered.

β_j is the generic *j*th parameter of the model.

Considering the zoning data described in Section 2 and through the Gretl software [20], different systematic utility functions were specified. Table 1 reports the best tested attributes and those that are statistically significant.

The attributes of the model that are statistically significant are the following:

- $Distance_{od}$ (km): it is measured considering the minimum path in street network between the university and the area where the residences are;
- Other $university_i$ (1,0): a Boolean variable that assumes value 1 if the residences are in the same zone where the other universities are (which are different from the technical and scientific departments discussed); it is 0 otherwise;
- Railway $accessibility_i$ (1,0): a Boolean variable that assumes value 1 if the walking distance between the residences area and the railway station is less than 900 m (this threshold was estimated);
- Constant-specific alternative: it considers other phenomenon that influences the users' choice but has not been explicitly taken into account through the other variable.

The values estimated were compared to the result of other studies (e.g. [21–25]).

One of the most interesting results is related to the values of railway accessibility: comparing the value of parameters estimated in the model, it is observed that to travel from a station at a distance less than 900 m (subway) is equivalent (in terms of perceived utility) to travel every day a distance greater than 28 km to reach the university.

Table 1: Model estimation results.

Parameter	Model
$distance_{od}$ [km]	–0.01
Std. Error/t-test	*0.01 to 2.60*
other university [1,0]	–0.35
Std. Error/t-test	*0.18 to 1.90*
railway accessibility [1,0]	0.28
Std. Error/t-test	*0.17 to 1.67*

4 CONCLUSION AND RESEARCH PERSPECTIVE

The aim of this paper was to estimate the effect of railway accessibility on the choice of university studies through the following:

a. The analysis of the geographical distribution of the residences of the students enrolled in the technical and scientific departments of the University of Naples Federico II.
b. The estimation, through a regression method, of the weight of a rail service attributes (railway accessibility) to the choice of university studies.

The geographic distribution of the students' residences shows that the students prefer to live in areas which are in close 'proximity' to the university. For example, 87% of students live within a distance of 50 km from the departments analysed against the 68% of inhabitants. In addition, mostly 50% of university students of the technical and scientific departments of the University of Naples Federico II travel a distance shorter than 1.5 km from the railway station. To evaluate the 'weight' of a specific railway accessibility attribute, a regression model was used. The result of the study shows the surprising role of railway accessibility in the choice of university studies: to travel to a station within 900 m from home is equivalent (in terms of perceived utility) to travel every day a distance greater than 28 km to reach the university.

This result shows that the public transport, and in particular the railway transport, influences not only the attendance of students for university activities, but also the choice of students in deciding which university to choose for their education.

The research prospective is to also consider the level of attributes of the railway services in the choice model (e.g. $time_{od}$, $cost_{od}$). In this way, the proposed model could be applied within the optimization planning procedures (e.g. [26]) in case of unexpected events too (e.g. [27, 28]).

The research prospects will include the following:

- extending this analysis to all departments of the University of Naples Federico II;
- evaluating the attributes that influence the choice of university studies (in particular, the 'weight' of specific attributes and the level of service of public transport) and discovering a discrete choice model.

REFERENCES

[1] Cascetta, E., Cartenì, A., Pagliara F. & Montanino, M., A new look at planning and designing transportation systems: a decision-making model based on cognitive rationality, stakeholder engagement and quantitative methods. *Transport Policy*, **38**, pp. 27–39, 2015. DOI: 10.1016/j.tranpol.2014.11.005.

[2] Cartenì, A., Urban sustainable mobility. Part 1: Rationality in transport planning. *Transport Problems*, **9(4)**, pp. 39–48, 2014.

[3] de Luca, S., Public engagement in strategic transportation planning: An analytic hierarchy process based approach. *Transport Policy*, **33**, pp. 110–124, 2014.DOI: 10.1016/j.tranpol.2014.03.002.

[4] Hansen, W.G., How accessibility shapes land-use. *Journal of the American Institute of Planners*, **25**, pp. 73–76, 1959. DOI: 10.1080/01944365908978307.

[5] Leonardi, G., Optimum facility location by accessibility maximizing. *Environment and Planning A*, **10**, pp. 1287–305, 1978. DOI: 10.1068/a101287.

[6] Cartenì, A., Accessibility indicators for freight transport terminals. *Arabian Journal for Science and Engineering*, **39**, pp. 7647–7660, 2014.DOI: 10.1007/s13369-014-1333-y.

[7] Cartenì, A., Galante G. & Henke I., The catchment area of high architectural railway stations: An application of the Graham scan algorithm. *WIT Transactions on the Built Environment*, **135**, pp. 463–474, 2014. DOI: 10.2495/CR140381.

[8] Cascetta, E., Cartenì, A. & Montanino, M., A behavioral model of accessibility based on the number of available opportunities. *Journal of Transport Geography,* **51**, pp. 45–58, 2016. DOI: 10.1016/j.jtrangeo.2015.11.002.

[9] Politecnico di Milano, *Mobility Management per il sistema universitario milanese,* 2007. http://progetti.unicatt.it/. Accessed September 2014.

[10] Università degli Studi di Pavia, *Gli spostamenti casa-università degli studenti dell'Università degli Studi di Pavia*, 2014. http://www-wp.unipv.it/. Accessed May 2015.

[11] Horstschraer, J., University rankings in action? The importance of rankings and an excellence competition for university choice of high-ability students. *Economics of Education Review*, **31**, pp. 1162–1176, 2012.

[12] Gürsesa, A., Büsra, K., Mindivana, F., Dogarb, Ç., Günes, K. & Yolcua H.H., Determining the relationship between students' choice of profession and mission and vision of their high school. *Procedia Social and Behavioral Sciences*, **15**, pp. 2595–2598, 2011. DOI: 10.1016/j.sbspro.2011.04.149.

[13] Cascetta, E. & Cartenì, A., The hedonic value of railways terminals. A quantitative analysis of the impact of stations quality on travellers behaviour. *Transportation Research Part A,* **61**, pp. 41–52, 2014.

[14] Cascetta, E. & Cartenì, A., A quality-based approach to public transportation planning: theory and a case study. *International Journal of Sustainable Transportation,* **8(1)**, pp. 84–106, 2014.

[15] Cascetta E., Cartenì, A. & Henke I., *Stations quality, aesthetics and attractiveness of rail transport: empirical evidence and mathematical models [Qualità delle stazioni, estetica e attrattività del trasporto ferroviario: evidenze empiriche e modelli matematici]*, Ingegneria Ferroviaria, 69(4), pp. 307–324, 2014.

[16] Cartenì, A., Galante G. & Henke I., An assessment of models accuracy in predicting railways traffic flows: a before and after study in Naples. *WIT Transactions on Ecology and the Environment*, **191,** pp. 783–794, 2014. DOI: 10.2495/SC140661.

[17] VI Rapporto dell'European Migration Network (Emn) Italia, *Gli studenti internazionali nelle università italiane: indagine empirica e approfondimenti*, http://www.yallaitalia.it, 2013. Accessed September 2014.

[18] Hosmer, D.W., Taber, S. & Lemeshow, S., The importance of assessing the fit of logistic regression models: a case study. *American Journal of Public Health*, **81(12)**, pp. 1630–1635, December 1991. DOI: 10.2105/AJPH.81.12.1630.

[19] Hilbe, J.M., *Logistic regression models*, CRC Press, 2009.

[20] Battaglia, F., *Metodi di previsione statistica*, Collana di Statistica e Probabilità Applicata, Springer Science & Business Media, 2010.

[21] de Luca, S. & Cartenì, A., A multi-scale modelling architecture for estimating of transport mode choice induced by a new railway connection: The Salerno-University of Salerno-Mercato San Severino Route [Un'architettura modellistica multi-scala per la stima delle ripartizioni modali indotte da un nuovo collegamento ferroviario: il caso studio della tratta Salerno-Università di Salerno-Mercato San Severino]. *Ingegneria Ferroviaria*, **68** (**5**), pp. 447–473, 2013.

[22] Cartenì, A., Updating demand vectors using traffic counts on congested networks: A real case application. *WIT Transactions on the Built Environment,* **96**, pp. 211–221, 2007. DOI: 10.2495/UT070211.

[23] Cartenì, A. & Punzo, V., Travel time cost functions for urban roads: A case study in Italy. *WIT Transactions on the Built Environment,* **96**, pp. 233–243, 2007. DOI: 10.2495/UT070231.

[24] de Luca, S. & Cantarella, G.E., Validation and comparison of choice models. *Travel Demand Management and Road User Pricing: Success, Failure and Feasibility*. New Delhi, India: Taylor & Francis, pp. 37–58, 2009.

[25] De Luca, S. & Papola, A., Evaluation of travel demand management policies in the urban area of Naples. *Advances in Transport*, **8**, pp. 185–194, 2001.

[26] Montella, B., D'Acierno, L. & Gallo, M., *A* multimodal approach for determining optimal public transport fares. *Journal of Applied Sciences,* **14**(**21**), pp. 2767–2781, 2014. DOI: 10.3923/jas.2014.2767.2781.

[27] D'Acierno, L., Placido, A., Botte, M. & Montella, B., A methodological approach for managing rail disruptions with different perspectives. *International Journal of Mathematical Models and Methods in Applied Sciences,* **10**, pp. 80–86, 2016.

[28] D'Acierno, L., Gallo, M., Montella, B. & Placido, A., The definition of a model framework for managing rail systems in the case of breakdowns. *Proceedings of IEEE ITSC 2013 – 16th International IEEE Annual Conference on Intelligent Transportation Systems*, The Hague, The Netherlands, pp. 1059–1064, 2013.

RELATIONSHIP BETWEEN ACCESSIBILITY IMPROVEMENT AND RESIDENTIAL PROPERTY APPRECIATION: AN OBSERVATION FROM SHANGHAI METRO

Q. WU & X. YE
Key Laboratory of Road and Traffic Engineering of Ministry of Education, Tongji University, China.

ABSTRACT
The costly construction and operation of urban rail transit have become escalating problems for cities worldwide, especially in developing countries. Reliable measures of residential property appreciation for urban rail transit can provide suggestions for policy-making of value capture to fund transit improvements. Using GIS techniques and residential property price data, the relationships between accessibility improvement value and residential property appreciation are analysed in Shanghai. The impacts of urban rail transit on residential property values are classified into traffic effect and agglomeration effect, both of which are measured by the accessibility improvement model. The results indicate that the goodness-of-fit of the model is greater than 93%. Traffic benefit is greater than agglomeration benefit in the suburb, which is completely different in the city centre. Furthermore, the residential property appreciation is about 5 times the accessibility improvement value per year. This study contributes to the evidence of capitalization impacts of public transit from a booming and transitional economy and provides suggestions for land use planning of areas surrounding stations.
Keywords: accessibility, agglomeration benefit, residential property appreciation, traffic benefit, transit-oriented development, urban rail transit.

1 INTRODUCTION

Traditional funding mechanisms, particularly for urban rail transit investments, have become increasingly inadequate, especially in China, a developing country. Because of the costly construction and operation, Chinese State Council had issued a notification to suspend all the underground rapid transit projects in the mid-1990s. Medda [1] presented that land value capture finance is often seen as an alternative method of revenue generation for urban rail transit systems. Although quantitative methods have been applied for nearly 20 years in China, the appreciation results according to different researchers are quite different, even for the same research object. Such differences may be due to the methods introduced from abroad which are not suitable for China. For example, the comparison method [2–4] can hardly be a fit in China because most cities have diverse land use and high-density land development. These land use patterns result in that it is extremely difficult to find a control area which has similar characteristics to its paired catchment area. The traffic cost method assumes that the city is mono-centric and the factors affecting residential property values are uniformly distributed while most Chinese cities are typically multi-centric or cluster-developed and have more than one Central Business District (CBD) in a city. The hedonic price model tries to estimate the implicit value of the determinant factors on the prices of heterogeneous goods [5, 6] by using the residential property transaction data and characteristics description, which are

DOI: 10.2495/TDI-V1-N3-348-358

easily obtained in many developed countries but difficult to be obtained currently in China. So a new approach to measure the impacts of urban rail transit on residential property values should be made. This paper considers the essential impact factors and quantifies them from the point of view that the main function of urban rail transit is to improve accessibility. Using GIS techniques and residential property price data, the relationships between accessibility improvement value and residential property appreciation are analysed in Shanghai.

2 ACCESSIBILITY IMPROVEMENT MODEL

2.1 Impact factor analysis

The primary function of urban rail transit is to improve travel accessibility around station areas. At the same time, it can promote land development, population gathering and job growth. Although these agglomeration benefits are some kind of economic phenomenon, the major reason is the convenient transportation from other origins to the stations by urban rail transit, which is called attract accessibility. Therefore, the impacts of urban rail transit on residential property values can be divided into two parts. One is travel benefit, called travel accessibility improvement. The other is agglomeration benefit, called attract accessibility improvement. However, rail transit may also bring negative impacts to property value [1, 7, 8]. Since the negative impacts are factored into the final transaction price, they are ignored in this article. The accessibility improvement model will be established as follows.

2.2 Modelling

No matter what the purpose of a trip, it will take a certain amount of cost. If the construction of a new transport facility reduces the original travel costs, the reduced part can be regarded as benefits brought by the new transport facility. Therefore, the accessibility improvement provided by urban rail transit is defined as the decrease of travel costs after the operation of urban rail transit. Since the most public transit trips within a city rely on bus transit before the construction of urban rail transit in China, the travel costs before the operation of urban rail transit are supposed to be the travel cost by bus transit. Although the bus transit also has slight impact on residential property values, it is not discussed in this study. For a generalized travel starting from a rail station, the generalized destinations are identified as areas within walking distance from other stations in the urban rail transit network of the city [9]. So the accessibility improvement is the weighted average of generalized travel cost differences between urban rail travels and bus travels starting from a rail station to areas within walking distance from other stations.

2.2.1 Travel accessibility improvement

From the residence to the starting rail station, shuttle transportation modes are generally required, such as walking, cycling, bus, taxi or private car in China. Because of the high coverage density of bus stops, the default shuttle transportation mode of bus is walking. So the travel accessibility improvement of a generalized travel from the residence to a generalized destination is the sum of the shuttle transportation accessibility improvement and the travel accessibility improvement excluding the shuttle transportation.

First, the shuttle transportation accessibility improvement expression is as follows:

$$\Delta C_i(r) = C_d(r) - \sum_{k=1}^{K} \gamma_k(r) \cdot C_{rk}(r) \quad , \tag{1}$$

where r is the network distance from the residence to the starting rail station. $C_d(r)$ is the generalized shuttle transportation cost from the residence to the starting bus stop. $\gamma_k(r)$ is the sharing rate of the shuttle transportation mode k (k =1, 2, 3… K, and K is the number of the shuttle transportation modes). $C_{rk}(r)$ is the generalized shuttle transportation cost from the residence to the starting rail station by mode k. The generalized shuttle transportation costs include fares, travel time and private transportation depreciation charges, all of which are converted to a unified monetary unit by the time value [9].

Second, the travel accessibility improvement excluding the shuttle transportation is calculated as follows:

$$\Delta C_i = \frac{\sum_{\substack{j=0 \\ j \neq i}}^{J} \omega_{ij1} \cdot (C_{ijd} - C_{ijr})}{J} = \frac{\sum_{\substack{j=0 \\ j \neq i}}^{J} \frac{\sum_{m=1}^{M} \alpha_m E_{jm}}{\sum_{m=1}^{M} \alpha_m E_{0m}} \cdot \frac{C_{i0r}}{C_{ijr}} \cdot (C_{ijd} - C_{ijr})}{J} \quad , \tag{2}$$

where ω_{ij1} is the travel importance weight from the rail station i to j (j = 0, 1, 2, …, J, and $j \neq i$, J is the number of urban rail transit stations within the city). C_{ijd} is the generalized travel cost from the bus stop near the rail station i to the bus stop near the rail station j by bus. C_{ijr} is the generalized travel cost from the rail station i to j by urban rail transit. E_{jm} is the number of job m in the impact range of the rail station j (m =1, 2, 3, …, M, and M is the number of job types). α_m is the number of attracting visitors per job m (obtained by survey). E_{0m} is the number of job m in the impact range of the centre station. C_{i0r} is the generalized travel cost from the rail station i to the centre station. The serial number of the centre station is 0. The generalized travel cost includes fares, travel time, transfer time and transfer punishment, all of which are converted to a unified monetary unit by the time value [9].

Assuming that the urban per capita daily travel frequency is n_{iy} in the impact range of the rail station i in year y, the travel accessibility improvement per capita in year y is as follows:

$$A_{iy}(r) = \left(C_{dy}(r) - \sum_{k=1}^{K} \gamma_{ky}(r) \cdot C_{rky}(r) + \frac{\sum_{\substack{j=0 \\ j \neq i}}^{J_y} \frac{\sum_{m=1}^{M} \alpha_{my} E_{jmy}}{\sum_{m=1}^{M} \alpha_{my} E_{0my}} \cdot \frac{C_{i0ry}}{C_{ijry}} \cdot (C_{ijdy} - C_{ijry})}{J_y} \right) \cdot n_{iy} \cdot t \quad , \tag{3}$$

where the indicators which have the subscript y are the corresponding values in year y; t is the service time of urban rail transit.

2.2.2 Attract accessibility improvement

The agglomeration benefits around urban rail transit stations are also attributed to convenient transportation from other origins to the stations by urban rail transit. Therefore, the attract accessibility improvement is the weighted average of generalized travel cost differences between urban rail travels and bus travels from other origin stations to the destination station i. Assuming that the urban per capita daily travel frequency is n_y in the city in year y, the attract accessibility improvement per capita in year y is as follows:

$$B_{iy} = \frac{\sum_{\substack{j=0 \\ j \neq i}}^{J_y} \omega_{ij2} \cdot (C_{ijdy} - C_{ijry})}{J_y} \cdot n_y \cdot t = \frac{\sum_{\substack{j=0 \\ j \neq i}}^{J_y} \frac{E_{iy}}{E_{0y}} \cdot \frac{N_{jy}}{N_{0y}} \cdot \frac{C_{i0ry}}{C_{ijry}} \cdot (C_{ijdy} - C_{ijry})}{J_y} \cdot n_y \cdot t, \tag{4}$$

where B_{iy} is the attract accessibility improvement per capita in year y. ω_{ij2} is the travel importance weight from the rail station j to i ($j = 0, 1, 2, \ldots, J$, and $j \neq i$). E_{iy} is the number of jobs in the impact range of the station i in year y. N_{jy}, N_{0y} are the resident population in the impact range of the rail station j and the centre station. J_y is the number of urban rail transit stations within the city in year y.

2.2.3 Accessibility improvement model

In summary, the accessibility improvement model is as follows:

$$D_{iy}(r) = A_{iy}(r) + B_{iy}. \tag{5}$$

Assuming that the residential area per capita in the impact range of the rail station i is s_{iy} in year y, the accessibility improvement by urban rail transit for per square metre of residential in the impact range of the rail station i is as follows:

$$\begin{aligned}\overline{D_{iy}(r)} &= \frac{D_{iy}(r)}{S_{iy}} \\ &= \left(C_{dy}(r) - \sum_{k=1}^{K} \gamma_{ky}(r) \cdot C_{rky}(r) + \frac{\sum_{\substack{j=0 \\ j \neq i}}^{J_y} \frac{\sum_{m=1}^{M} \alpha_{my} E_{jmy}}{\sum_{m=1}^{M} \alpha_{my} E_{0my}} \cdot \frac{C_{i0ry}}{C_{ijry}} \cdot (C_{ijdy} - C_{ijry})}{J_y}\right) \cdot \frac{n_{iy}}{S_{iy}} \cdot t \\ &+ \frac{\sum_{\substack{j=0 \\ j \neq i}}^{J_y} \frac{E_{iy}}{E_{0y}} \cdot \frac{N_{jy}}{N_{0y}} \cdot \frac{C_{i0ry}}{C_{ijry}} \cdot (C_{ijdy} - C_{ijry})}{J_y} \cdot \frac{n_y}{S_{iy}} \cdot t\end{aligned} \tag{6}$$

3 RESIDENTIAL PROPERTY APPRECIATION CALCULATION METHOD

Although the benefits associated near a rail station and other benefits are not easy to distinguish, Wu and Ye [10] designed a method to separate the benefits of urban rail transit from other benefits. The specific process is shown as the following.

The characteristics of residential property are generally divided into physical characteristics, neighbourhood characteristics and location characteristics [7, 8, 11, 12], which are the main factors affecting residential property values. The impact scope of urban rail station on residential property value is about 500 to 2000 m radius around a station, depending on the location of the station in the city and the size of the urban area. Within the same impact scope, most neighbourhood characteristics and location characteristics are generally similar. If the physical characteristics are controlled to be the same, the difference between the prices of residential property in different places within the same impact scope should only be caused by the distance to the nearest rail station.

Within the impact scope of a rail station, residential districts which have the similar physical characteristics are chosen as the research sample. A scatterplot about the average price per square metre and the distance to the station is drawn. Then a quadratic curve is adopted to fit the scatterplot. The inflection point of the price curve is the border of the impact scope, where the residential property appreciation is considered to be zero. The residential property appreciation at a certain place within the impact scope is calculated as follows:

$$\Delta P_{il} = P_{il} - P_{i0}, \tag{7}$$

where ΔP_{il} is the average appreciation per square metre of the residential district l within the impact scope of the station i ($l = 1, 2, 3, \ldots, L$, and L is the sample set size for the station i). P_{il} is the average price per square metre of the residential district l within the impact scope of the station i. P_{i0} is the average price per square metre of the residential district on the border of the impact scope. If the calculated result is positive, it means the impact of urban rail transit on residential property value is positive; otherwise, the impact is negative.

4 STUDY AREA AND DATA

The study area is the city of Shanghai, located in the Yangtze River Delta in East China. Shanghai is one of the global financial centres, with a population of more than 25 million as of 2014. As of the end of June 2014, 14 rail transit lines have been operated in Shanghai (excluding the maglev line). Shanghai Metro Line 11 crosses from the CBD to the

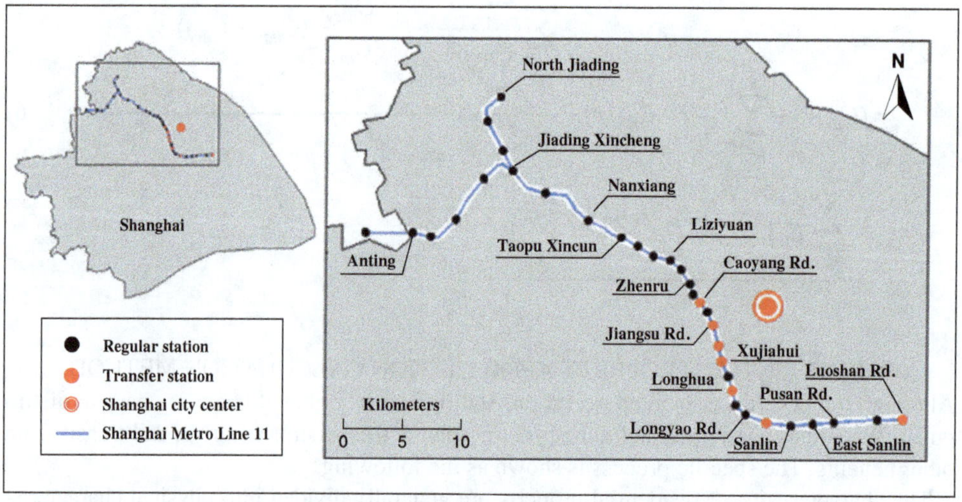

Figure 1: Shanghai Metro Line 11.

north-western outskirts of Shanghai. To ensure that the impact scopes of adjacent stations do not overlap and the residential property price data within the impact scope is rich and evenly distributed, 16 newly built stations of Line 11 between 2009 and 2013 are taken into consideration as a case study (see Fig. 1).

Data for this study are gathered from various sources and all of them are from 2014 to maintain consistency. The accessibility improvement data include information of rail travel, bus travel, population and employment. Rail travel information comes from the Shanghai Metro official web site; bus travel information comes from the Shanghai GIS map with the city's bus transit network loaded; population and employment information come from Shanghai Statistical Yearbook and field survey. The residential price data come from various online real estate platforms and include information of average price per square metre and network distance to the nearest station for 2,236 residential districts which have similar main physical characteristics.

The parameters that cannot be obtained by actual survey will be referred to previous research literature and reports. The waiting time is set as half of the departure interval in the peak period; the time value is set as 20 yuan per hour referring to the income level per capita in Shanghai (Shanghai Statistical Yearbook, 2013); the daily travel frequency per capita is set as 2 referring to the report on the fourth comprehensive traffic survey in Shanghai (2010); the residential area per capita is set as 30 m^2 referring to Statistical Bulletin of Urban Housing Overview (2003) and Shanghai Statistical Yearbook (2013); the penalty coefficient of the shuttle transportation time, waiting time and transfer time is set as 1.5 compared to the time on the subway or in the bus.

5 RESULTS AND DISCUSSION

By eqns (3–7), the travel accessibility improvement average per square metre $\overline{A_i}$, the attract accessibility improvement average per square metre $\overline{B_i}$ and the residential property average appreciation per square metre $\overline{\Delta P_i}$ within the impact scope of each station could be calculated. The results and analysis are shown in Table 1. T_i is the travel time by urban rail transit from the station i to the city centre station. $\overline{D_i}$ is the sum of $\overline{A_i}$ and $\overline{B_i}$; H_i is the ratio of $\overline{A_i}$ and $\overline{B_i}$.

It can be seen from Table 1 that both of the residential property average appreciations and the accessibility improvement average values show the trend that they are larger in the suburb but smaller in the city centre in Shanghai. The relationship between the travel accessibility improvement and the attract accessibility improvement will be discussed, as well as the relationship between the accessibility improvement and the residential property average appreciation.

5.1 Relationship between travel accessibility improvement and attract accessibility improvement

It can be seen from Table 1 that H_i increases significantly with the increase of travel time to the city centre station (see Fig. 2). In the station areas away from the city centre (Anting, North Jiading, Jiading Xincheng, Nanxiang, Taopu Xincun, Sanlin, East Sanlin, Pusan Rd. and Luoshan Rd.), the ratio is 1.5–2, suggesting that the travel benefit is 1.5–2 times the agglomeration benefit. In the station areas close to the city centre (Liziyuan, Zhenru, Caoyang Rd., Jiangsu Rd., Xujiahui, Longhua and Longyao Rd.), the ratio is less than 1, suggesting that the travel benefit is less than the agglomeration benefit.

The quality of public transportation in the suburb is generally worse than that in the city centre in Chinese cities. After the construction of the urban rail line, the transportation around the suburban station is improved significantly. So the travel benefit is more than the agglomeration benefit

Table 1: Results and analysis of the accessibility improvement and the residential property average appreciation.

Station name	T_i (min)	$\overline{A_i}$ (yuan m^{-2})	$\overline{B_i}$ (yuan m^{-2})	$\overline{\Delta P_i}$ /(yuan m^{-2})	$\overline{D_i}$ (yuan m^{-2})	H_i
Anting	65	334.3	175.7	2417.9	510.0	1.902
North Jiading	60	336.9	172.2	2395.7	509.1	1.956
Jiading Xincheng	50	368.0	189.8	2827.0	557.8	1.939
Nanxiang	41	233.9	124.6	2087.3	358.5	1.877
Taopu Xincun	37	182.8	101.2	1314.2	284.0	1.805
Liziyuan	29	153.4	172.0	1928.0	325.4	0.892
Zhenru	24	120.5	140.6	1369.6	261.1	0.857
Caoyang Rd.	19	141.2	224.5	1820.6	365.7	0.629
Jiangsu Rd.	8	56.2	91.9	1102.9	148.1	0.611
Xujiahui	12	91.7	147.1	1232.1	238.8	0.623
Longhua	24	128.3	147.6	1555.3	275.9	0.869
Longyao Rd.	26	207.7	223.4	2063.4	431.1	0.930
Sanlin	28	247.3	164.9	2286.3	412.2	1.5
East Sanlin	30	252.0	168.0	2092.3	420.0	1.5
Pusan Rd.	32	251.5	167.7	2188.4	419.2	1.5
Luoshan Rd.	40	414.1	276.1	3773.0	690.2	1.5

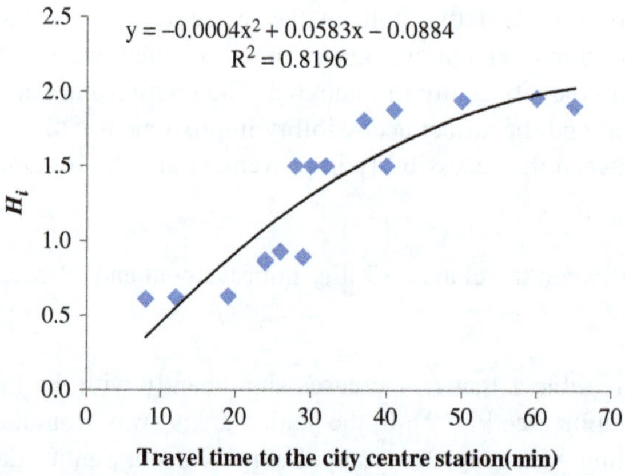

Figure 2: Relationship between H_i and the travel time to the city centre station.

in the suburban station areas. On the other hand, the introduction of the urban rail line attracts more travellers over a long distance to the station areas close to the city centre on employment and consumption. So the agglomeration benefit is more than the travel benefit in the urban station areas. That is the reason why the relationship between the travel benefit and the agglomeration benefit in the suburban station area and in the urban station area is distinct. Further explanations to this phenomenon will be made by analysing representative stations as below.

5.1.1 Luoshan Rd. station

Luoshan Rd. station is in the suburban area of Shanghai, and the linear distance to the city centre is about 15 km. Before the opening of the rail station, public transportation was very poor (see Fig. 3).

After the opening of Luoshan Rd. station, public transportation is improved significantly. There are 9 bus stops in the range of 500 m around Luoshan Rd. station, which is convenient for travellers to transfer from bus to subway (see Fig. 4).

(a) Bus stop not in use around the station. (b) Land use around the station.

Figure 3: Public transportation and land use before the opening of Luoshan Rd. station.

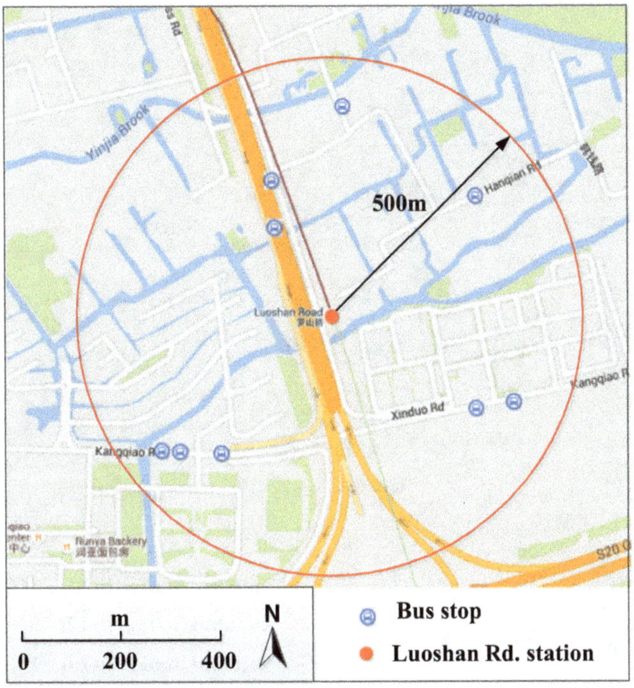

Figure 4: Bus stops in the range of 500 m around Luoshan Rd. rail station.

5.1.2 Xujiahui station

Xujiahui station is a three-line transfer station of Line 1, Line 9 and Line 11, located in the Shanghai core area. Xujiahui is one of the four commercial circles in Shanghai. Agglomeration benefits are strengthened significantly due to the opening of rail lines, which can be seen in Fig. 5.

(a) Xujiahui core area in the 1980s. (b) Xujiahui core area in 21st century.

Fiugre 5: Xujiahui core area development after the opening of rail lines [13].

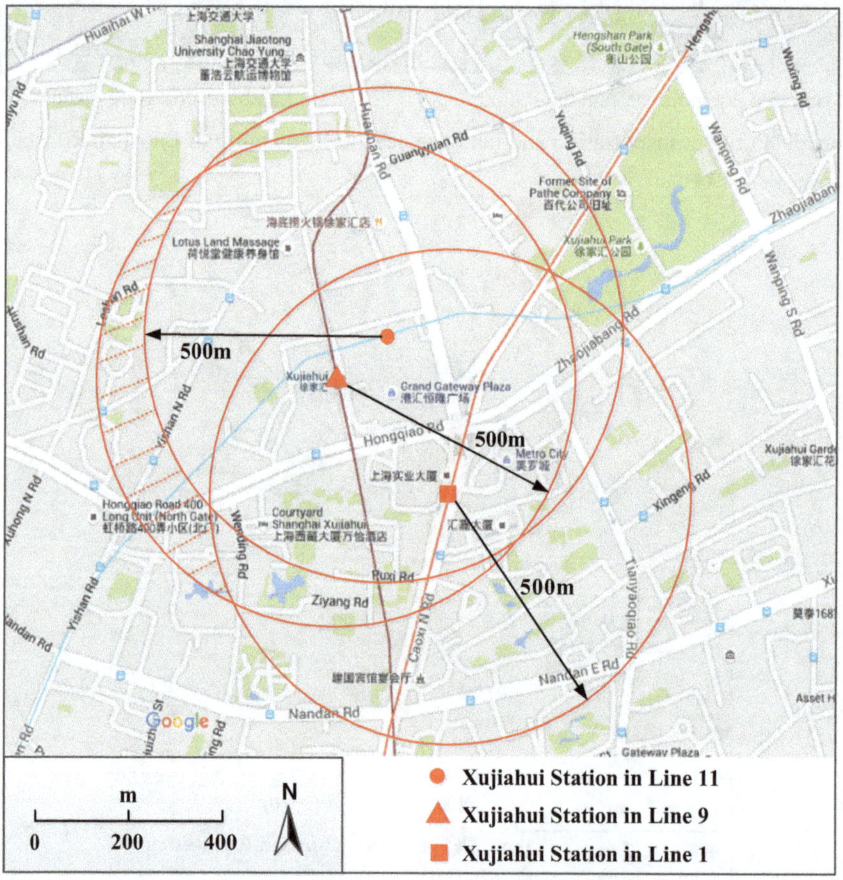

Figure 6: 500 m walking distance around Xujiahui station.

Line 1 and Line 9 were opened in 1993 and 2007, respectively. Line 11 was opened in 2009 and brought convenience for travellers from Jiading district to the city core area. The opening of Line 11 also expanded the scope of the 500 m walking distance around Xujiahui station, as the shaded part shows in Fig. 6.

Through the analysis given earlier, the traffic benefit is greater than the agglomeration benefit in suburban area, which is completely different in the city centre. Therefore, the suburban station should be considered to strengthen the traffic function and improve the shuttle transportation while the urban station should be considered to connect with the surrounding commercial building and provide convenience for travellers when planning urban rail transit stations.

5.2 Relationship between accessibility improvement and residential property appreciation

The accessibility improvement average value is significantly associated with the residential property average appreciation (see Fig. 7).

The fitting function coefficient is 4.77, close to 5, suggesting that the residential property appreciation is about 5 times the accessibility improvement value per year. The goodness-of-fit of the model is greater than 0.93. In theory, the fitting line should be through the origin of coordinates. In fact, the intercept is 176.63 in this case study because of bias. But compared to the appreciations level, the intercept is close to zero.

6 CONCLUSIONS

The introduction of urban rail transit benefits to the accessibility and land development of the city. One of the most significant benefits is the impact on property values. A new approach which researches the impacts of urban rail transit on residential property values has been proposed here. The impacts of urban rail transit on residential property values have been classified into traffic effects and agglomeration effects, which are measured by the travel accessibility improvement and the attract accessibility improvement, respectively. A case on 16 stations of Shanghai Metro Line 11 has been studied to verify the accessibility improvement model, whose goodness-of-fit is greater than 93%. Both of the residential property average appreciation and the accessibility improvement average value show the trend that they are

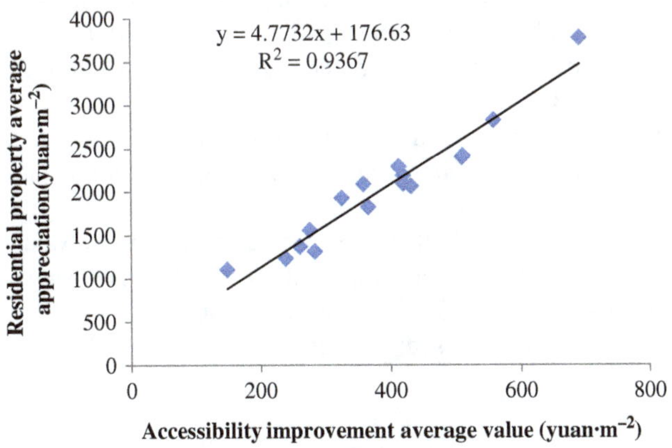

Figure 7: Relationship between the accessibility improvement average value and the residential property average appreciation per square metre.

larger in the suburb but smaller in the city centre. Traffic benefit is greater than agglomeration benefit in the suburb, which is completely different in the city centre. Furthermore, the residential property appreciation is about 5 times the accessibility improvement value per year.

It should be noted that the research findings are subject to the specific selected data samples and only reflect the case of Shanghai Metro in the given period of examination. Future studies will consider a more detailed and generalized analysis of such phenomenon for other cities in developing countries like China, as property value impacts of urban rail transit consist of the basis for value capture, providing finance in the development of new rail lines.

REFERENCES

[1] Medda, F., Land value capture finance for transport accessibility: a review. *Journal of Transport Geography*, **25**, pp. 154–161, 2012. DOI: 10.1016/j.jtrangeo.2012.07.013

[2] Cervero, R. & Landis, J., Assessing the impacts of urban rail transit on local real estate markets. *Transportation Research A*, **27**, pp. 13–22, 1993. DOI: 10.1016/0965-8564(93)90013-B.

[3] Du, H. & Mulley, C., The short-term land value impacts of urban rail transit: quantitative evidence from Sunderland, UK. *Land Use Policy*, **24(1)**, pp. 223–233, 2007. DOI: 10.1016/j.landusepol.2005.12.003.

[4] Pagliara, F. & Papa, E., Urban rail systems investments: an analysis of the impacts on property values and residents' location. *Journal of Transport Geography*, **19(2)**, pp. 200–211, 2011. DOI: 10.1016/j.jtrangeo.2010.02.006.

[5] Lancaster, K.J., A new approach to consumer theory. *Journal of Political Economy*, **74(2)**, pp. 132–157, 1966. DOI: 10.1086/259131.

[6] Rosen, S., Hedonic prices and implicit markets: product differentiation in pure competition. *Journal of Political Economy*, **82(1)**, pp. 34–55, 1974. DOI: 10.1086/260169.

[7] Bowes, D.R. & Keith, R.I., Identifying the Impacts of Rail Transit Stations on Residential Property Values. *Journal of Urban Economics*, **50(1)**, pp. 1–25, 2001. DOI: 10.1006/juec.2001.2214.

[8] Seo, K., Golub, A. & Kuby, M., Combined impacts of highways and light rail transit on residential property values: a spatial hedonic price model for Phoenix, Arizona. *Journal of Transport Geography*, **41(0)**, pp. 53–62, 2014. DOI: 10.1016/j.jtrangeo.2014.08.003.

[9] Wu, Q., Ye, X. & Lin, X., Attraction area model for urban rail transit stations. *Journal of Tongji University (Natural Science)*, **42(7)**, pp. 1058–1063, 2014.

[10] Wu, Q. & Ye, X., Analysis of relationships between urban rail transit accessibility improvement and residential property appreciations. *Journal of Tongji University (Natural Science)*, **44(6)**, pp. 899–906, 2016.

[11] Dube, J., Theriault, M. & Des, R.F., *Commuter rail accessibility and house values: The case of the Montreal South Shore*, Canada, 1992–2009. *Transportation Research Part A–Policy and Practice*, **54**, pp. 49–66, 2013.

[12] Kay, A.I., Noland, R.B. & DiPetrillo, S., Residential property valuations near transit stations with transit-oriented development. *Journal of Transport Geography*, **39**, pp. 131–140, 2014. DOI: 10.1016/j.jtrangeo.2014.06.017.

[13] Image, www.baidu.com. Accessed: May 26, 2016.

OPTIMIZATION SYSTEM OF RESERVED/NON-RESERVED SEATING PLANS FOR IMPROVING CONVENIENCE AND REVENUE OF INTERCITY EXPRESS TRAINS

S. NAKAGAWA, M. SHIBATA, & N. FUKASAWA
Signalling and Transport Information Technology Division, Railway Technical Research Institute, Japan.

ABSTRACT

In Japan, seating plans of reserved/non-reserved seats in intercity express trains do not necessarily correspond to passenger demands which may vary depending on many factors such as calendar days and operating sections. This mismatch possibly causes passengers' inconvenience due to passenger congestions on board, and railway operators can lose revenue. Optimizing seating plans is expected to be an effective solution for these problems, where predicting both potential passenger demands and reflections to the potential demand caused by the passenger congestions is necessary.

For establishing the prediction methods, we conducted some demand analyses and some questionnaire surveys. According to these results, we establish demand estimation models of reserved/non-reserved seats which can consider potential demands and behaviour changes if the passengers face congestion. We developed a prototype of seating plans optimization system applying these models, simulation of passengers' behaviour and genetic algorithm. This paper describes these models, the optimization system and the result of applying this system to some existing intercity express trains as a case study. We got the optimized seating plan which is expected to improve both of the passengers' convenience and the railway operators' revenue.

Keywords: congestion, genetic algorithm, intercity trains, potential demands estimation, revenue management, seat reservation, second best choice behaviour.

1 INTRODUCTION

In Japan, seating plans of reserved/non-reserved seats in intercity express trains do not necessarily correspond to passenger demands, because the demand is more flexible than the seating plans (Minami et al. [1]). This problem possibly causes passengers' inconvenience, and railway operators can lose revenue. We suppose optimizing seating plans is expected to be an effective solution, as described in Shibata et al. [2, 3].

This paper describes the seating plans optimization system we have developed and the models involved in the system, which estimate potential demands and passengers' behaviour. By conducting a case study, we can show that the optimized seating plan improves both the passengers' convenience and the railway operators' revenue.

DOI: 10.2495/TDI-V1-N3-359-370

1.1 Abbreviations and terminologies

In this article, we use the following abbreviations:

RS(s) – reserved seat(s).
NRS(s) – non-reserved seat(s).
ICT(s) – intercity express train(s).
LF – load factor.

We also use the terminology 'OD'. This is the combination of 'origin' and 'destination' stations of passengers' travel.

2 ESTABLISHING POTENTIAL PASSENGER DEMANDS ESTIMATION MODEL

2.1 Existing revealed demand data

To estimate the potential demand, we have to know the revealed demands of the target ICTs. We can get three kinds of the revealed demand data as described in Table 1. We use these data to calibrate the potential passenger demands estimation model.

(1) Sales results accumulated in seat reservation system

All RSs are sold through the seat reservation system. We can get sales results of RSs for all days and trains from the system. By using these data, we can observe all passengers' OD. We have to consider that the sales results do not necessarily correspond to the revealed demands; the tickets may not necessarily be used in accordance with their validity. We also have to consider that we cannot get the sales results of NRSs from these data.

(2) OD survey for ICTs

For some ICTs, OD survey has been conducted to gain the OD data. Usually this survey is carried out in some weekdays per year, by collecting used tickets.

From this survey, we can observe OD data of not only RSs passengers but also NRSs passengers with relatively higher accuracy. On the other hand, we cannot grasp short-term change (e.g. seasonal change) of the demand from this survey, because this survey is annual.

(3) Visual count of ridership

The train conductors of ICTs have to report the number of ridership in predetermined target sections. They walk through the train and generally count the passengers in each seat class. If the target section consists of some station-to-station sections and some passengers get on board or drop off within the target section, the record should be the maximum number of ridership through the target section. The most important feature of these data is that they are recorded through all days, all trains and all of the seat classes. By analysing them, therefore, we can recognize the changes of the demands depending on the season, the day of the week, etc. On the other hand, we have to consider some defects of them – their accuracy is relatively low compared to other data and they are not OD data.

2.2 An online survey

To establish the potential passenger demands estimation model for NRS, we conducted an online survey. In this survey, we asked how passengers behave when they want to board NRS

Table 1: Existing real demand data.

	(1) Sales results recorded in seat reservation system	(2) OD survey for ICTs	(3) Visual count of ridership
Method to get the data	From the seat reservation system	Collecting used tickets	Visual count by the train conductors
Type of the data	OD	OD	Ridership on target sections
Date	All days	Target date of the survey	All days
Seat type	RS and first-class seats	All types	All types
Features	OD data of RS are available	OD data of NRS are available	Demand data on all days and of all seat types are available
Demerits	Do not always correspond with revealed demands	Cannot grasp short-term change of the demand from this data	Less accuracy than other data

and face passenger congestion. The outline of the survey is described in Table 2. We also asked how passengers choose their seat class to board and the impact of the price difference between RS and NRS (Nakagawa et al. [4]).

The results are shown in Table 3. Most of the passengers want to get a seat in NRS, although there are only few seats available in NRS. On the other hand, if there is no seat available (i.e. they have to be standing in deck or aisle), 35% of them avoid boarding the NRS.

Table 2: Outline of the online survey.

Target of the survey	Persons who: – belong to the age group between the age of a high school student and a 69 years old – live in Japan (excluding Okinawa Pref.) – have experiences to get on the ICT (not including Shinkansen super express) operated by JR (Japan Railways) in last one year
Period of the survey	From 30 January to 13 February 2014
Number of answerer	Target: 8,130 Valid response: 6,036 (response rate: 74.2%)

Table 3: Answers for 'How do you behave when you want to board NRS and face passenger congestion?.'

	Do not mind	Mind a little but not feeling uncomfortable	Board the ICT although feeling uncomfortable	Feeling uncomfortable and do not board the ICT
(1)	3,758	1,616	562	100
(2)	3,357	1,854	628	197
(3)	3,134	1,989	743	170

(continued)

Table 3: continued

	Do not mind	Mind a little but not feeling uncomfortable	Board the ICT although feeling uncomfortable	Feeling uncomfortable and do not board the ICT
(4)	832	714	2,388	2,102
(5)	501	350	1,989	3,196

(1) Both the window seat and the aisle seats are available (for all empty seats, and adjacent, front and rear seats are occupied)

(2) There are some window seats available but the aisle seats are fully occupied (ditto)

(3) There are some aisle seats available but the window seats are fully occupied (ditto)

(4) Cannot be seated due to full occupancy (but no one stands in the deck)

(5) Cannot be seated due to full occupancy and the deck is crowded

2.3 THE POTENTIAL PASSENGER DEMANDS ESTIMATION MODEL

We established the potential passenger demands estimation model from the data described above. The outline of the model is shown in Fig. 1. This model estimates the potential OD data of some ICTs on a particular date from the revealed demand data of the ICTs. Because there are few revealed OD data for NRS, we tried to estimate the potential OD data indirectly; i.e. this model estimates the ridership on some sections and converts them into OD data.

(1) Calculate the average of the counts of ridership reported on the same date over the past years, where we consider the structure of calendar date. This is treated as the revealed ridership on the section on the target date and train.
(2) From the estimated ridership calculated in (1), we estimate the potential ridership, by using the equation describing the relationship between the potential demands and the revealed demands.
(3) Convert the potential ridership calculated in (2) into the potential OD data of the target date and train, by applying the result of OD survey for target ICT as a OD pattern data.

The detail of each method is as follows.

2.3.1 Relationship between the potential and revealed demands (for NRS)

We tried to formulate the relationship between the revealed ridership (actual demands) and the number of the passengers who avoid to board the ICTs due to passenger congestion (lost

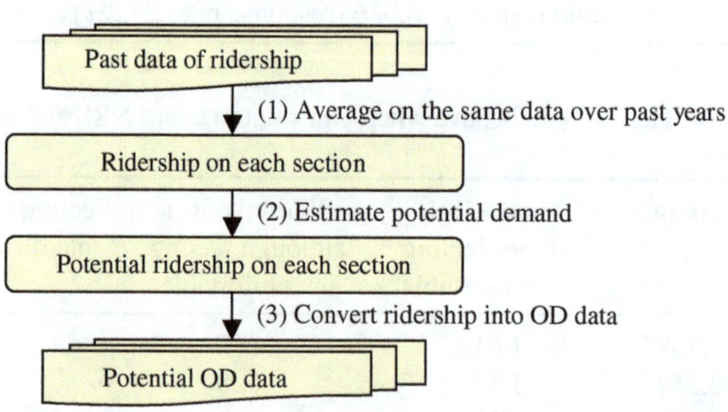

Figure 1: Outline of the potential passenger demands estimation model.

demands), by utilizing the result of the online survey described in Section 2.2. First, we assume that the LF is 75% if there are some vacant seats – either window seat or aisle seat. This is because usually half of the seats in the ICTs are window seats and the rest of the half is aisle seats. Then, we assume that when the LF is between 75% and 100%, behaviour of ones who intend to get on is average of the second and third rows in Table 3. Under this assumption, the ratio between passengers becoming the revealed demands (i.e. who take the ICT) and passengers becoming the lost demands (i.e. who abort taking the ICT) is calculated as below:

$$(3357+1854+628+3134+1989+743) : (197+170) = 11705 : 367$$

Hence, when the LF is between 75% and 100%, the lost demands are estimated to be 367/11705 of the LF minus 75%.

In the same way, when the LF is over 100%, the ratio is as below:

$$(832+714+2388) : 2102 = 3934 : 2102$$

Therefore, when the LF is over 100%, the lost demands are estimated to be the summation of 2102/3934 of the LF minus 100% and the lost demands are generated if the LF is between 75% and 100%.

By the estimation described above, we can formulate the relationship between the potential LF and revealed LF of NRS. The result is shown in Fig. 2. Note that we consider that the lost demands are not generated when the LF is less than 75%.

2.3.2 Relationship between the potential and revealed demands (for RS)

To estimate the potential demands of RS is more difficult than that of NRS, for the reasons described below.

(1) The LF of RS, the number of seats reserved divided by the number of RS, never exceeds 100%. For this reason, we cannot conclude that 'the ICT whose revealed LF is high has high potential demands' with absolute certainty.
(2) Sometimes the seat reservation request is rejected though the LF is less than 100%. This is because the request is accepted only when there is a seat available through the requested OD. Therefore, we cannot understand whether the rejection of the seat reservation request occurred or not from the LF.

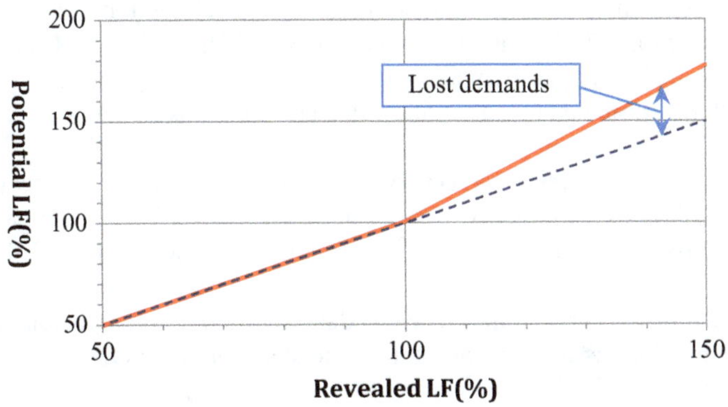

Figure 2: The relationship between the potential and revealed LF of NRS.

(3) Most of the rejections for seat reservation requests are not recorded in the seat reservation system due to its specifications. In addition, the records of the rejections do not necessarily represent the passengers' initial choice. For these reasons, we cannot observe how many rejections of the seat reservation request have truly occurred from the seat reservation system.

We thought about a method to estimate the potential demands of RS from the revealed demand data considering these problems. First, while there is the problem (1) shown above, we decided to formulate the relationship between the potential and revealed LF of RS, which is similar to that of NRS. It is because we have no other data from which we can estimate the volume of potential demands. Next, we assumed that the rejection for the seat reservation occurred only when the LF is over 85%, as a result of an interview survey for a railway operator. Furthermore, we analysed some sales results of ICT recorded in the seat reservation system and noticed that most of the RSs are reserved on the day of operation and the rate of reservation on the previous days of operation rarely exceeds 85%. From these situations and knowledge, we supposed some assumptions described below can be applied.

(a) When the revealed LF of RS is under 85%, no rejection for reservation request has occurred. When the LF is over 85%, some rejection has occurred at a constant rate for the excess of 85%. In other words, the equations described below are satisfied, where y is the potential LF and x is the revealed LF:

$$y = x \quad (\text{if } x <= 0.85)$$
$$y = x + p(x - 0.85) \quad (\text{if } x > 0.85;\ p \text{ is a coefficient})$$

(b) The rejection for reservation occurs only on the day of operation.
(c) There is linear relationship between the rate of reservation on the previous day of operation and the potential LF.

With these assumptions, we conducted a linear regression analysis between the rate of reservation on the previous day of the operation and the revealed LF (from assumption (a), this is equal to the potential LF) for every ICT whose LF is under 85%. Then we calculated the coefficient p to fit the regression line with the ICTs whose LF is over 85%. Using the obtained value of p, relational expression between the potential LF and revealed LF has been calculated.

2.3.3 A method to get the OD data from the ridership on each section

As described in Section 2.1, there are few actual measured OD data for NRS. For this reason, we established a method to get the potential OD data by adjusting the past real OD data to correspond with the estimated potential ridership. The outline is as below.

(1) Consider the previous result of OD survey (or the result of an extra survey, if the target date is a holiday) of the target ICT as the basic OD data. We note that we describe the number of passengers who take the ICT from station i to station j in this basic OD data as t_{ij}.
(2) For every station appeared in the basic OD data, we define the coefficient a_i. We assume that the number of passengers from station i to station j in the OD data is calculated by $t_{ij} * a_i$ (as shown in Fig. 3).

Basic OD data		Get off at					
		Sta. 1	Sta. 2	Sta. 3	Sta. 4	Sta. 5	Sta. 6
Get on at	Sta. 1		10	20	50	20	30
	Sta. 2			10	40	10	20
	Sta. 3				30	10	20
	Sta. 4					30	20
	Sta. 5						10
	Sta. 6						

After multiplying the coefficients		Get off at					
		Sta. 1	Sta. 2	Sta. 3	Sta. 4	Sta. 5	Sta. 6
Get on at	Sta. 1		$10a_1$	$20a_1$	$50a_1$	$20a_1$	$30a_1$
	Sta. 2			$10a_2$	$40a_2$	$10a_2$	$20a_2$
	Sta. 3				$30a_3$	$10a_3$	$20a_3$
	Sta. 4					$30a_4$	$20a_4$
	Sta. 5						$10a_5$
	Sta. 6						

Figure 3: An example of applying the converting method.

(3) Convert the OD data assumed in (2) into the ridership on every section. Then calculate the coefficient a_i which minimizes the square error between the converted ridership and the potential ridership. The potential OD data are calculated by using this a_i. For example, when the OD data assumed in (2) are as given in the table in Fig. 3 and the potential ridership in the section between station 3 and station 4 is 200, this method calculates a_i to fit $100a_1 + 70a_2 + 60a_3$ to 200.

In other words, this method stands on the assumption that the ratio of the passengers who get off at each station to who get on a certain station is constant. Under this assumption, this method calculates the potential OD data by adjusting the basic OD data to suit the estimated potential ridership which is the input of the method.

3 ESTABLISHING SECOND BEST CHOICE BEHAVIOUR MODEL

To utilize the potential demands to determine how the transport service is, we have to consider how the customers think when they faced congestion. In other words, we have to understand how the potential demands change to revealed demands. To measure this, we carried out a questionnaire survey for the passengers of some ICTs. In this survey, we observed their choice preference if they could not take a seat/they could not get a reservation. For RS passengers, the question was 'How did you behave if you couldn't buy a ticket for RS due to full occupancy?' For NRS passengers, the question was 'How did you behave if you faced passenger congestion of NRS when you got on board?' The outline of the survey is described in Table 4.

The result is shown in Table 5. As for RS passengers, about half of them would try to reserve RS of other trains. About 40% would decide to board on NRS. Only 4% answered that they would avoid boarding the ICT.

Though they had to stand in the deck or the aisle, 63% of the NRS passengers would choose to board the NRS. In other words, 37% of them would avoid boarding the NRS. This corresponds well to the result of the other online survey described in Section 2.2. 25% of the passengers would wait for the next train: 8% would try to reserve RS, and 17% would wait in line for NRS.

Table 4: Outline of the onboard survey.

Target of the survey	Persons who: – board the target ICT on the sections where the visual count of ridership is recorded – seat in RS or NRS – are not young children
Period of the survey	17 and 18 January 2015
Target ICT	10 trains each day
Number of answerer	Distributed: 4,194 Valid response: 3905 (response rate: 93.1%)

Table 5: The passengers' choice behaviour if they could not take the seat they were seating.

	Passengers of RS	Passengers of NRS
Reserve RS of earlier service	261	–
Reserve RS of latter service	291	209
Board NRS of earlier service	6	–
Board NRS of this train	447	1,748
Board NRS of latter service	15	470
Abort taking ICT	41	93
Others/invalid answers	83	241

In addition, we conducted chi-squared tests for these results to check whether the attributions of the passengers and trains significantly affect the choice preference or not. The results of these tests imply that the choice preference is significantly different by some factors; for RS passengers, their age and number of the accompanying persons; and for NRS passengers, their age and OD.

From these tests, we decided to establish various second best choice behaviour models depending on these factors. As an example, in Fig. 4, we show the model for the NRSs passengers who are in their thirties travelling long-distance trip. NRS passengers have four alternatives as a second best choice as shown in Fig. 4. RS passengers have six alternatives, including 'switch to the RS of earlier service' and 'switch to the NRS of earlier service'.

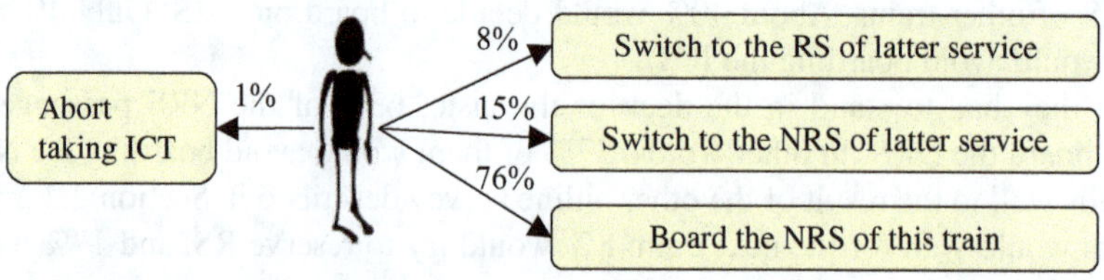

Figure 4: An example of second best choice behaviour model.

4 THE DEVELOPED SYSTEM AND THE OPTIMIZATION ALGORITHM

We developed the seating plans optimization system by using the models described above. The outline of the algorithm is shown in Fig. 5. This system is based on the genetic algorithm, whose solutions are combinations of seating plans of each target ICT and sections on the target date. In this system, to evaluate the solution, we have to consider the relationship between the potential demands and the solution, especially the effects on the demands caused by a certain seating plan. Therefore, this system involves the simulation of selling process of RS before departure and seating behaviour of NRS when a certain ICT arrives, which is called 'S&S simulation' hereafter.

The calculation flow of the system is described as below.

(1) Estimate the potential OD of the target date and trains from the real demand data in the past, applying the potential passenger demands estimation models described in Section 2.
(2) Generate various seating plans as initial solutions of the genetic algorithm.
(3) For each of the solutions generated in (2), conduct S&S simulation using the potential OD estimated in (1). This simulation represents the selling process of RS from the launch day. The outline is as follows.
(3–1) We assume that the number of passengers who wanted to reserve RS every day follows the sale curve estimated by the revealed demand data. Based on this assumption, the system implements the simulation, where RSs are reserved gradually on a first-come-first-served basis. When a passenger cannot reserve the RS due to full occupancy in this simulation, he/she changes his/her choice in accordance with the second best choice behaviour model. We note that if he/she chooses to reserve another train's RS but again faces full occupancy, he/she again changes his/her choice.
(3–2) When the train's departure time comes in the simulation, the passengers of NRS, including those who decide to take NRS in (3–1), begin to try to take a seat. We assume that they make a queue on the platform and take a seat in random order if there is a NRS available. When the passenger cannot take a seat in the NRS due to full occupancy, he/she changes his/her choice in accordance with the second best choice behaviour model.

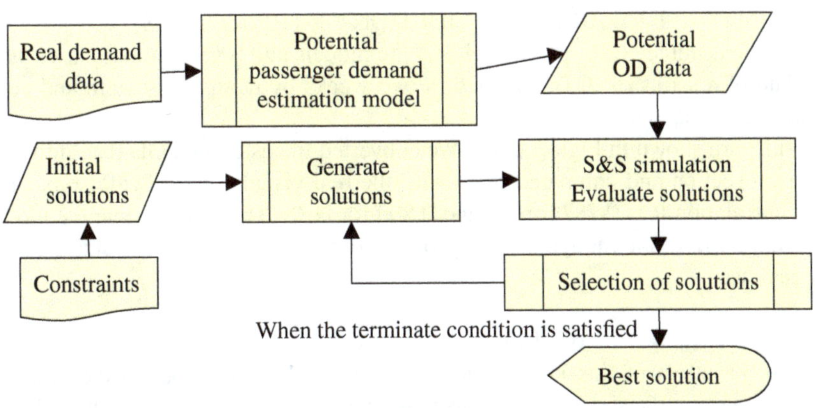

Figure 5: The outline of the seating plan optimization algorithm.

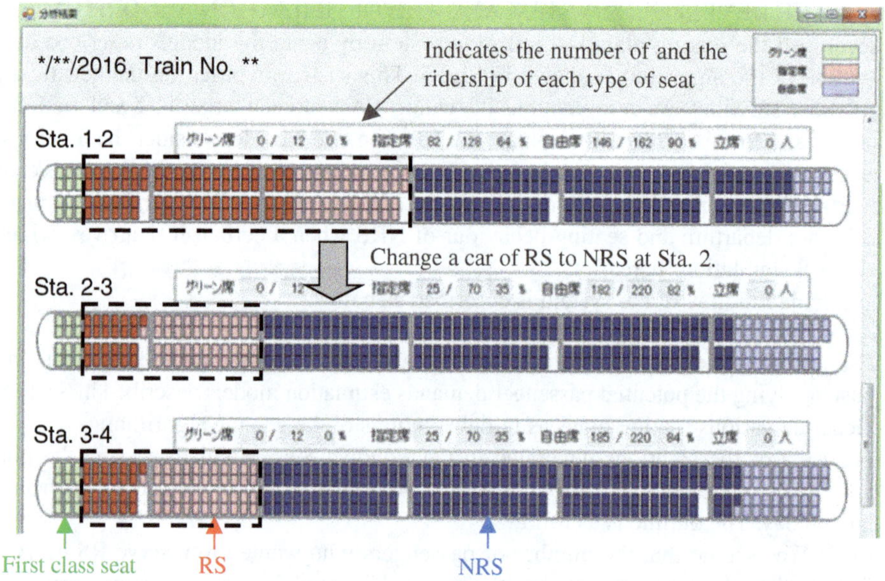

Figure 6: An example of seat map representing optimized seating plan.

(4) Evaluate the solution considering the OD data which are the result of S&S simulation derived using procedure (3) and calculate the evaluated indices of each solution as an objective function. Based on the genetic algorithm, the system generates and selects the solutions, using a certain index which is set in advance. We can choose a certain objective function among four types of evaluation indices: the number of passengers per kilometre who take ICT but cannot seat (minimize), the difference between the LF of RS and NRS (minimize), the LF (maximize), and the total revenue (maximize).
(5) When the condition for termination defined in advance (e.g. the limit of calculation time) is satisfied, this system terminates the continuous calculations and outputs the best solution.

The solution is displayed in the form of seat map, as shown in Fig. 6.

5 RESULT OF VALIDATION OF THE MODELS

To validate the accuracy of re-productivity of the models and S&S simulation, we estimated the OD data of a series of ICTs operated in two weeks by using the system and compared them with real demand data.

The results are shown in Figs 7 and 8. We converted the estimated OD data into the ridership on each section and compared them with the real visual count of ridership. The coefficient of correlation R is 0.787 for RS and 0.800 for NRS. These imply that the models and S&S simulation have an enough accuracy of re-productivity and these seem to be suitable for practical use.

6 RESULT OF A CASE STUDY

To evaluate the effect of this kind of flexible seating plans, we applied the seating plans optimization system to a series of ICTs operated in a certain line, as a case study. We treated two types of evaluation indices, as described in Table 6, and conducted the calculation for

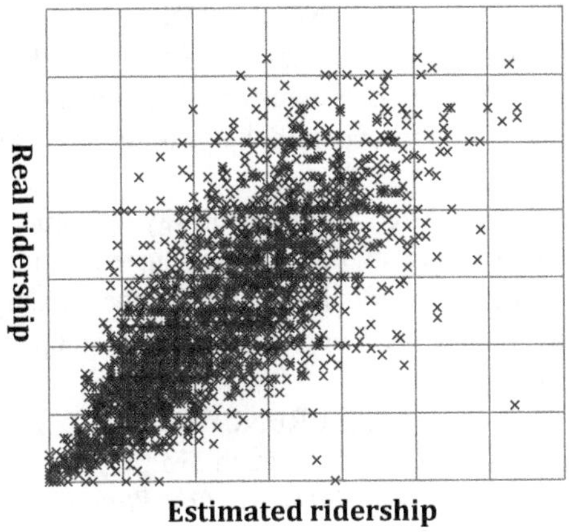

Figure 7: Validation result of accuracy (for RS).

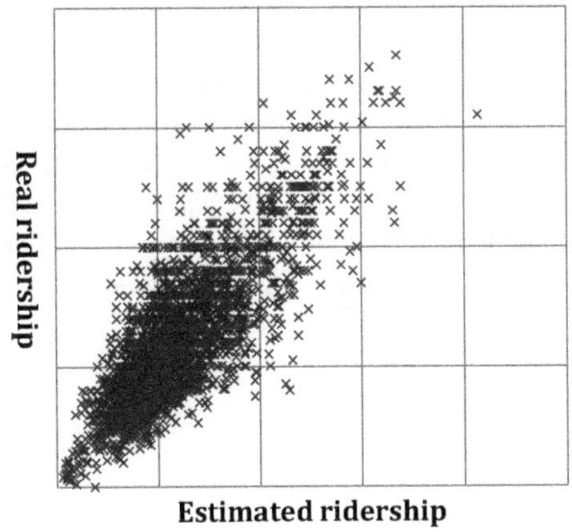

Figure 8: Validation result of accuracy (for NRS).

two days (a certain weekday and a holiday.) We also calculated the evaluation indices of the current seating plan and compared them with the indices of optimized seating plans. The results are shown in Table 6. These suggest that the optimization result possibly decreases the passengers per –kilometre (P-K) who take an ICT without taking a seat and increases the revenue.

Table 6: The effect of optimizing seating plans.

	Minimize P-K*		Maximize total revenue	
Date	P-K	Total revenue	P-K	Total revenue
A weekday	–74.1%**	+0.1%	–40.0%	+1.3%
A holiday	–60.1%	+0.1%	–9.2%	+1.1%

* P-K means 'passengers per kilometre who take an ICT but cannot take a seat'.

** Percentage in this table shows the results of comparison with the evaluation indices of current seating plan.

7 CONCLUSION

We established the potential passenger demands estimation model and the second best choice behaviour model and developed a seating plans optimization system for ICTs. This system can generate solutions which can improve both the passengers' convenience and the railway operators' revenue.

This system can be effectively utilized when the railway operators consider their ICTs' seating plan. While there are some further studies to be conducted about the train operation practically, e.g. how to show whether the seat is RS or NRS to the passengers, we aim to carry out this effective flexible seating plan for ICT operations with the cooperation of ICT operators.

REFERENCES

[1] Minami, K., Terabe, S., Ieda, H. & Mizuguchi, M., Optimization of seat allocation on inter-city transport. *Proceedings of Infrastructure Planning*, Vol. 27, CD-ROM, 2003. (in Japanese)

[2] Shibata, M., Terabe, S. & Uchiyama, H., An effectiveness of flexible seat class assignment on intercity express trains. *Transport Policy Studies' Review*, **13(1)**, pp. 2–13, 2010. (in Japanese)

[3] Shibata, M., Terabe, S.& Uchiyama, H., Optimization system for train set planning of inter-city rapid trains with the application of the flexible seat class assignment approach. *Proceedings of the 9th World Conference on Railway Research*, Paris, 2011.

[4] Nakagawa, S., Shibata, M., Ozaki, N., Fukasawa, N. & Suzuki, T., A study on demand of limited express trains to optimize allocation of reserved/non-reserved seats. *Journal of Railway Engineering, Japan Society of Civil Engineers*, **18**, pp. 179–186, 2014. (in Japanese)

A SOFTWARE ARCHITECTURE FOR AUTONOMOUS MAINTENANCE SCHEDULING: SCENARIOS FOR UK AND EUROPEAN RAIL

CHRIS TURNER[1], PRITHYUKSHAA THOPPUR RAVI[1], ASHUTOSH TIWARI[1], ANDREW STARR[1], & KEVIN BLACKTOP[2]
[1]Manufacturing and Materials Department, Cranfield University, UK.
[2]Network Rail, The Quadrant, UK.

ABSTRACT
A new era of automation in rail has begun offering developments in the operation and maintenance of industry standard systems. This article documents the development of an architecture and range of scenarios for an autonomous system for rail maintenance planning and scheduling. The Unified Modelling Language (UML) has been utilized to visualize and validate the design of the prototype. A model for information exchange between prototype components and related maintenance planning systems is proposed in this article. Putting forward an architecture and set of usage mode scenarios for the proposed system, this article outlines and validates a viable platform for autonomous planning and scheduling in rail.
Keywords: decision support systems, rail planning and scheduling, software architecture.

1 INTRODUCTION

According to the Office of Rail Regulation (ORR) [1], for the period 2013–14 total rail industry expenditure in the UK was £12.7 bn, of which £6.2 bn (49%) was incurred in operating rail infrastructure alone. A significant percentage of these costs were incurred through maintenance of railway infrastructure. Railway infrastructure state may be affected by a range of factors such as ‘the track geometry, topography, geology and weather conditions’ [2]. Maintenance plans must be devised and the risk of failures forecasted in order to maintain safe operation with high levels of service availability.

With a significant increase in railway traffic in the UK expected over the coming years the scheduling and distribution of maintenance (and possession) time is increasingly challenging. Hence ‘An integrated maintenance software framework’ is essential to streamline future maintenance activities based on factors of cost, urgency and the flagging of issues sensed by intelligent assets and infrastructure. These factors, especially the rise in the use of sensors, are driving a need for the automated planning of maintenance tasks within rail. With this need comes the potential for autonomous scheduling.

1.1 The AUTONOM project

The examination of autonomous planning and scheduling for rail maintenance is a core component of the AUTONOM project (integrated through life support for high-value systems) [3] sponsored by the Engineering and Physical Sciences Research Council (EPSRC), a UK government–backed major research funder and UK rail infrastructure provider Network Rail

DOI: 10.2495/TDI-V1-N3-371-381

(Network Rail is the organization that is responsible for maintaining and developing the UK rail infrastructure including signalling, bridges, tunnels, level crossings, viaducts and 17 key stations within the country) [4]. Encompassing areas such as sensor fusion (collection and analysis of data streams from rail vehicles and infrastructure) and cost analysis (costs and benefits in maintenance decisions) the development of a detailed architecture, encompassing scenarios of operation, was necessary. The AUTONOM system is comprised of four modules:

- WP1: Integration – integrates system functionality and acts as a user 'dashboard'
- WP2: Sensor Fusion – gathers data from data stores and live feeds, provides an analysis of asset degradation trends and alerts to flag areas in need of urgent maintenance
- WP3: Planning and Scheduling – optimally schedules maintenance jobs based on importance and cost
- WP4: Costing – determines the cost and benefit of different maintenance jobs

While other notation sets such as Business Process Modelling Notation (BPMN) and Flowcharts [5] have been used in enterprise level projects, the Unified Modelling Language (UML) is still the accepted modelling standard. Examples of UML use in rail are in evidence through works such as Berkenkotter *et al.* [6, 7]. The modelling standard provides a set of notations designed to support various domain specialisms and stages involved in the engineering of software systems.

2 RELEVANT RESEARCH

Planning in railway operations is a complex task, due to the large solution space this work is normally partitioned into several problems that are solved sequentially [8]. Klabes [9] presents a novel framework detailing the different planning processes involved:

- Network planning: determines the detailed layout of the railway infrastructure
- Line planning: determines the lines and the frequency of train operation on them
- Timetabling: governs the arrival and departure times of trains at train stations
- Capacity allocation: including insertion of train path requests into the working train timetable
- Vehicle planning: rolling stock assignment
- Crew planning: crew rostering and staff planning
- Re-scheduling: controlling the movements of trains during operation

To date, many studies have examined the problem of maintenance scheduling for railway systems. The maintenance concept can be categorized into three groups: corrective maintenance; periodic maintenance; predictive maintenance [10]. Though works on autonomous operation and scheduling in rail are limited, one particular study by Dadashi et al. [11] suggests an intelligent infrastructure to move from find and fix to predict and prevent. Schlake *et al.* [12] have conducted research into the trackside monitoring of rail vehicles, an essential development to enable predictive maintenance. Their work [12] examined the economic impact of train delays and the effect of introducing lean production methods in passenger and freight railway operations to improve rail vehicle maintenance and monitoring procedures.

Automated planners have been investigated by authors such as Cresswell *et al.* [13] and Fernandez *et al.* [14]. Such planners require action models described using languages such as the Planning Domain Definition Language (PDDL) [15].

Apart from such standardized data formats, a semantic model with a high level of 'structured interoperability' between information systems is required to correctly combine and manage complex scheduling information. Verstichel *et al.* [16] describe both UML and OWL as methods of semantically describing models. Marcano *et al.* [17] while stating that UML offers a standard systems modelling notation also acknowledge that it lacks formal semantics to model safety critical scenarios.

In summary, the literature suggests that interactions and message passing are modelled either using sequence diagrams or interaction diagrams, the former being more prevalent. As far as message exchange is concerned, numerous research articles point towards an XML-based schema due to its simplicity and dual advantage of being both machine and human readable.

3 METHODOLOGY

The methodology followed for this research is detailed in Fig. 1. As shown in Fig. 1 the background study phase consisted of two sub-tasks, namely to study relevant literature for modelling in the context of railways and to analyse the input and outputs between the systems/modules within the maintenance framework. The literature review performed helped to identify the UML approaches present. The model and document phases dealt with modelling the interactions and message passing between the aforementioned modules. UML was identified from literature to be the most suitable visual modelling language for the case study. It was decided at this stage that the standard notation set of UML should be used in this research due to its familiarity and acceptability to both software developers and the rail industry (with the proviso that such models could be transformed to newer more rail specific standards, as they gain in popularity and use, in the future). The validation phase involved identifying operational scenarios from domain experts, pertaining to this modelling environment.

4 SCENARIOS FOR AUTONOMOUS PLANNING AND SCHEDULING

As previously mentioned in this article the use of UML as a notation for the description of software including complex railway information systems is quite established, it is by far the most widely accepted modelling language in the software engineering community. In the development of the scenario set supporting the AUTONOM project the process of modelling is divided into three phases, namely scenario specification phase, architecture engineering phase and interaction design phase.

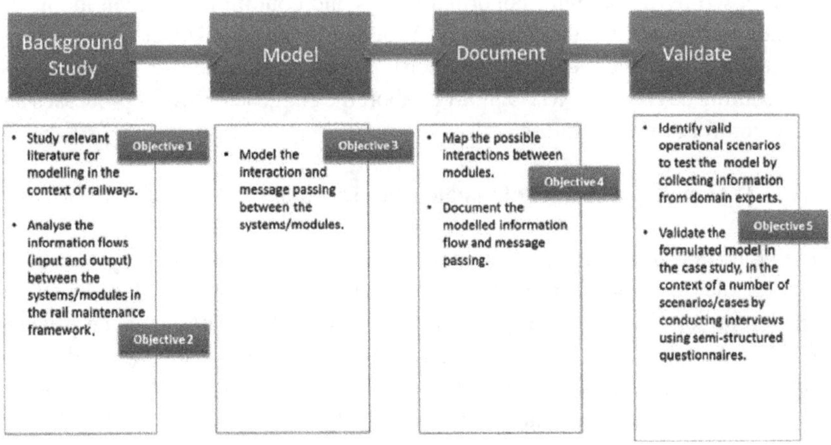

Figure 1: Methodology employed.

4.1 Scenario descriptions

In order to completely specify and model the requirements of the AUTONOM system, it is essential to understand the scenarios that it can work under. With careful analysis and discussions with the stakeholders, the following scenarios were identified:

a. amanual production of maintenance schedules
b. bsemi-automated production of maintenance schedules
c. cautomated production of maintenance schedules
d. dautomated/person-in-loop production of maintenance schedules

Scenario 1: Manual production of maintenance schedules
The first scenario deals with the production of schedules for further processing from existing Network Rail systems; a 'person in the loop' manually prompts AUTONOM to start running. In this scenario, the data is static in nature and comes from a database store at Network Rail (NR). AUTONOM needs to identify manual alterations and produce schedules that will be stored in the database shared with all NR's systems. The decision dashboard is generated and displayed to user. The scenario ends with user having to manually select one of the many scheduling choices from the dashboard.
Scenario 2: Semi-automated production of maintenance schedules
The semi-automated use of AUTONOM involves the generation of automatic alerts to trigger actions within the system. The alert is generated after studying the live data streams supplied by NR systems. Communication protocols at both NR and AUTONOM ends need to be devised to respond to such alerts. As with Scenario 1 the ultimate decision of schedule selection is manual.
Scenario 3: Automated production of maintenance schedules
In this scenario a user is not required as an autonomous decision engine is used to select a schedule as against manual selection in Scenario 2. In this scenario the system is fully autonomous, notifying NR systems of updated maintenance schedules when available, based on alerts raised by the sensor fusion module.
Scenario 4: Automated/person-in-loop production of maintenance schedules
Automated/person-in-loop is an extension of Scenario 3 where a user can veto autonomous decisions proposed by the system; abnormal values and conditions are then monitored and communicated to the user via the dashboard. When such situations occur, an alert is generated which requires user intervention for a decision to be made. The user can either continue with the scheduling advised by AUTONOM or abort the suggested flow of processes and take over manual control of the operations.

4.2 Scenario 1: Manual production of maintenance schedules

There are two actors in this use case, namely User and the AUTONOM system itself. The user clicks on the dashboard to update the workflow log. This workflow log is used to identify different activities that are performed as a part of the maintenance process. The AUTONOM system reads the static data, listens for alterations in state (raised by the user starting the system), generates the decision dashboard, creates schedules and parses the workflow log. In addition a user may wish to manually import data from a live Hadoop big data store.

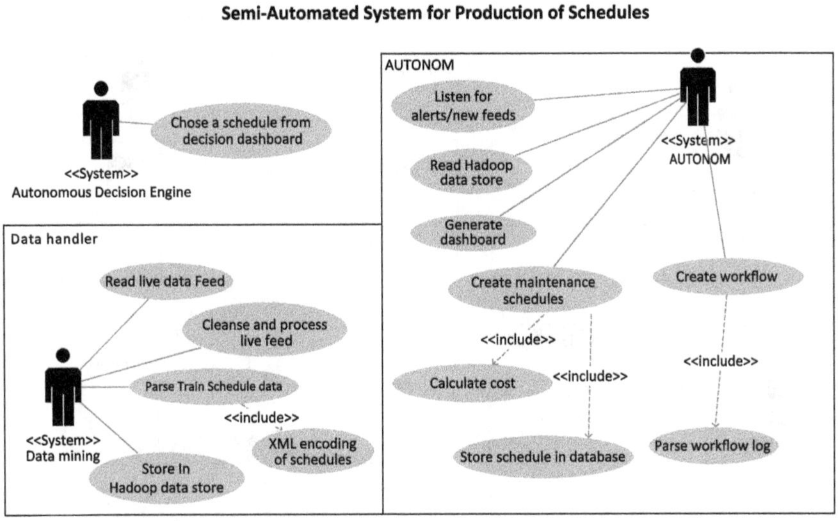

Figure 2: Use case diagram for semi-automated system for production of schedules.

4.3 Scenario 2: semi-automated

There are three actors in this use case. The additional actor is provided by the 'data handler' 'live data streams'. In this scenario the user has one task, that of selection of a schedule from the decision dashboard returned by AUTONOM. In this scenario AUTONOM listens for alterations that can be triggered by an incoming data feed as against the manual mode in Fig. 2. The processes such as 'Generate Dashboard' and 'Create Schedules' are similar to the manual mode of operation. In this scenario the Hadoop big data store is utilized. Creation of workflow including parsing is performed at the AUTONOM end. The data handler reads the live data feeds. It also creates an XML encoding of the existing schedules and stores both the schedules and cleansed data in the Hadoop data store.

4.4 Scenario 3: automated

The use case diagram (shown in Fig. 3) details the system working in fully autonomous mode. In Fig. 3 it can be seen that the user has been removed from the loop.

4.5 Scenario 4: automated /person-in-loop

In this scenario the user is placed back into the diagram for this scenario. The actors AUTONOM system, data handler and autonomous decision engine execute the same set of processes as found in Fig. 3. The autonomous decision engine apart from choosing a schedule also checks for abnormalities and notifies the user of an unexpected event. The user is simply tasked to act on such alerts by approving the course of action suggested by AUTONOM.

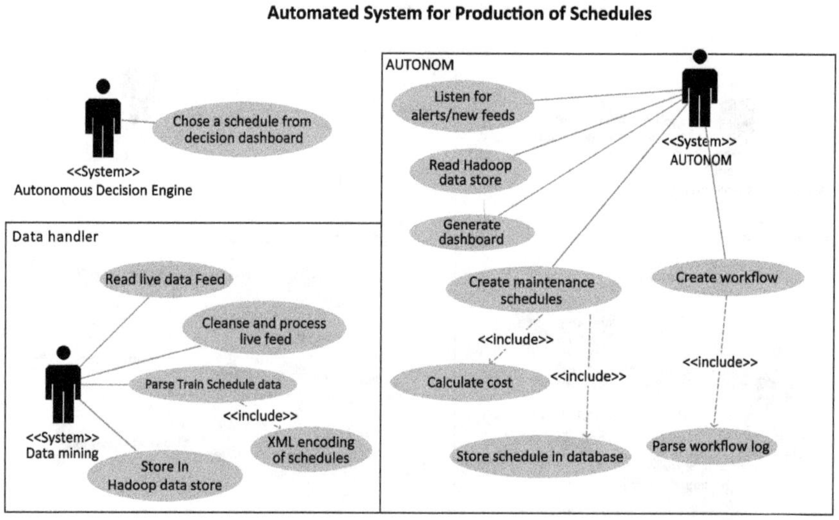

Figure 3: Use case for automated system for production of schedules.

4.6 Sequence diagrams of the use cases

The case scenarios can be modelled as a sequence diagram depicting the messages exchanged as in Appendix 1 (showing the sequence for the automated scenario). The initial objects involved are the user, AUTONOM prototype, static data store, the database (to store the schedules), NR systems and a 'workflow handler' which is again a part of AUTONOM. An additional component in this diagram is the Hadoop data store (in the manual scenario it is an inactive entity due to the use of static data). With the fully automated production of schedules, the AUTONOM prototype along with NR systems listens for alerts (or new data feeds). Meanwhile the 'live data handler' works with the live data streams, cleanses live data and parses the data on schedule (to represent schedules as XML). Once the data is stored in the Hadoop data store by 'live data handler', a notification is received by the AUTONOM software of an incoming feed. The production of maintenance schedules is commenced by reading the Hadoop data store. A set of schedules and dashboard of possible schedules are produced.

In the fully autonomous scenario the user remains inactive until and unless the 'autonomous decision engine' creates an alert for the user to handle an emergency situation.

4.7 Sequence diagram for whole AUTONOM project

The next step is to model the information exchange between the AUTONOM modules described in Section 1.1, this is shown in Fig. 4. Apart from these four main components, there exists another object: the Web Application. This application is used to mirror the dashboard displayed on desktop application through a web browser, thereby enabling portability across several devices.

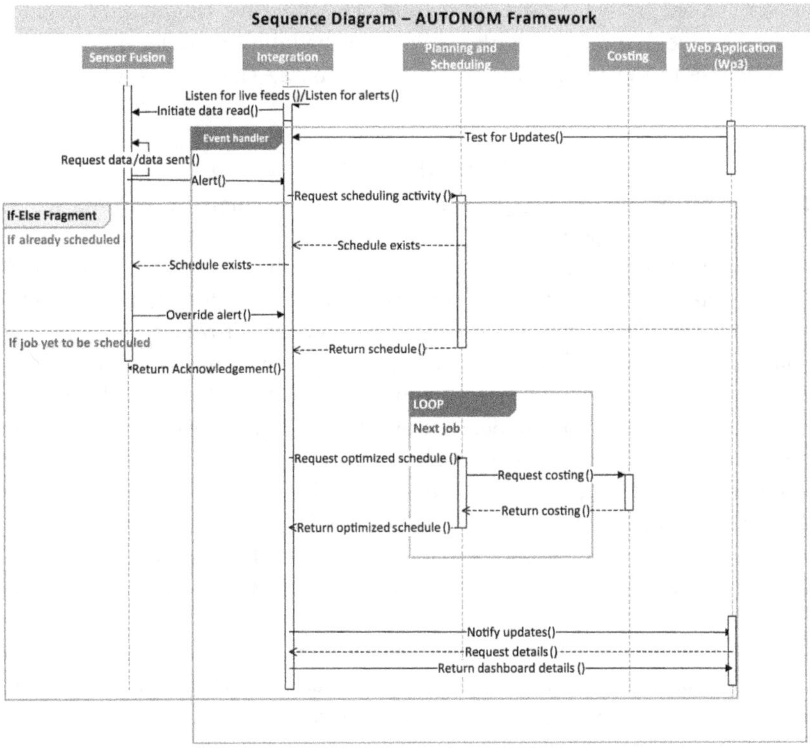

Figure 4: Sequence diagram for the AUTONOM framework.

There are two UML frame elements being used in this sequence diagram: one to denote an alternative fragment and another to denote a loop fragment. The process begins with the 'integration module' listening for alerts or new feeds. When a new data stream is detected, the integration module directs the sensor fusion module to read data from the corresponding data store (could be static data or a live feed via the Hadoop data store). The data is read and processed by the sensor fusion work package; it then generates an alert when the data values display the presence of a fault. The sensor fusion module creates an alert and directs integration to schedule a maintenance activity to manage the fault detected. The integration module subsequently requests the scheduling of the activity. The planning and scheduling module either returns a schedule or messages integration about the existence of that particular activity in the schedule. In the case of the schedule already being present, the sensor fusion overrides the alert. Alternatively, the integration module sends an acknowledgement to the senor fusion module. The model element 'Alternative Segment' is used in order to visually represent the two aforementioned mutually exclusive logic flows. Once the planning and scheduling module receives a request to produce an optimized schedule, it triggers the costing module to calculate the schedule's cost. This action is completed for numerous combinations of schedules, for each of which the cost value is obtained from the costing module. This iterative fragment is denoted by a 'loop' segment in the diagram. The planning and scheduling module, finally, passes the optimized schedule to the integration module dashboard. The Web Application mirrors this action.

5 VALIDATION

In the validation of the architecture a semi-structured questionnaire was drafted and the insights of eight industry experts sought. This questionnaire was used to judge the models based on various parameters such as completeness, simplicity, clarity and accuracy. The user community chosen for this validation process were industry experts with experience in planning and scheduling and the design and development of prototypes. The questionnaire for this study consisted of seven parts, beginning with an introduction a second section stated details of the case study, and the subsequent sections listed: UML models to be evaluated; questions on scenario modelling; use case diagrams; AUTONOM component interaction; sequence diagrams. The message passing aspect was tested using questions under Section 6 followed by questions about the overall model. For each of the questions, the experts were expected to rate the model (from a rating scale of 1, low, to 5, high). The overall feedback from the eight interviews conducted showed parameters such as accuracy, simplicity and completeness scored more than 4.5. The clarity of the models was evaluated to be 4.3 out of 5.

The main changes made in response to expert comments included the insertion of a sub-task within `create maintenance schedules', called `calculate cost', in the use case diagrams and improvements in the way the data store was represented in the sequence diagrams. The data stores are presented as hollow rectangles whereas the elements representing functions are shown as solid coloured figures.

6 DISCUSSION AND CONCLUSIONS

This article has now presented the UML modelling pertaining to AUTONOM. The study includes the high level design (HLD) of the prototype with this article focussing on the communication between identified modules in the AUTONOM framework. Scenario-based modelling best describes the information flow between the different modules. UML provides a unique bridge for developers and business users, permitting efficient communication that supports the implementation of software. In the further development of this research the following suggestions for future work need to be taken into account: An XML-based message framework should be employed to specify generic templates of various messages that could be exchanged. Description of constraints is an essential step in introducing railway terminology into the UML profile. UML standard supports a notation, the Object Constraint Language (OCL) for specification of constraints. In this study, scenarios were used as the foundation for requirement specification. Use case models derived from the identified scenarios project a black box representation of the system. The use of sequence diagrams with use cases allow for clarity in software functionality. Together, this set of UML diagrams will act as a standardized documentation of the system and will be used to communicate system specifications to the implementation team.

REFERENCES

[1] GB rail industry financial information 2013–14 (Office of Rail Regulation), available at http://orr.gov.uk/__data/assets/pdf_file/0005/16997/gb-rail-industry-financials-2013-14.pdf (accessed 4 February 2016).

[2] Cores, F., Caceres, N., Benitez, F. G., Escriba, S. & Jimenez Redondo, N., A logical framework and integrated architecture for the rail maintenance automation. In European Transport Conference 2013, 2013/9/30 to 2013/10/2, Frankfurt.

[3] AUTONOM (integrated through life support for high-value systems), available at https://www.cranfield.ac.uk/research/research-activity/current-projects/research-projects/autonom.html (accessed 4 February 2016).

[4] Turner, C., Tiwari, A., Starr, A. & Blacktop, K., A review of key planning and scheduling in the rail industry in Europe and UK. *Proceedings of the Institution of Mechanical Engineers, Part F: Journal of Rail and Rapid Transit*, **230**(3), 2015. DOI: 10.1177/0954409714565654.

[5] Mary, S.R. & Rodrigues, P., Software architecture-evolution and evaluation. *International Journal of Advanced Computer Science and Applications* **3**(8), pp. 82–88, 2012. DOI: 10.14569/IJACSA.2012.030814.

[6] Berkenkötter, K., Bisanz, S., Hannemann, U. & Peleska, J., Executable hybrid UML and its application to train control systems. *Integration of Software Specification Techniques for Applications in Engineering*, eds. H. Ehrig, W. Damm, M. Große-Rhode, W. Reif, E. Schnieder & Westkamper, Springer Verlag: Berlin, Germany, pp. 145–173, 2004.

[7] Berkenkötter, K. & Hannemann, U., Modeling the railway control domain rigorously with a uml 2.0 profile. *Computer Safety, Reliability, and Security (LNCS – 4166)*, ed. J. Górski, Springer, Heidelberg: Springer, pp. 398–411, 2006.

[8] Caprara, A., Kroon, L., Monaci, M., Peeters, M. & Toth, P., Passenger railway optimization. *Handbooks in Operations Research and Management Science* **14**, pp. 129–187, 2007. DOI: 10.1016/S0927-0507(06)14003-7.

[9] Klabes, S.G., Algorithmic railway capacity allocation in a competitive European railway market, Doctoral dissertation, PhD thesis, RWTH Aachen, 2010.

[10] Camci, F. & Chinnam, R.B., *Process Monitoring, Diagnostics and Prognostics in Machining Processes*, Saarbrücken, Germany: LAP Lambert Academic Publishing, 2010.

[11] Dadashi, N., Wilson, J.R., Sharples, S., Golightly, D. & Clarke, T., A framework of data processing for decision making in railway intelligent infrastructure. *2011 IEEE First International Multi-Disciplinary Conference on Cognitive Methods in Situation Awareness and Decision Support (CogSIMA)*, Miami Beach, FL, IEEE, pp. 276–283, 2011.

[12] Schlake, B.W., Barkan, C.P.L. & Riley Edwards, J., Train delay and economic impact of in-service failures of railroad rolling stock. *Journal of the Transportation Research Board*, **2261**, pp. 124–133, 2011. DOI: 10.3141/2261-14.

[13] Cresswell, S.N., McCluskey, T.L. & West, M.M., Acquiring planning domain models using LOCM. *Knowledge Engineering Review*, **28**, pp. 195–213, 2013. DOI: 10.1017/S0269888912000422.

[14] Fernández, S., De La Rosa, T., Fernández, F., Suárez, R., Ortiz, J., Borrajo, D. & Manzano, D., Using automated planning for improving data mining processes. *Knowledge Engineering Review*, **28**, pp. 157–173, 2013. DOI: 10.1017/S0269888912000409.

[15] McDermott, D., The AIPS'98 Planning Competition Committee. PDDL–the planning domain definition language. Tech. rep., available at http://www.cs.yale.edu/homes/dvm/ (accessed 18 September 2015).

[16] Verstichel, S., Ongenae, F., Loeve, L., Vermeulen, F., Dings, P., Dhoedt, B. & De Turck, F., Efficient data integration in the railway domain through an ontology-based methodology. *Transportation Research Part C: Emerging Technologies*, **19(4)**, pp. 617–643, 2011. DOI: 10.1016/j.trc.2010.10.003.
[17] Marcano, R. & Levy, N., Using B formal specifications for analysis and verification of UML/OCL models. In Workshop on consistency problems in UML-based software development. *5th International Conference on the Unified Modeling Language*, Dresden, Germany, pp. 91–105, 2002.

APPENDIX 1: SEQUENCE DIAGRAM FOR FULLY AUTOMATED PRODUCTION OF SCHEDULES

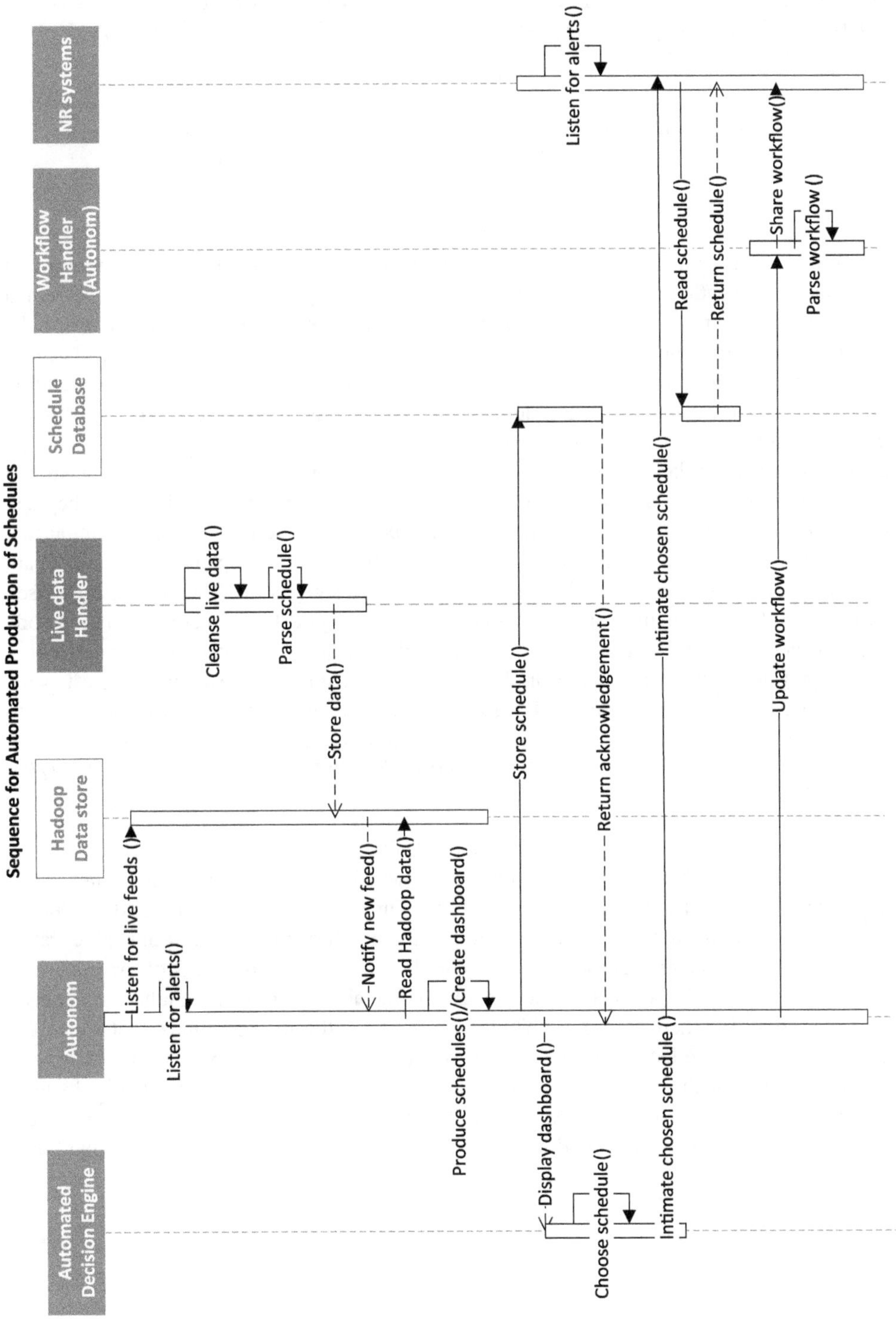

INTERACTIVE TIMETABLING SYSTEM WITH FUNCTION OF ESTIMATING TRAIN CONGESTION ON TIMETABLE PLANNED

DAISUKE TATSUI[1], TAKETOSHI KUNIMATSU[1] & TAKASHI SAKAGUCHI[2]
[1]Railway Technical Research Institute, Japan.
[2]West Japan Railway, Japan.

ABSTRACT

It is important for railway operators to make suitable timetables. On the assumption that passengers' train paths are invariant even if the timetable changes, the timetables are determined according to the various statistical data under the current timetable. In planning the timetable, it is difficult to grasp the change of passengers' paths due to timetable modifications because it is too complex. In this paper, we propose a framework of timetabling with due consideration of the path change and propose a practical timetabling system which can quickly estimate the congestion of each train by using passengers' origin–destination (OD) data collected by automatic ticket checkers. With the system, timetable planners can interactively make and modify timetables by trial and error while confirming the congestion, and finally, they can reach the most preferable one. In order to realize such an interactive system, it is important to develop a fast estimation algorithm of train congestion. We developed a new shortest path search algorithm to determine which trains each passenger gets on. The algorithm devised based on the Dijkstra method has two features. First, the shortest path search from each node to all the destination stations in the composed graph network is executed only once. Second, overlapping the path searches are omitted using stored information of the shortest paths which have already been searched. By applying this algorithm, it took only about 10 s to estimate the train congestion under a timetable of the whole day on an urban commuter line.

Keywords: Dijkstra method, estimation of congestion, interactive system, passengers' path, shortest path search, timetabling.

1 INTRODUCTION

For railway companies, it is necessary to prepare timetables so that they provide enough transportation capacity to satisfy passengers' demands. Currently, a software dedicated for timetabling is used, but timetabling itself is not automated. Therefore, the quality of timetables relies on the timetable planners' experience and skill. The planners conduct lots of investigations under current timetables and determine a modified timetable based on the results, their experiences and so on. But there are two problems with respect to this method. One is that timetables they prepare are not reflecting passengers' real demands. The other problem is that the timetable planners assume that passengers' train paths will not change in case the timetable changes.

Especially during the rush time when many people use the railway for commuting, there are many people who give up boarding on the heavily congested trains and get on other trains

DOI: 10.2495/TDI-V1-N3-382-391

which have space for them to get on. It means that the passengers choose trains against their wills. Therefore, it is difficult to grasp all the passengers' real demands by the investigations under current timetable.

Also, passengers choose trains based on the given timetable in accordance with the purposes of trips. If they need to reach the destination station as soon as possible, they choose the earliest trains regardless of train congestion. If the timetable changes, train paths of these passengers may change because the earliest train will change according to the train reordering. But the changes of passengers' train path are so complex that the timetable planner has difficulty in grasping them by the investigation under current timetable.

So, as shown in the article, we have built a new timetabling system which can estimate train congestion by using passengers' OD data collected by the automatic ticket checkers. In order to reflect passengers' real demands, the system supposes that all the passengers get on the trains with the aim to reach the destination station as soon as possible. The system calculates passengers' train path under planned timetables, based on passengers' real demand.

The system has three features as follows. First, the timetable planners can estimate congestion of each train at sight by fast calculation algorithm of train congestion. Second, they can grasp congestion of each train visually and intuitively by the coloured train lines on train diagram. The colour is determined based on the estimated train congestion. Third, on the diagram which is coloured according to train congestion, they can modify timetable properly by using mouse operation such as dragging, dropping, copying and pasting.

At the same time, in this article, we propose a workflow which helps the timetable planners make the timetables more efficiently. In the workflow, the timetable planners can prepare an appropriate timetable interactively by repeating modification of the timetable and estimation of the train congestion.

2 PROBLEMS RELATED WITH PREPARATION OF TIMETABLE

2.1 Workflow of timetabling and problems associated with timetabling

Timetables are determined based on a comprehensive view of management policy, passengers' demands on the lines, passengers' volume under current timetable, and the constraints of resources required for operating trains such as the number of the rolling stocks and crews.

For timetabling, the timetable planners use the dedicated software, which enables them to conduct their task very efficiently. When they input departure time at the starting station, train line is drawn automatically by calculating the arrival time and departure time at intermediate stations. Moreover, it has a function that allows them to make cyclic pattern timetables by copying the existing train lines. However, it is impossible to propose appropriate number of trains, the number of vehicles of each train or the operating sections of each train so as to meet passengers' demands.

In order to help the timetable planners make decisions, many investigations about transportation situation are conducted. In these investigations, investigators on platforms count and record the number of passengers on each train. Also, in order to grasp the transportation situation, train operating companies have begun to estimate the number of passengers on each car based on its weight and passengers' trip data stored at automatic ticket gate (Kubota et al. [1]).

However, even if the investigation results clarify the problems of the current timetables such as existence of the heavily congested trains, it will not be obvious whether the modified timetable becomes more convenient for passengers, because it is difficult to grasp real passengers' demand from the results of executed investigations under current timetable, and it is also difficult to grasp all the passengers' train choice change caused by timetable modification. For example, suppose that 4 rapid trains and 8 local trains are operated every hour. Timetable planners cannot grasp train congestion change in case where the number of rapid train changes from 4 to 6, and that of local train changes from 8 to 6, or in the case where the places where the rapid trains overtake the local trains change. If the congestion of rapid trains drops due to increase in the number of rapid trains, there is possibility that passengers getting on a local train because of heavy congestion under current timetable may shift to a rapid train.

2.2 Related research

Many researches on estimating train congestion under planned timetable have been executed, making use of passengers' OD data obtained from the automatic ticket gate. For example, we built a prototype 'Multi-Agent Train Simulator with Intelligent Information Exchange' which has a function by which passengers' flow and delay time of each train are estimated simultaneously by using the data and timetable data (Kunimatsu et al. [2]). Also, Nagasaki et al. [3] proposed a fast-estimating algorithm of passengers' flow under a given timetable by preliminary search in a reverse direction (a direction going back in time).

However, in these researches, the timetable planners have to make a timetable for estimating the train congestion every time. So it is beyond the scope of their researches to consider their actual work flow with which they prepare timetables by trial and error based on the estimated train congestion in the stage of planning and making timetables. They conduct various modifications to the current timetable one by one and finally reach target timetable. Especially, when they prepare dozens of candidate timetables because of large-scale modification of current timetable, it is not realistic to prepare relevant data and estimate train congestion for all the candidate timetables. Therefore, it is necessary to unite the function of preparing timetable with that of estimating train congestion so that they can make lots of timetables and examine which one is the best. Then, the timetable planners can estimate train congestion and grasp them promptly every time they modify the timetable.

On the other hand, as for research on visualization, Inagawa et al. [4] proposed a coloured diagram in order to grasp the result of evaluation of timetables easily. In the research, at first, actual result operation data of some selected days are collected from the programmed train control system, and statistics of delay of departure and arrival time at each station are obtained. Then, train lines on the diagram are coloured according to the delay results. This method enables the timetable planners to grasp and specify easily where delays occur, but it is impossible to grasp the change of train congestion when the timetable changes because the method is based on the actual result.

2.3 Requirements for timetabling system

Based on the discussion above, in this research, we propose an interactive timetabling system which fulfils three requirements as follows. First, the system can estimate train congestion quickly under modified timetable. Second, the timetable planners can grasp promptly train congestion easily and visually. Third, the timetable planners can modify the timetable instantly with confirmation of the estimated train congestion.

3 INTERACTIVE TIMETABLING SYSTEM WITH FUNCTION OF ESTIMATING TRAIN CONGESTION

3.1 Proposed work flow of preparing timetables

Considering the requirements mentioned above, we propose a work flow of preparing timetables with the use of the system we built, as shown in Fig. 1.

First, the timetable planner inputs the initial draft timetable data into the system (e.g. the current timetable data). After that, estimation of train congestion is conducted under the input timetable according to user command. The calculation is performed based on the input timetable data, passengers' demands of the target line (origin stations, destination stations, time zones when passengers appear at origin stations, the number of passengers. These data are created by converting those obtained by automatic ticket gate.), and infrastructure data (order of stations, track layout, platform layout at each station). Then, the result is shown as a coloured diagram in which each train is coloured according to the train congestion so that the timetable planners can grasp train congestion easily (see Fig. 2). Therefore, in case the timetable planners can find some congested trains, they modify the timetable on the coloured diagram.

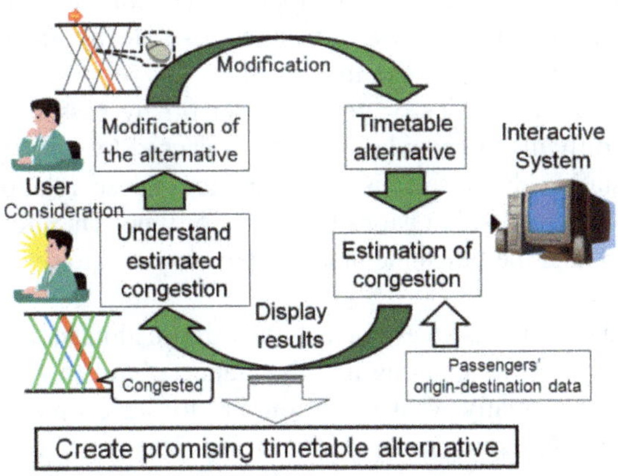

Figure 1: Process of interactive timetabling incorporating estimation of train congestion.

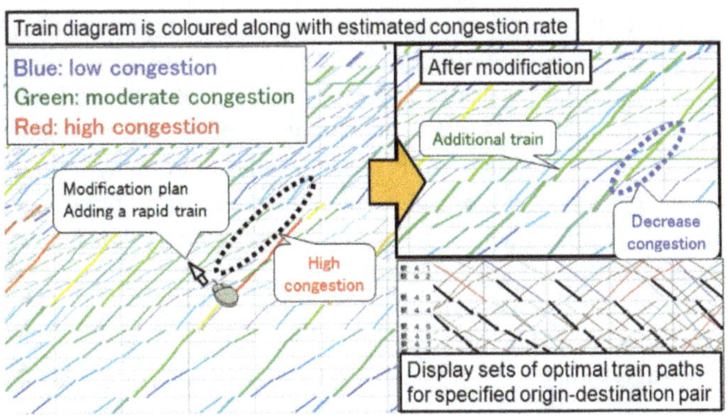

Figure 2: An example of system interfaces for timetabling.

After modification of the timetable, if the timetable planner executes estimation again, then the result of re-estimated train congestion under the modified timetable is shown as the coloured diagram. As a next step, the timetable planners compare train congestion before and after the modification of the timetable, and they can confirm whether the modification is appropriate. In this way, by repeating modification of the timetables and confirmation of the train congestion under the modified timetable, the timetable planners can find proper modification with trial and error by which the train congestion can be adjusted. Then, they can finally prepare the most suitable timetable.

If this work flow is executed, it is possible both to improve timetable quality based on the estimation of the train congestion and to make timetabling more efficient. In order to achieve this, the interactive system we built needs to have fast algorithm for estimating train congestion.

3.2 Presupposition of this research

Considering the purpose of the system (i.e. prior evaluation of the revised timetable), estimation of train congestion is conducted on the following three presuppositions.

First, passengers' demand is invariant even if a timetable changes. Passengers' demand will change if a new line is constructed, the mutual direct operation starts or the required time is shortened by vehicle speed-up. However, such a large-scale modification is not conducted frequently, so it is out of consideration in this research.

Second, delay caused by troubles or getting on or off by passengers at each station is not considered. It is difficult to consider train delay caused by getting on or off trains by passengers at each station, because it is necessary to analyse relationship between train congestion and dwelling time which varies by station or time. Therefore, in this research, it is out of consideration, and the timetable planners are supposed to set enough dwelling time so as not to cause delay.

Third, passengers choose trains so as to reach their destination station as soon as possible. It is based on the thought of timetabling that the planners have to grasp passengers' needs for rapid train which they really want to get on and to make timetables to have enough transportation capacity. For example, as shown in Fig. 3, if the ratio of the number of the rapid trains to that of the local trains is 1 to 1, train congestion of the local trains is found to be 120% and that of the rapid trains is found to be 150%. But the train congestion is calculated assuming the inclusion of the passengers who avoid getting on rapid train because of heavy congestion, which amounts to 30% of the total amount of the passengers. The problem is that these passengers are invisible to the investigators. Therefore, even if train congestion of the rapid trains is higher than those of the local trains, there is possibility that passengers' demand for rapid trains is not so huge as to operate rapid trains and local trains with the ratio of 2 to 1. Under the same situation, if passengers can get on any train they like without avoiding rapid trains of heavy congestion, train congestion of the rapid trains and the local trains is estimated to be 180% and 90%, respectively. Accordingly, it is better to operate rapid trains and local trains with the ratio of 2 to 1 in order to meet passengers' actual demands. To obtain an idea to conduct such a modification, it is necessary to modify the timetables according to passengers' actual demands for each train without considering passengers who avoid heavily congested trains.

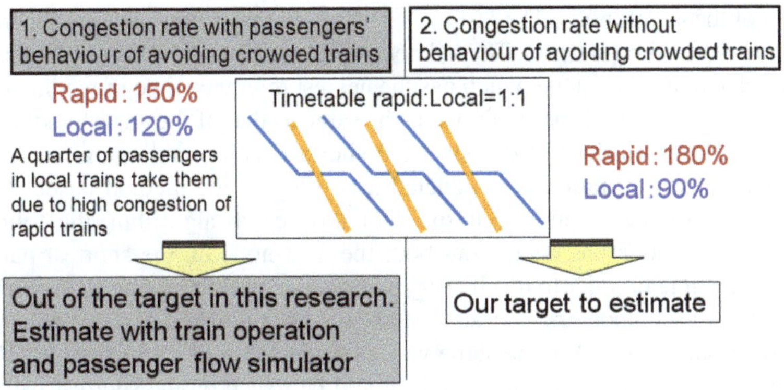

Figure 3: Estimation reflecting passengers' real demands.

4 ALGORITHM FOR FAST ESTIMATION OF TRAIN CONGESTION

4.1 Method of estimating train congestion

In this system, estimation of train congestion is conducted by the following steps.

First, passengers' appearances at each station are created based on passengers' demands data. When passengers' data based on leaving time from the destination stations are available, they have to be converted to the format of passengers' demands data described in Section 3.2 according to the required time from the origin stations to the destination stations. Based on the data and Poisson distribution, appearance time of each passenger is determined by stochastic method. For further details, see Kunimatsu et al. [2].

Second, the graph network which represents the timetable is constructed. The graph network has nodes (including information of reference time and platform) and arcs as follows.

- Nodes

There are three types of nodes. The first is a train arrival node that represents each train arrival at each station. Here, the reference time is the arrival time, and the reference platform is where the train arrives and departs. The second is a train departure node that represents each train departure at each station. Here, the reference time is the departure time, and the reference platform is where the train arrives and departs. The last is a stay node that represents passengers stay on the platform in order to wait for the next train. Each departure node has one stay node. Also, the reference time is equal to the departure time plus Δ which means minute time smaller than the minimal time of timetable (e.g. 1 s). The reference platform is the same as that for the departure node.

- Arcs

There are three kinds of arc, and the cost of each kind of arc is equal to the difference between the reference times of the nodes regardless of their kinds. The first one is a train arc that represents a train running between stations or dwelling at stations. The second one is a transfer arc that represents passengers' change of train. A tail node of transfer arc is an arrival node, and a head node of it is a corresponding stay node which has the nearest reference time immediately after the arrival node, considering the transfer time. Here, the reference platform of the stay node is different from that of the arrival node, but the station of the stay node is the same as that of the arrival node. The last one is a stay arc that represents passengers' waiting for next train on the same platform. The tail node of the stay arc is the stay node, and the head of it is a corresponding stay node which has the nearest reference time immediately after the arrival node. The platform of the stay node is the same as that of the arrival node.

After creating a graph described above, the shortest path search from every arrival or departure node to every station is conducted according to the method based on Dijkstra method. In order to accelerate search speed of this search algorithm, the following are devised. First, each node has data about both the next node on the shortest path and the shortest required time in order to reach every station. Second, according to the arrival or the departure time of each node, the shortest path search is conducted in a reverse direction (a direction going back in time) in the same way as Nagasaki et al. [3] proposed. Third, both the shortest required time and an arc which should be traced on the shortest path from the node are recorded or updated as the data of the shortest path, according to the sum of the time data of another node linked by the arc and the cost of the arc. Moreover, using the stored shortest path data, the calculation of re-search is accelerated by omitting overlapping path search (see Section 4.2).

Based on the result of the shortest path search, the number of passengers who appear at the origin station at the time zone between nodes is assigned to each node, classifying passengers according to their destination stations.

Finally, the number of passengers is assigned to each train and section according to both the number of passengers and the shortest path data. By accumulating them, the total number of passengers of each train in each section is calculated.

4.2 Algorithm for fast re-estimation

In order to speed up the re-estimating train congestion after the first estimation, the path search which has no possibility that train paths will change by timetable modification is omitted.

In this research, we propose a method by which re-search of the shortest path is conducted only for the node which may have changed the shortest path according to updated information of the structure change of the graph network. The following process is performed every time the path search is conducted and the timetable is modified.

After the path search described in Section 4.1 is conducted, the data of the shortest path at each node are retained. The data include the next node on the shortest path and the shortest required time in order to reach every station.

On the other hand, according to timetable modification, the structure of the graph network is updated, and changing the data of the shortest path of each node is recorded with the method described in Section 4.2.1. By means of this method, all the changes of the graph structure after the last path search are stored.

When research is carried out after timetable modification, data of the shortest path are updated with the method described in Section 4.2.2 based on the stored data described above.

4.2.1 Timetable modification and updated data of the network configuration

In each operation of modification of a timetable, data of network structure are updated, and list of nodes necessary for checking is updated by the following method. As for the operations which are not described in this section, they can be substituted by cancelling and adding train to the parts in need of correction.

Let N_0 be a set of network nodes before modification; A_0, a set of arcs before modification; a_s, the tail of arc a; a_d, for the head of arc a; a_c, cost of arc before modification.

When the user cancels trains (including shortening of operating section) and changes the train operation from stopping to passing, all the nodes and arcs related to the train, the station and the section are deleted. Let N_- be a set of deleted nodes, and A_- be a set of deleted arcs.

In the case where the user adds trains (including extension of operating section) and changes the train operation from passing to stopping, all the nodes and arcs related to the train, the station and the section are added. Let N_+ be a set of added nodes.

If timetable is modified according to N' (a set of node which satisfies $N' = N_0 - N_- + N_+$), the graph structure after the modification is created by the method described in Section 4.1. Let A' be a set of arcs of the graph corresponding to the modified timetable.

As for recording or storing in the list of nodes necessary for checking, four kinds of the nodes are all recorded and stored in the list as described in Fig. 4. They are added nodes (N_+), data of the tail nodes of the added arcs which belong to not A_0 but A', data of the tail nodes of the deleted arcs which belong to not A' but A_0 in the case where the nodes are not deleted or belong to N', and data of the tail nodes of the arcs whose cost changed in the case where the arcs belong to both A_0 and A' and the a_c is different from a'_c. The data recorded and stored are a set of [node n, destination station s], and the recorded data are the one about every destination station from the nodes.

4.2.2 Updating the shortest path data using network updated data

For the graph network of the modified timetable, determining whether cost of the shortest path has changed and updating the data of the shortest path are conducted with the use of the list necessary for checking by the following steps until the list becomes empty.

First, the node which has the last reference time in the list is considered as a target. Then, the node and the destination stations are deleted from the list. Here, the order of destination stations is not considered.

In the case where there is neither the target node nor the data of the shortest path (the next node on the path, the shortest required time), the path search for the destination stations and data updating are conducted by the method described in Section 4.1. In addition, all tail nodes of all the arcs whose head are the target node and the destination station are added to the list; then the first step is conducted again.

If the node u and the destination station s have already had data of the shortest path, confirming whether it is optimal is conducted with the following method. Here, let C_{xy} be the cost from the target node x to the node y; C'_{us}, the cost recorded in the shortest path data; and v_0, the next node in it.

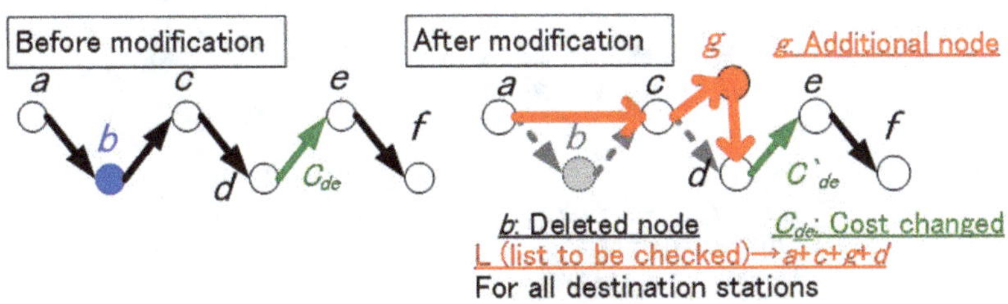

Figure 4: Process of specifying nodes to be checked.

1. Minimum cost C^*_{us} and v^* which minimizes the cost are obtained by calculating $C_{uv} + C_{vs}$ for every arc.
2. If $v_0 = v^*$ and $C_{uv0} + C_{v0s} = C^*_{us} = C'_{us}$, then the information is not changed (Fig. 5 (case 1)).
3. If v_0 is not in the changed graph and $C'_{us} = C^*_{us}$, or $v_0 \neq v^*$ and $C_{uv0} + C_{v0s} > C'_{us} = C^*_{us}$, then the next node is changed from v_0 to v^* because the minimum cost is the same but the next node is changed (Fig. 5 (case 2)).
4. If $v_0 = v^*$ and $C_{uv0} + C_{vos} = C^*_{us} \neq C'_{us}$, then the minimum cost is changed from C'_{us} to C^*_{us} because the path is the same but the cost is changed. In addition, each tail node of each arc whose head is the node u is added to the list necessary for checking with respect to the destination station s (Fig. 5 (case 3)).
5. If v_0 is not in the changed graph and $C'_{us} \neq C^*_{us}$, or $v_0 \neq v^*$ and $C'_{us} \neq C^*_{us}$, then the next node is changed from v_0 to v^*, and the minimum cost is changed from C'_{us} to C^*_{us}. In addition, the each tail node of each arc of which the head is node u is added to the list necessary for checking with respect to the destination station s (Fig. 5 (case 4)).
6. The first step of this section is conducted again. Target node is determined.

In this way, both confirmation of the possibility that the shortest path from each nodes is changed by timetable modification and path change operation in case where the shortest path is changed are conducted at the same time by propagation on the graph. This enables us to narrow the re-search width without giving any influence on the optimality of the shortest path. For example, if local change of the network structure such as reordering of one train is only considered, it is possible to shorten the calculation time drastically.

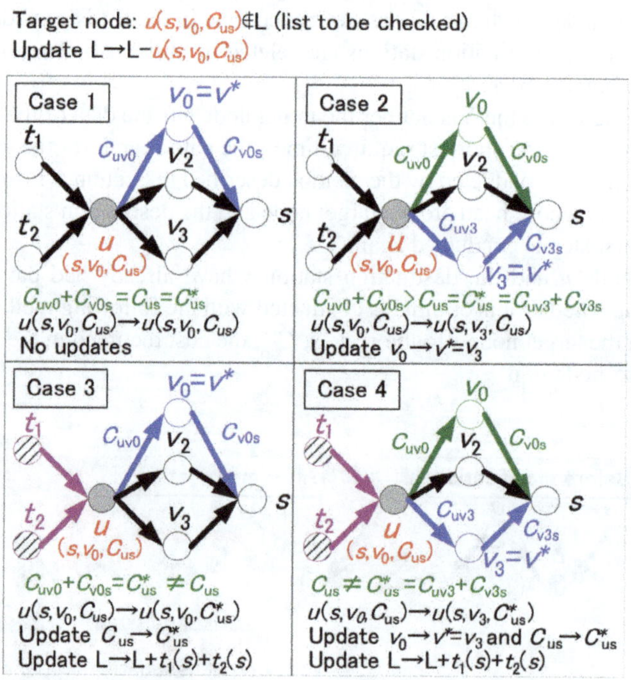

Figure 5: Process of updating the optimal path for the destination.

5 EVALUATION OF A PROTOTYPE BUILT FOR AN ACTUAL COMMUTER LINE

In order to evaluate the system, we built a prototype for an urban commuter line and measured estimating time for the current timetable and re-estimating time when the train order between one rapid train and a following train is changed at 20 stations.

The target line includes 103 stations, the number of passengers per day is 1,600,000 and the number of trains per day is 2,319. The measurement of the estimation time was executed on a PC equipped with a 2.4 GHz Pentium Core i7 processor, 8 GB memory and Windows 7 64 bit.

As a result, it took approximately 10 s for the estimation of train congestion under current timetable, and 7 s for the estimation under modified timetable. If the algorithm which was proposed by Kunimatsu et al. [2] is applied, the estimating time is approximately 180 s. So the estimating time of the algorithm we propose is considered sufficiently practical.

Moreover, comparing the estimating time under the current timetable and with that under the modified timetable, we confirmed the effect of the proposed fast algorithm for re-estimating train congestion in the case where the shortest path search for the nodes is not conducted if there is no possibility that the path of the node does not change even in case of modification of timetable.

However, it is early to conclude that the calculation speed is fast enough to use the system as an interactive system. As a future work, it is necessary to realize faster calculation (e.g. algorithm for summing up the passengers).

6 CONCLUSION

In this research, we built the prototype interactive timetabling system in order to realize raising both the quality of timetabling work and effectiveness of it. Specifically, we designed user-friendly interface and fast algorithm of estimating train congestion. As for fast-estimating algorithm of train congestion, it took 10 s in order to estimate current timetable on the actual train line, and it took only 7 s in order to estimate train congestion under a modified timetable when the train order between one rapid train and a following train is changed at 20 stations. As a result, effectiveness of fast re-estimation algorithm of train congestion is confirmed.

As for the future work, it is necessary to consider faster algorithm and to verify the usability of the system in case of applying it to various complex train lines.

REFERENCES

[1] Kubota, T., Morita, Y., Ota, M., Furuya, S. & Ieda, H., Analysis of rate of train overcrowding by stored data at ticket gates and load weight units, Infrastructure planning view. *Japan Society of Civil Engineers*, **25**, pp. 641–646, 2008. (in Japanese)

[2] Kunimatsu, T., Hirai, C. & Tomii, N., Evaluation of timetables by estimating passengers' personal disutility using micro-simulation. *Proceedings of the 3th International Seminar on Railway Operations Modelling and Analysis* (RailZurich2009), 2009.

[3] Nagasaki, Y., Asuka, M. & Komaya, K., A fast method for estimating railway passenger flow. *Computers in Railways* **X**, pp. 179–187, 2006. DOI: 10.2495/CR060181.

[4] Inagawa, M., Tomii, N. & Ushida, K., Visualization of train operation record data. *J-Rail*, **16**, pp. 745–748, 2009. (in Japanese)

MAXIMIZING FREIGHT TRAFFIC BY RE-ROUTING

C. MEIRICH AND N. NIEßEN
Institute of Transport Science, RWTH Aachen University, Germany.

ABSTRACT
This article describes how to maximize the number of freight trains through global networks. Nowadays, the capacity of railway lines and also of railway nodes can be calculated by using analytic algorithms. Currently there is no generally accepted method to allocate the overall capacity of lines and nodes of such networks by using analytic algorithms. However, to maximize the number of train runs, the rail passenger service will be fixed on the scheduled train course so that the rail freight service can use the remaining capacity on a planned or re-routed train course. Existing or detected bottlenecks could be eliminated by means of sensible re-routing due to optimization, which will reveal the best train paths through the network. The article concludes with an illustrative computation for a generic railway sub-network to exhibit how the optimization is working.
Keywords: assessment of railway infrastructure, linear programming, network optimization, railway capacity, routing.

1 INTRODUCTION

It is undeniable that the rail traffic forecast will rise in Europe, so that in the strategic network planning it will be more and more important to optimize the capacity available throughout networks. Currently, it is only possible to calculate the capacity separately for railway lines and railway nodes by using analytical methods, especially the principles of queuing theory. Apportion of the capacity for lines and nodes to describe a required capacity for networks has not been possible hitherto. One of the reasons for this condition is the differing assumptions used for the calculations. The importance of facilitating such apportionment so as to justify suitable means of developing and maintaining the network in longer-term planning will nevertheless grow in the years ahead.

In addition to maximizing the number of trains through a network, the procedure is designed to serve as a basis for identifying and reducing bottlenecks.

The present work aims to use the capacity available on lines or nodes in a macroscopic network to calculate the maximum number of trains through such a railway network. Moreover, possible train courses can be used to re-route (especially) freight trains through the network to maximize the given capacities. Therefore, the important assumption in this article is that the capacities of lines and nodes can be applied to each other.

2 RESEARCH PROGRESS TO DATE

Set out in the following section is the current state of scientific progress as well as the point of departure for the network-wide optimization of capacity. It includes the general description of railway networks, capacity and how the capacity can be calculated.

DOI: 10.2495/TDI-V1-N3-392-402

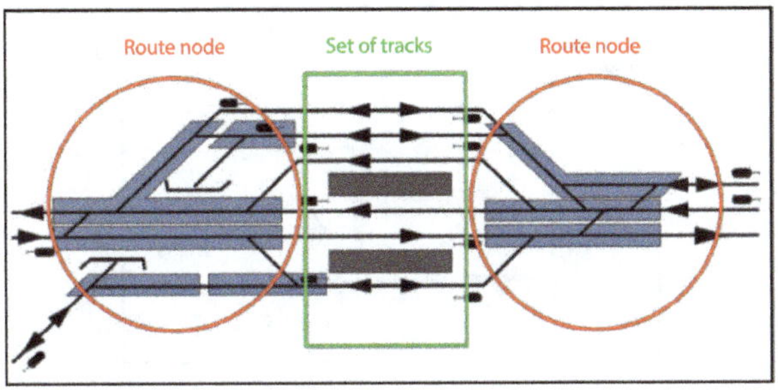

Figure 1: Route node and set of tracks.

2.1 Lines, nodes and set of tracks

A railway network can be divided into individual infrastructure elements that are explained in greater detail below and basically comprises nodes and lines.

Nodes are deemed to be either stations linking at least two lines with one another or junctions. Furthermore, a node can be divided into waiting positions (set of tracks) and route nodes (cf. Fig. 1).

The set of tracks could be subdivided into platforms, through and passing tracks as well as handling tracks, which are used as (scheduled or unscheduled) stopping or waiting positions. In addition, all tracks in a set of tracks have to substitute each other. If there are some tracks which could not be collated, it is possible to set up an additional set of tracks in the same node.

Between two nodes, 'line' is the term given to the running path. Up to this, intermediate stations can be part of a line so that only the major nodes in a network have to be set up in detail for a macroscopic approach. It is the major nodes, however, that from the start and end of the line.

2.2 Capacity of transport systems

The capacity of transport systems is generally taken to mean the ability to meet demand for the conveyance of persons, goods or information to a desired level of quality.

Hansen and Pachl define the capacity of infrastructure as follows:

The maximum number of trains that may in theory be operated over a defined part of the infrastructure over a certain period of time, through this limiting value is unlikely to be reached in practice [1].

The transport system's actual loading is compared and contrasted with a basic quality indicator as a rule. The higher the loading, the sooner the queue lengthens and hence the sooner the waiting times arise. This correlation between traffic units and waiting time is shown in Fig. 2.

The theoretical capacity of an infrastructure element is equitable with the maximum possible number of train runs n_{max}. In this case the performance graph in Fig. 2 will rise to infinite congestions because of an infinite waiting time and infinite delays. In the practical applications, the theoretical capacity is a value that is of no relevance. Normally the quality available (waiting times, knock-on delay) with a given level of service (LoS) is used to describe the dimensioning railway infrastructure. To establish the (economical) optimum loading, the number of train runs n_{opt} is used at which congestion is at a defined level so that

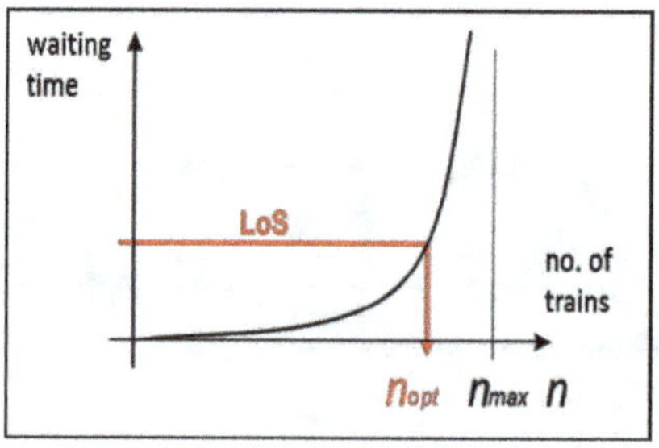

Figure 2: Correlation between loading and quality.

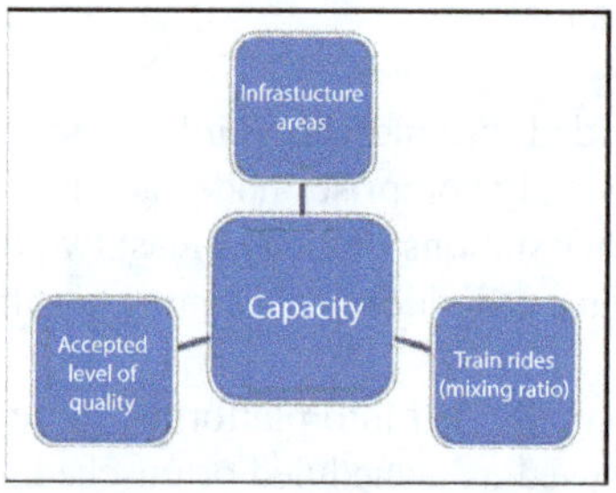

Figure 3: Factors impacting on capacity on the railways.

both infrastructure managers as well as train operating companies will be marketed with most profit. This optimum number of trains is thus referred to as 'practical capacity' [2].

To determine the capacity, the infrastructure area and its properties, the period under review, the operating schedule (train rides as well as the mixing ratio) and at least the quality of traffic have to be defined (cf. Fig. 3).

Having regard to a running path, a train move is defined as a path which includes minimum two railway nodes. Furthermore an infrastructure area is set up by the current infrastructure elements like set of tracks, route nodes or lines.

The introduced factors are not the only ones which have a direct influence on the capacity. In addition, attention needs to be paid when studying capacity of a plethora of operational requirements, rescheduling strategies, speeds, length of block sections, automatic train control systems and signalling facilities. These influences are considered by minimum headway times (cf. Hansen & Pachl [1]). Using analytical methods, minimum headway times are an establishing approach to the applicable technical standard.

2.3 Procedure for establishing capacity

To calculate the capacity of a railway infrastructure there are various procedures. But overall, most procedures have to map the infrastructure and train characteristics as well as the specific national signalling systems, the operating process and input factors.

Literature on the subject summarizes the approaches available in differing ways. A very good overview of the establishing approaches is given by Pouryousef et al. [3]. Abril et al. [4] divide these approaches into analytical, optimization or simulation methods. Hansen and Pachl [1] advocate a breakdown into analytical and simulation procedures, which is the most frequently used approach in the literature. A further option involves a division into timetable-based procedures and these are not underpinned by any precise timetable (Sameni et al. [5]).

In addition, the capacity can be calculated by using compilatory procedures. One of the common approaches draws on the blocking-time theory (cf. Happel [6]). To establish the capacity by the UIC (Code 406 Capacity) the stepped blocking-time series have to be compiled as soon as possible without producing conflicts. To this end, the compensated and non-compensated consumed time is used to determine the concatenated track-occupation ratio. Placing this concatenated track-occupation ratio in relation to quality the potential capacity can be established (cf. [7]).

A further method to calculate the capacity is by using simulation. With the aid of ensuring parameters the behaviour of a system can be measured and replicated. Normally this method focuses on delay development here. Simulation can be subdivided into synchronous and asynchronous simulations, where the first group only allows to review the running of traffic, however (cf. Pachl [8]). But overall, to get the most benefit of both simulation methods, these are often used in a hybrid formation (cf. Kuckelberg et al. [9]).

Another procedure is to calculate capacity using analytical procedures. The railway system is viewed with the aid of queuing theory approaches to this end. For example, railway lines and railway nodes are modelled as (multi-channel) queuing systems. For the service system, train moves are analogous with the demands. The service times are described by the minimum headway times for different sequence-of-trains scenarios. Former information about queuing theory in the field of railway systems are given in Kleinrock and Gail [10], Gnedenko and König [11] and Fischer and Hertel [12]. Once again the upshot of establishing capacity analytically is that the possible number of train moves for an infrastructure section under review is output as a ratio of a predefined LoS.

Furthermore, there are some published approaches to assess the overall capacity of a railway network. First of all, the macroscopic software tool NEMO (traffic and re-apportionment model) could be named, which contains a network modelling algorithm to lead trains through a network to assess the profitability (Prinz et al. [13]).

With interest focusing on a direct correlation between capacity and quality in a railway network, analytical methods are most suitable for the procedure presented here.

3 MODE OF PROCEDURE AND METHOD

There follows a general course of action for the optimization targeted.

3.1 Process

The general concept presented in this section is shown in Fig. 4. First of all the required input data have to be set up. A sub-network and a level of detail of this network have to be defined. Furthermore, the minimum headway times and in addition the capacity of infrastructure elements cannot be calculated without train runs, so a reference operating programme with (pattern-)trains has to be appointed. As an additional input factor the mixing ratio can be calculated using the reference operating programme.

The next step is to detect the running paths of the passenger trains. Here are different constrains like frequency of stops, so there is only limited scope for conducting running-path searches in rail passenger traffic. As a consequence the passenger trains will be fixed and the number of them (n_p) will be counted to reduce the capacity limits in the optimization. After that, a running path search of rail freight service through the network by defining the

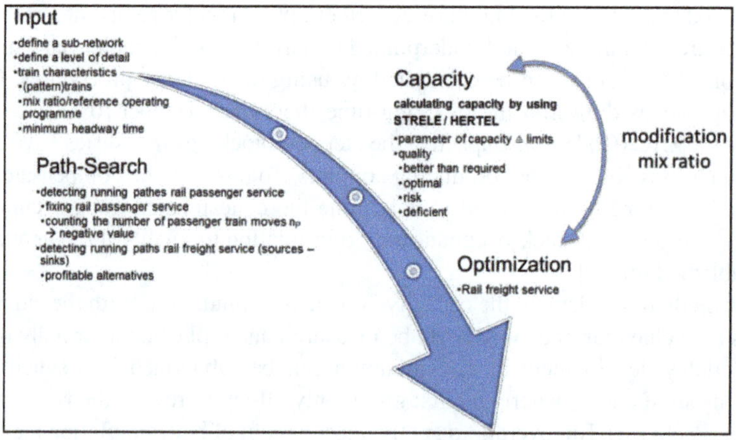

Figure 4: Optimization flowchart.

sources and sinks of the trains has to be conducted. Various route-search procedures exist, which establish the shortest or fastest link between two points in a network. One of the most well-known is the Dijkstra algorithm [14]. It is necessary here, however, to output all possible routes below the highest boundary condition. For example, if there is a wrong traction unit (diesel/electric) it is impossible for this train to use the proposed alternative. All routes in the process have to be acyclical; hence there is no duplicate negotiation of nodes or lines.

Schwanhäuser [15] has explained how to calculate the capacity of lines and route nodes by using the STRELE formula (also in Nießen [2]) which is archived in current EDP-based procedures [16]. To calculate the set of tracks capacity the method of HERTEL can be used with the assumption that there will be an intensive disturbed operation [2, 12]. The capacity of each infrastructure also can be calculated by using the method of UIC Code 406 [7]. By using predefined levels of service, it is possible to calculate the capacity of each infrastructure segment.

To maximize the number of freight trains through the network, an objective function has to be determined:

$$\max c^T \cdot n_T \tag{1}$$

where

n_T is the vector of the number of train moves and $n_T \geq 0$

c is the vector for the cost for weighting running paths and $0 < c \leq 1$

The possible running paths are the variables in this function. For weighting this function the constant c is used and for example it depends on the shortest path by length or time (Radtke [17]).

An upper limit C_{max} and, where applicable, a lower limit C_{min} for capacity are required as input parameters for each infrastructure element. For simplicity's sake, computations are confined to the upper limit C_{max}.

It is to ensure that the number of trains is lower than or equal to the calculated upper limits C_{max} for each infrastructure element (line, set of track, route node).

One of the following conditions accordingly is applied as an equality constraint for each infrastructure element.

Lines: $$n_{T,L_{i,j}} \leq C_{\max,L} \tag{2}$$

Set of tracks: $$n_{T,SoT_i} \leq C_{\max,SoT} \tag{3}$$

Route nodes: $$n_{T,RN_{i,r}} \leq C_{\max,RN} \tag{4}$$

Here $n_{T,L_{i,j}}$ is the number of trains on the line from note i to node j; n_{T,SoT_i} is the number of trains in the set of tracks of the node i. $n_{T,RN_{i,r}}$ is the number of trains on the route node r of the node i.

As already pointed out, if there are any passenger train moves on the infrastructure elements the limit of capacity has to be reduced due to the number of passenger trains n_p on this element (cf. Fig. 4).

One of the important assumptions in this article is that all capacities of lines, route nodes or set of tracks can be applied to each other.

With this background, the actual optimization is performed by a commercial solver. Using the previous steps the maximum number of train runs on a sub-network by forming a set of linear equations (cf. eqns (1)–(4)) can be calculated.

Furthermore, the solver also outlines the selected train paths which are used. It is additionally possible to output the loading for each individual infrastructure element. Finally, the new mixing ratio on the infrastructure elements can be calculated.

3.2 Altering the mixing ratio and adapting capacity

As illustrated in Fig. 4, train moves or, respectively, the mixing ratio has a significant impact on capacity so that there is an iteration loop. The reason is, if the number of freight trains on an infrastructure element will increase due to the optimization and there also are passenger train runs the mixing ratio will change. And in conclusion, it has to be checked if the capacity limits have to be adjusted.

Figure 5 shows illustrative existing capacity by percentage variations in the mixing ratio divided into long-distance passenger service ($p_{\text{long-dist.}}$) and local passenger service (p_{local}), arrived for a generic line. Computations were conducted for each interpolation node (at 20% stages). As an evidence for this, the missing percentages up to 100% are the freight trains. For example if $p_{\text{long-dist.}}$ is 20% and p_{local} is 30% there are 50% rail freight service. Only the marked surface is important, because the sum of the percentages cannot be over 100%.

Up to this, after changing the number of freight trains it has to be checked which new capacity depending on the optimized values is on the infrastructure elements now. There are two obvious methods. On the one hand, the capacity could be calculated again by using the methods pointed out in Sections 2.3 and 3.1. On the other hand, it can be interpolated for every mixing ratio and the adjusted capacity can be read out as shown in Fig. 5. Currently a mixed-linear approach is used to include a linkage between long-distance and regional traffic.

It becomes apparent that mixing fast and slow services yields differing capacity curves. With freight and local-passenger traffic travelling at comparatively homogeneous speeds, this curve is higher than the curve for the applicable comparison of these types of traffic with long-distance passenger traffic.

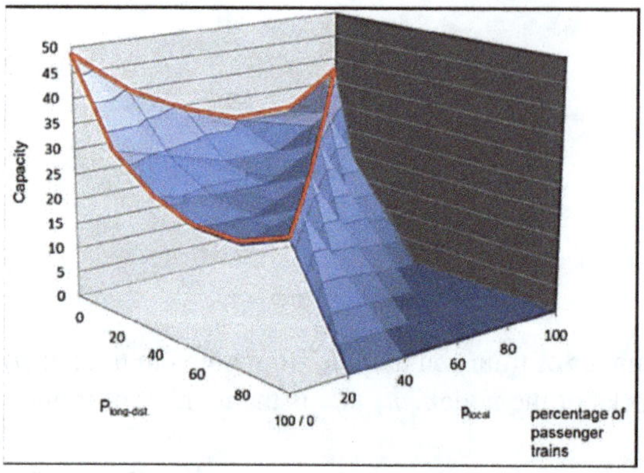

Figure 5: Illustrative representation of output capacity for given mixing ratios.

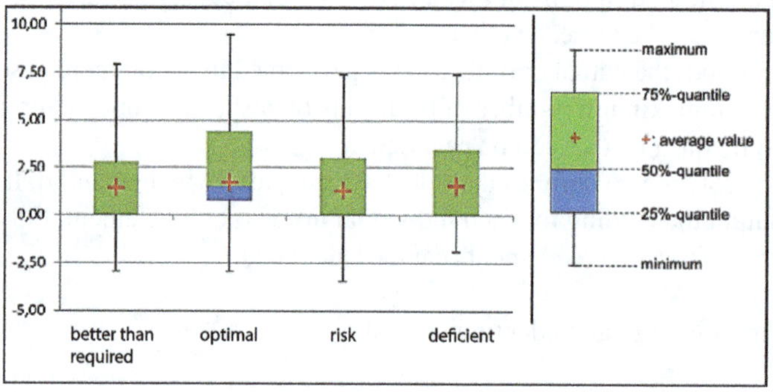

Figure 6: Representation of percentage deviations.

3.3 Deviation between capacity interpolated and calculated

The percentage deviation between the values interpolated and exact values calculated using LUKS® software (Studies into the Output Capacity of Nodes and Lines) was established illustratively for four levels of service, those being 'better than required', 'optimum', 'risk' and 'deficient'. Figure 6 portrays these deviations with the aid of boxplots. The average level of deviation for all interpolation nodes amounts to <2%. No deviations exceed 10%. Though the results could be refined by adding further interpolation nodes, this constitutes an adequate degree of accuracy for the course of action targeted.

3.4 Evaluation of results

Once optimization has been concluded, the results need to be evaluated. The number of possible train runs is composed by the number of freight trains and the base loading (passenger trains). As outlined in Section 3.1, the chosen freight train running paths in the network could be read out, even if these are not the shortest running paths.

As another result, the loading of each infrastructure element can be calculated by knowing the limits of capacity as well as the actually consumed capacity. Applying recognized indicators like quality and track-occupation ratio, congested infrastructure can be detected and a comparison between different infrastructure and/or operating-schedule variants can be made.

By comparing and gauging the number of train paths in the sub-network with the aid of length or run-timing factors a further indicator could be set out. Therefore, longer-term expansions, redesign or downsizing of a network depending on the cost/benefit ratios will be possible by considering the whole networks.

4 ILLUSTRATIVE COMPUTATION

The methods and theories presented thus far are elucidated in greater depth with the aid of an illustrative computation in this section. To simplify this example how many freight trains can run through a network without any passenger trains should be analysed. In addition the train paths will be chosen by the optimizer on detected point-to-point routes.

A small model network consisting of seven nodes linked by the lines represented in Fig. 7 was set up as an illustrative network. The respective running times and distances between and within nodes are further posited as being known. The sources and the attendant sinks for the freight traffic were taken as follows:

- Point-to-point route 1: from node 1 to node 7
- Point-to-point route 2: from node 4 to node 3

The number of train moves is to be conducted in this sub-network on the point-to-point routes detailed above, with various alternatives being predefined for each point-to-point route (cf. Table 1).

In this example, four alternatives are listed for each of two point-to-point routes; hence eight variables are used in the objective function. For simplicity's sake, the indices for the

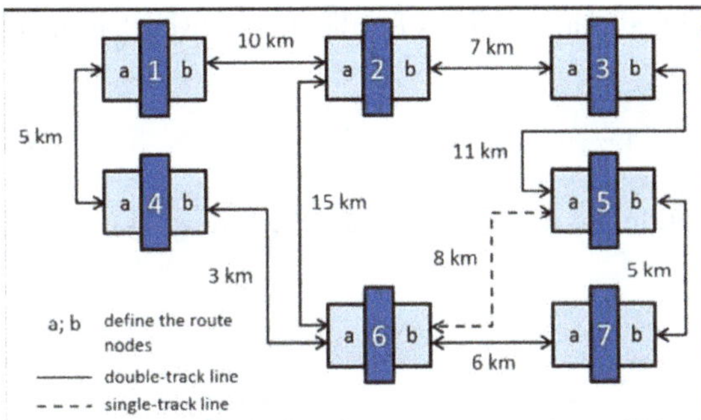

Capacity Lines			Capacity nodes			
			Set of tracks		Route Node	
$L_{i,j}$	$L_{j,i}$	$C_{L,i,j}$	i	$C_{SoT,i}$	$C_{RNi,a}$	$C_{RNi,b}$
1,2	2,1	33	1	50	58	48
2,3	3,2	15	2	80	90	92
1,4	4,1	28	3	75	70	80
3,5	5,3	40	4	55	54	53
2,6	6,2	27	5	95	100	100
4,6	6,4	28	6	60	65	58
5,7	7,5	36	7	72	75	72
6,7	7,6	20				
5,6		19				

Figure 7: Illustrative network with given capacity limits.

Table 1: Running paths of point-to-point routes for freight traffic.

Point-to-point route 1		Point-to-point route 2	
Alternative	Nodes	Alternative	Nodes
1	1,4,6,7	1	4,1,2,3
2	1,2,3,5,7	2	4,6,2,3
3	1,2,6,7,	3	4,6,5,3
		4	4,6,1,5,3

point-to-point route (1 or 2) and the number of the alternative (1; 2; 3 or 4) are assigned to each variable (cf. Table 1).

The weighted target function (cf. eqn (1)) has to be set out by using the given distances between the different nodes (cf. Fig. 7), the length of the route nodes (here simplified to 0.5 km) as well as the length of the set of tracks (here simplified to 1 km).

For each alternative (cf. Table 1) the distance between source and attendant sink can be calculated by adding the distances of the lines, the route nodes and the set of tracks. Furthermore, the shortest train course of every point-to-point route can be appointed and due to this the ratio for the alternative to this shortest train course can be calculated. Because of the maximization the reciprocal will be needed so that the shortest path of the point-to-point route will have the highest weight (equal to 1.0). The other weights will be less than 1.0. Using eqn (1) the weighted objective function is given as

$$\max c^T \cdot n_T = n_{T1,1} + 0.667\, n_{T1,2} + 0.737\, n_{T1,3} + 0.966\, n_{T2,1} + 0.718\, n_{T2,2} + 0.778\, n_{T2,3} + n_{T2,4} \quad (5)$$

$n_{T_{n,m}}$ is vector of the number of train moves on the point-to-point route n and the alternative m (cf. Table 1).

It needs to be ensured for all equality constraints that the sum of all train moves over the respective infrastructure element is lower than the capacity ceiling for this element (cf. eqns (2)–(4)). In the cause of enhanced assimilation, one equality constraint per infrastructure element is exemplified hereafter.

If the line from node 1 to node 2 is considered, the following equality constraint needs to be observed:

$$n_{T1,2} + n_{T1,3} + n_{T2,1} \le C_{L_{1,2}} \quad (6)$$

The set of tracks in node 5 thus contains all variables for the train moves that run through this node:

$$n_{T1,2} + n_{T2,3} + n_{T2,4} \le C_{SoT_5} \quad (7)$$

With regard to the left-hand station throat in the same node, the following condition is to be met:

$$n_{T1,2} + 2 \cdot n_{T2,3} + n_{T2,4} \le C_{RN_{5,a}} \quad (8)$$

Table 2: Results of optimization (consumed capacity).

Point-to-point route		Lines				Set of Tracks		Route Nodes			
$n_{T1,1}$	19	$L_{1,2}$	26	$L_{6,7}$	20	SoT_1	26	$RN_{1,a}$	26	$RN_{1,b}$	26
$n_{T1,2}$	0	$L_{2,3}$	15	$L_{4,1}$	26	SoT_2	26	$RN_{2,a}$	37	$RN_{2,b}$	15
$n_{T1,3}$	0	$L_{3,5}$	15	$L_{6,2}$	0	SoT_3	34	$RN_{3,a}$	15	$RN_{3,b}$	34
$n_{T2,1}$	9	$L_{2,6}$	11	$L_{5,3}$	19	SoT_4	54	$RN_{4,a}$	26	$RN_{4,b}$	28
$n_{T2,2}$	0	$L_{4,6}$	28	$L_{7,5}$	0	SoT_5	34	$RN_{5,a}$	53	$RN_{5,b}$	0
$n_{T2,3}$	11	$L_{5,7}$	0	$L_{5,6}$	19	SoT_6	39	$RN_{6,a}$	39	$RN_{6,b}$	39
$n_{T2,4}$	15					SoT_7	20	$RN_{7,a}$	20	$RN_{7,b}$	0

In eqns (6) to (8) the right side is the existing upper limit of capacity for lines (L), set of tracks (SoT) and route nodes (RN).

A further special feature that can be mapped when taking this course of action is the reversing of train moves in a station. The additional condition is that there must be scope for double negotiation of a station throat (RN) (cf. eqn (8)).

This is thus a course of action that enables all equality constraints to be imposed for lines, sets of tracks and route nodes.

After the solver has calculated the results the following output information can be derived (Table 2). The predefined ceilings for the infrastructure show that a total of 44 freight train moves are possible in the sub-network (sum of point-to-point route; cf. Table 2). If the solver does not find a solution, there is no possibility to route another freight train through the network. The second and third alternative from point-to-point route 1 as well as the second alternative from point-to-point route (cf. Table 1).

If there were any passenger trains in the sub-network, it must be checked how the limits of the capacity will be changed. This course of action is to be verified following optimization for each line, set of tracks and route node. A process of iteration with the new ceiling is to be launched should limiting factors be encountered. Analysis is then to be conducted in such an instance to find out into whether a solution to the equation problem tends towards one value or else towards the degree of accuracy following which iteration can be ended.

5 SUMMARY AND OUTLOOK

Establishing network-wide capacity constitutes a key developmental goal of strategic network planning. It is possible by addressing the capacity of railway lines and railway nodes to effect the meaningful optimization by automated means of running paths and capacity consumption for entire networks.

To reach this goal for a net-wide optimization a macroscopic network can be described by a set of linear equations. With the aid of a solver the maximum number of trains in this network can be determined. Furthermore, a running-path search for freight trains through the network is possible by knowing the sources and the appropriate sinks of them.

Optimization provides data on possible train paths within the sub-network. Building upon these it is possible to evaluate train-path consumption, the loading of individual infrastructure elements and the most worthwhile running paths through the sub-network. There is scope for comparing differing infrastructure or operating-schedule scenarios, thus allowing pronouncements to be made regarding current or future bottlenecks and their elimination or prevention.

The approach presented is capable of pointing up the direct impact that expansion, redesign or downsizing measures have on a railway network's total output capacity. Depending on this fact a benefit/cost ratio could be used to measure the effects of synergy of the whole network. For example, if there is a local expansion on a line an improvement of capacity will be recognizable but there are circumstances, such as when an adjoining node limits the number of train moves, under which it will not benefit the network as a whole.

REFERENCES

[1] Hansen, Ingo A. & Pachl, Jörn, eds., *Railway Timetabling & Operations*. 2nd revised and extended edition. Hamburg: Eurailpress, 2014.

[2] Nießen, Nils, Queueing. *Railway Timetabling & Operations*, eds. A. Hansen, Ingo, Jörn Pachl, 2nd revised and extended edition. Hamburg: Eurailpress, pp. 117–131, 2014.

[3] Pouryousef, Hamed, Lautala, Pasi, & White, Thomas, Railroad capacity tools and methodologies in the U.S. and Europe. *Journal of Modern Transportation*, **23**(**1**), pp. 30–42, 2015.

[4] Abril, M., Barber, F., Ingolotti, L., Salido, M.A., Tormos, P., Lova, A., An assessment of railway capacity. *Transportation Research Part E: Logistics and Transportation Review*, **44**(**5**), pp. 774–806, 2008.

[5] Sameni, Melody Khadem, Dingler, Mark, Preston, Jhon M., & Barkan, Christopher, P.L., Profit-generating capacity for a freight railroad. *TRB 90th Annual Meeting*, TRB: Washington, DC, 2011.

[6] Happel, O. Sperrzeiten als Grundlage der Fahrplankonstruktion. *Eisenbahntechnische Rundschau*, **8**(H.2), pp. 79–80, 1959.

[7] Union of Railways, UIC Code 406, 1st edition. Paris, 2004.

[8] Pachl, Jörn. *Railway Operation and Control*. VTD Rail Pub.: Mountlake Terrace, WA, 2009.

[9] Kuckelberg, Alexander, Janecek, David, & Niessen, Nils, Grundlagen zur Simulation der Fahrplanerstellung und Betriebsabwicklung. *Eisenbahntechnsiche Rundschau* **7+8**, pp. 50–55, 2013.

[10] Kleinrock, Leonard & Gail, Richard. *Queueing Systems: Problems and Solutions*. Wiley: New York, 1996.

[11] Gnedenko, B.W. & König, D., *Handbuch der Bedienungstheorie Vols I and II*. Akademie-Verlag Berlin: Berlin, 1983.

[12] Fischer, Klaus & Hertel, Günter, *Bedienungsprozesse im Transportwesen : Grundlagen und Anwendungen der Bedienungstheorie. 1. Aufl.* Transpress-Verl.-Ges.: Berlin, 1990.

[13] Prinz, Robert, Sewcyk, Bernd, & Kettner, Michael, NEMO – Netz-Evaluations-Modell bei der ÖBB. *Eisenbahntechnsiche Rundschau*, 50 (3), pp. 117–121, 2001.

[14] Dijkstra, E.W., A note on two problems in connexion with graphs. *Numerische Mathematik*, **1**(**1**), pp. 269–271, 1959.

[15] Schwanhäusser, Wulf, *Die Bemessung der Pufferzeiten im Fahrplangefüge der Eisenbahn*. Schriftenreihe des Verkehrswissenschaftlichen Instituts der RWTH Aachen H., 20, 1974.

[16] Janecek, David & Weymann, Frédéric, LUKS – Analysis of lines and junctions. *Proceedings of the 12th World Conference on Transport Research (WCTR)*, Lissabon, Session C2.15, 11–15 July 2010.

[17] Radtke, Alfons, *EDV-Verfahren zur Modellierung des Eisenbahnbetriebs*. Hanover: Institut für Verkehrswesen, Eisenbahnbau und -betrieb Universität Hanover; Eurailpress, 2005.

TRAIN TIMETABLE RESCHEDULING GENERATION BASED ON VEHICLE TYPE AND TRAIN ROUTE COMBINATIONS

T. KATORI & T. IZUMI
Nihon University, Japan.

ABSTRACT

Train rescheduling means a transient situation to correct a train diagram in a suspended state due to traffic accidents or disasters. Automatic (or half manual) rescheduling has been studied, and such previous research has shown promising results.

In this paper, we describe a train timetable rescheduling method. In Japanese urban areas, some private companies operate on each other's tracks. If vehicle types have limitations due to ground facilities or company rules, the vehicle has to be operated under these limitations even when running on a rescheduled timetable. Even if the ground facilities of different companies have uniform conditions, local and rapid trains must be operated in a distinct manner.

Therefore, we suggest a rescheduling method. With this method, each vehicle type and its vehicle routes based on the track layout are registered, and rescheduling diagrams are composed with the route combinations.

Important conditions to decide the combinations are vehicle location at the operation resumption time and the introduction of same-type vehicles at an originally unscheduled timing. We compare the traffic effects for some combinations of the latter situation where originally unscheduled same-type vehicles are introduced for rescheduling. The evaluation values are average headway time and its standard deviation at all stations on the timetable.

We apply our rescheduling method to a theoretical line and timetable modelled on existing urban lines in Japan where trains go and come back on double tracks, and indicate the efficacy of our rescheduling method.

Keywords: route combinations, train rescheduling, vehicle type.

1 INTRODUCTION

Railway transport is generally considered safe and punctual, but sometimes the transport cannot operate on the scheduled timetable due to traffic accidents or disasters. Train rescheduling means a transient situation to correct a suspended state. When a traffic accident suddenly happens, a rescheduling timetable has to be generated quickly, within 5 to 10 min. Train rescheduling means swapping the departure order and changing the turning or terminal station. Here, a short delay of a few minutes which can be solved by a shorter turning time at the terminal station is not included.

In Japan, a rescheduling timetable used to be generated by expert operators, but the number of operators was decreased due to management streamlining, and now this work has to be processed automatically. Computer processor performance improved and processors are able to solve realistic size problems [1, 2]. Many railway companies use programmed traffic control systems, and the timetable data are kept and processed on such computer systems. Therefore, automatic rescheduling became fit for computers, because manual converting is not needed [3].

DOI: 10.2495/TDI-V1-N3-403-413

However, companies running on through other company facilities to accommodate passengers face issues such as running area restrictions for different vehicle types. This limitation makes rescheduling difficult [4].

In this paper, we describe how to generate train rescheduling timetables considering vehicle-type operation. The rescheduled timetable has to satisfy two limitations, namely certain track areas are bound to certain vehicle types, and trains cannot be positioned at the same location at the same time. To solve these limitations, the track layout is managed by dividing the time interval system into data blocks. Train routes are registered based on the tracks between the start and terminal stations based on running time. The rescheduling timetable is composed with combinations of these routes.

This rescheduling procedure has four parameters: traffic accident happening time, communication time for all trains, resumption time and complete recovery time. We compare some rescheduled timetables based on variations of the above parameters. In this paper, we adopt only turning operation on double tracks as is usual in Japanese urban areas.

2 ASSUMPTIONS AND DATA TYPES FOR GENERATING TRAIN RESCHEDULING

2.1 Assumptions for generating train rescheduling

Train rescheduling timetables are generated under the following assumptions and modelling:

- Train operation unit time is 30 s. This means train locations are renewed every 30 s.
- Minimum headway time between trains is 2 min.
- Minimum dwell time is 5 min at turning operation and 30 s at the other middle stations.
- Running time between stations is the same on the scheduled timetable as on the rescheduling timetable. The train stops at the station only, does not stop between stations and does not operate at slow speed. In reality, trains may run at a slower speed than normal, but on the rescheduling plan, trains run on standard operation time only.
- All vehicles have the same driving performance. (On Japanese urban lines, most trains use the same vehicle type.)
- Trains are never partitioned or combined.

2.2 Data types and format

2.2.1 Track layout data

The track layout is divided by data blocks based on running time, and the blocks have ID numbers. In this study, the unit time is 30 s as indicated in Section 2.1. When the running time between stations is 2 min, this area is divided into four blocks. For crossover, 30 s are allotted, even if the train can run the distance in a shorter time. So, the track layout is expressed by blocks considering both time and distance. These block sections differ from real signalling sections, because this model is a theoretically simple model. If the unit time is shortened by 1 or 5 s, the sections can express a more detailed solution than signalling sections.

Figure 1 shows an example of a block division for a sample track layout and the data concerning the block number and the location from the terminal.

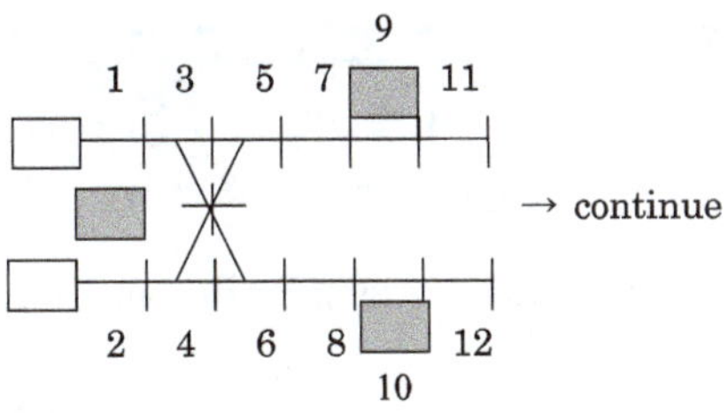

Figure 1: An example of a block division for track layout. (▩ : platform).

Table 1: Data form for a train schedule.

Vehicle type	Time (min) Train No.\	0	0.5	1	1.5	2	2.5	
1	0	1	3	5	7	9	9	…
2	1	16	14	12	10	10	8	…
1	2	61	59	58	56	54	54	…
2	3	63	63	63	65	67	67	…

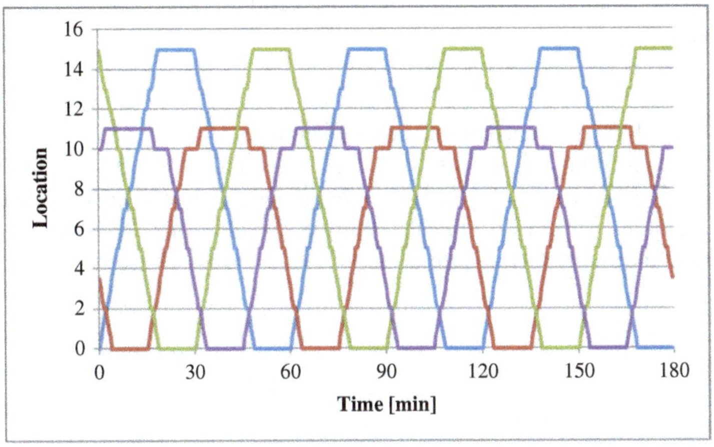

Figure 2: Sample of a train schedule diagram.

2.2.2 Train diagram data

Train diagram data express a row of track layout block numbers and train location versus time. Each vehicle operates differently, and the vehicles continue to have location data after turning at the terminal station.

Table 1 shows a data form for a train schedule, and Fig. 2 is a train diagram based on Table 1.

2.2.3 Route data

Route data are already combined in the scheduled operation. Therefore, these data are only needed for rescheduling.

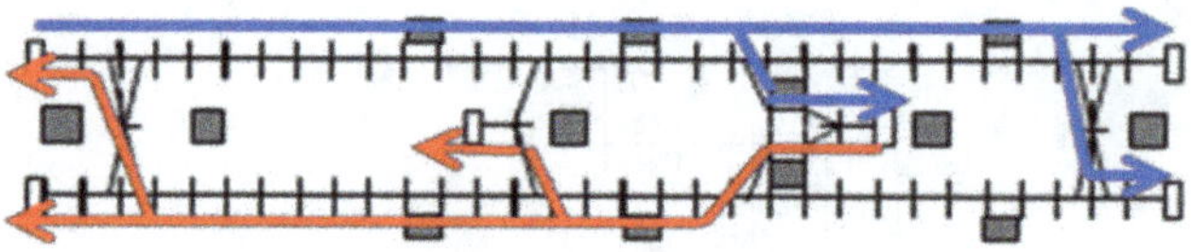

Figure 3: Some examples of route sets.

Route data contain information about possible routes, required time and possible vehicles for a particular route between the start and terminal stations.

Figure 3 shows some examples of route sets. Blue arrows indicate routes from the upper left track, and the red ones from the middle right track. All possible routes are registered per vehicle type, between all turning tracks.

3 HOW TO GENERATE A TRAIN RESCHEDULING TIMETABLE CONSIDERING VEHICLE TYPES AND ROUTE COMBINATIONS

3.1 Parameters for rescheduling

The following four parameters have to be provided to generate a timetable:

- time and location of the traffic accident
- time needed to communicate the traffic accident to all trains
- resumption time
- complete recovery time.

Figure 4 shows these parameters on the train diagram.

The time needed to communicate the traffic accident to all trains means the delay that is required till all other trains are informed. The train in the accident stops soon, but other trains continue to run on the scheduled timetable until they receive the accident information. All trains stop after this communication time.

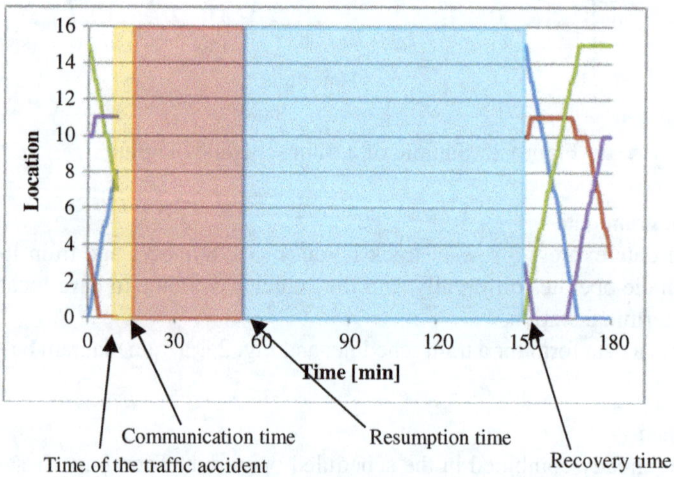

Figure 4: Rescheduling parameters.

Resumption time can set different times for the accident area and different areas and considers turning operation on the part of the line.

Complete recovery time means the end of the rescheduling operation time. The rescheduling timetable is generated between the resumption and complete recovery time.

3.2 Rescheduling timetable generation

3.2.1 Pre-processing

All trains stop at the accident time or the communication time. The stopping location is the nearest stopping station for each train.

Trains are operated by the rescheduling timetable after the resumption time.

First, new destination stations are redefined for all trains. As an experimental rule, the new destination is the original destination of the train, but if there are no tracks available when the train is to reach its original destination, the destination station changes to a closer station with a possibility to turn. If the new destination is closer than original one with a possibility to turn, trains cannot operate because the number of turning equipment is less than the number of trains. The decision of a new destination station is a very important condition, because this condition defines the rescheduling solution quality.

Also, each train is not necessarily located at its start station on the scheduled timetable at the complete recovery time. Therefore, the origin station and starting time are defined for running trains at recovery time.

Figure 5 shows pre-processed results for Fig. 4. Train diagrams are added until the destination station after the resumption time and from the origin station before complete recovery time.

A rescheduling timetable is calculated between the new origin station at the starting time and the destination at the complete recovery time for each train. This time is called rescheduling time.

3.2.2 Vehicle type and operation

If each vehicle is not on the location of the original scheduling diagram at the complete recovery time, there is no problem if a vehicle of the same type operates at that location and direction. Therefore, same vehicle-type connection at the complete recovery time is also an important condition.

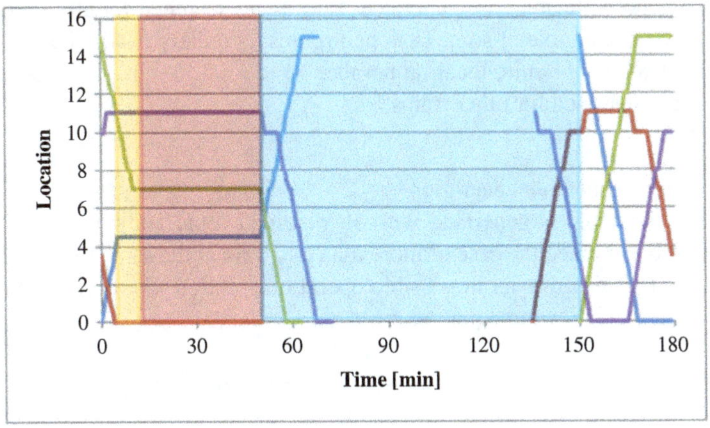

Figure 5: A result of pre-processing.

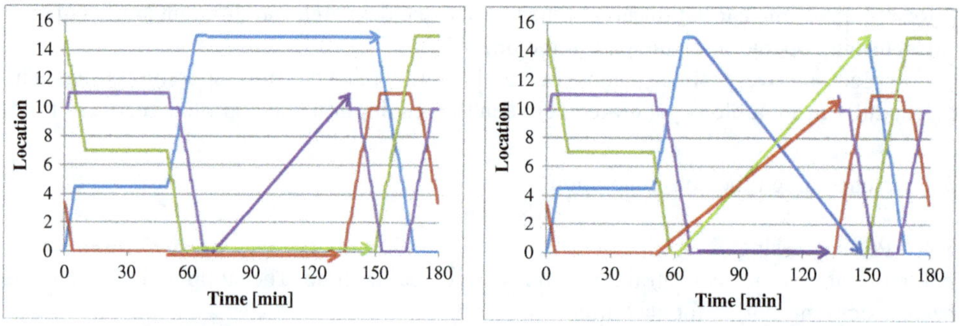

(a) No vehicle change (b) Vehicle change

Figure 6: Connection for vehicle type on operation.

In this paper, this is also set as a parameter, and the vehicle connection effect is compared.

Figure 6 shows some connections for vehicle operation. Figure 6(a) indicates the trains connecting with the original operation, and Fig. 6(b) indicates a substitution between same-type vehicles.

3.2.3 Route generation

All possible train routes between any turning station during rescheduling time are investigated. For this procedure, route data in Section 2.2.3 are used, and all possible route combinations are investigated recursively and saved. This investigation uses the start and terminal track numbers and required time. The total required time should be shorter than the rescheduling one. This investigation can be considered a knapsack problem with a limitation, where the terminal track number g_i in route R_i must be the start track number s_{i+1} in next route R_{i+1}.

If the rescheduling time is *rtime*, this relation gives the inequality

$$\Sigma R_i < rtime\ . \tag{1}$$

Because the route data include the possible vehicle types for a specific route, only route combinations for possible areas are generated. Routes are searched for all running trains.

Figure 7 shows an example of some searched routes. Numbers 1–4 in the figure are routes on the diagram from $t = 68$ min, location number 15 to $t = 150$ min, same location number. Similar routes are searched for other trains.

3.2.4 Rescheduling timetable generation

A rescheduling timetable is generated with all obtained route combinations for all trains. These combinations include the case of more than one train at the same time and on the same track. Such case is rejected as an impossible operation solution. For each train *i* and *j*, location *L* in time *t*, if the following equation gives

$$L_{it} = L_{jt} \qquad (i \neq j), \tag{2}$$

then a backtrack procedure is adopted, and computation time decreases. This way, crossing hindrance is also avoided. When trains are not located on the same track, the route combinations that do not maintain the minimum headway time are rejected.

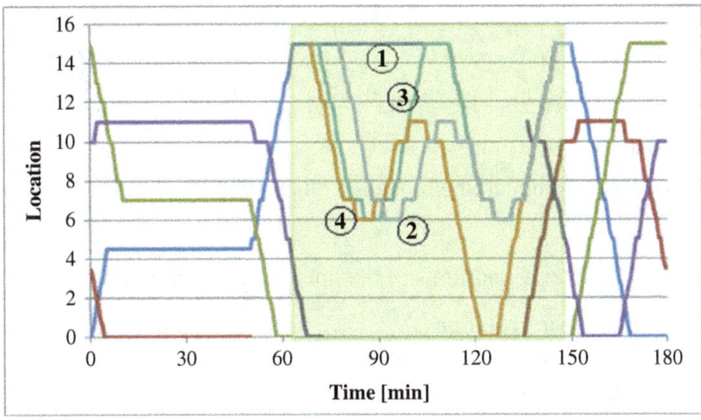

Figure 7: A route tree example (from terminal to terminal).

3.2.5 Evaluation value

The quality of the generated rescheduling timetable is evaluated, and the best solution is adopted. The evaluation value is the total product at station *i* on average headway time ave_i and standard deviation s_i for the same operating direction. A shorter average headway time at any station results in a large number of trains operations, which is convenient for the passenger. Additionally, a smaller standard deviation at the station means a uniform density operation. Both the smaller elements are good, and the evaluation value is given in eqn (3).

$$E = \Sigma\left(ave_i * s_i\right) (\text{min}^2). \tag{3}$$

The shorter the average headway time becomes, the higher the number of train, but distribution per an hour cannot be decided only by this. Therefore, standard deviation is used to uniformly distribute the train headway time and ensure even distribution of passenger numbers on the trains.

Under the current combination of parameters, several rescheduling timetables are possible. However, the smallest evaluation value is adopted as the solution.

4 RESULTS AND DISCUSSION

4.1 Model line for this study

The rescheduling procedure is applied to a line modelled on existing urban lines, and a rescheduling timetable is generated.

Figure 8 shows the line modelled on existing lines, and Table 2 shows the conditions of the line under consideration.

On this model line, two vehicle types operate. One type can operate on the whole line, and the other can run between station numbers 0 and 5. This condition means, for example, that direct current is available between station numbers 0 and 5, but alternative current or diesel operation is used between station numbers 5 and 8.

Figure 9 (= Fig. 2) shows the scheduled diagram for the model line. Blue and green vehicle types can run on the whole line, but the red and purple ones are limited and can only run until the middle turning station. Only local trains operate on this line. Trains running till station 5 and trains running till station 8 operate alternatively every 15 min. A rescheduling timetable is generated for this line and diagram by changing the parameters.

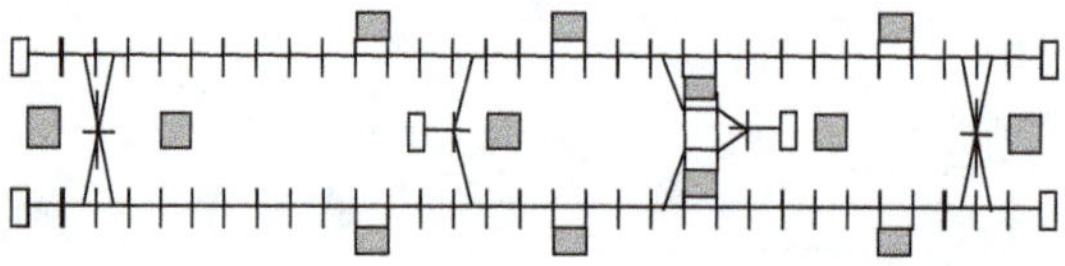

Figure 8: Theoretical line modelled on existing line (▩ : station).

Table 2: Conditions of the line under consideration.

Number of stations	9 stations
Running time (non-stop)	15 min
Total number of blocks	68
Number of routes	18
Total length	16.2 km

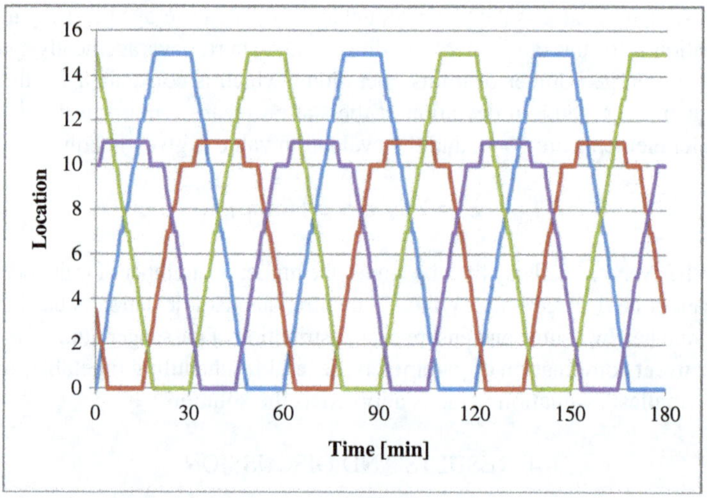

Figure 9: Originally scheduled train diagram (= Fig. 2).

4.2 Results

On the scheduled timetable, a traffic accident happens 5 min after the train departure. Communication time is set to take 5 min, resumption time is 40 min, and complete recovery operation time is set to vary. All parameter conditions and their evaluation values are shown in Table 3.

Figures 10(a), 10(b) and 10(e) are similar to Fig. 6(a), i.e. they connect to the original vehicle operation, and Figs. 10(c), 10(d) and 10(f) are similar to Fig. 6(b), i.e. they connect to same-type vehicles in different operations. In addition, Figs. 10(e) and 10(f) indicate waiting solutions where the train waits until the cycle repeats. These solutions can possibly not be called rescheduling. For Figs. 10(e) and 10(f), the resumption time is set at 30 min because setting it at 40 min resulted in the same solutions.

Figure 10 shows a sample of the rescheduled diagram under these conditions.

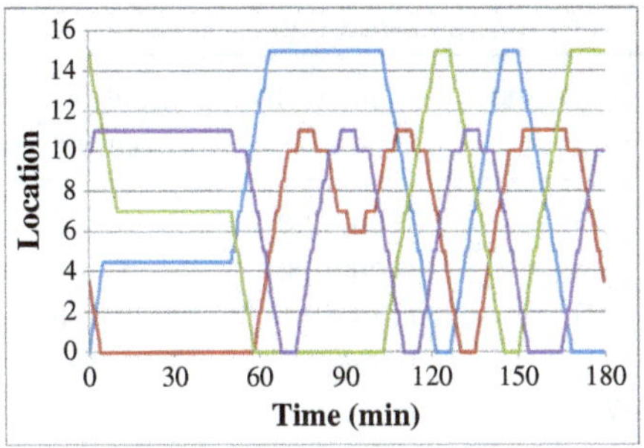

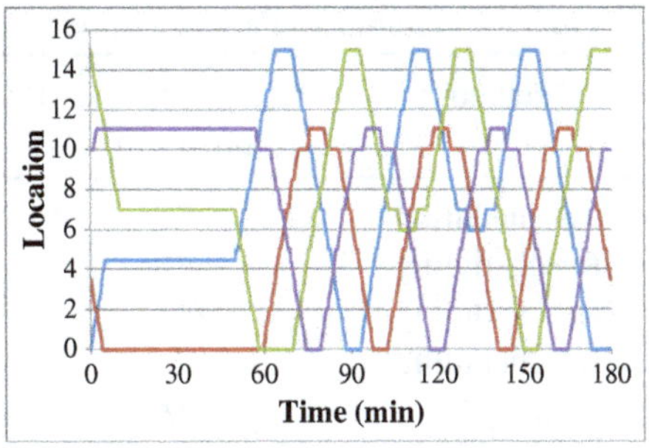

(a) Rescheduling without vehicle change. (Recovery time = 150(min))

(b) Rescheduling without vehicle change. (Recovery time = 178(min))

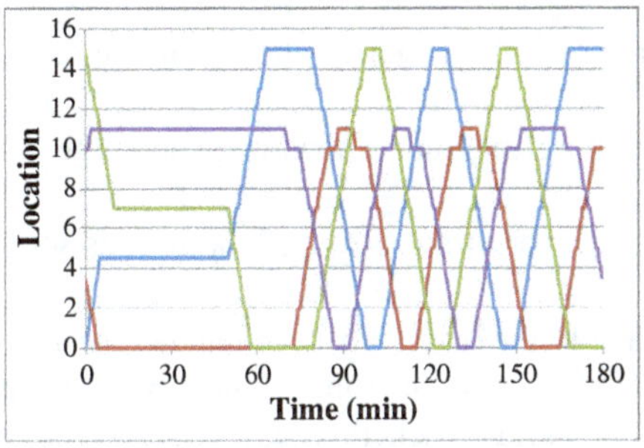

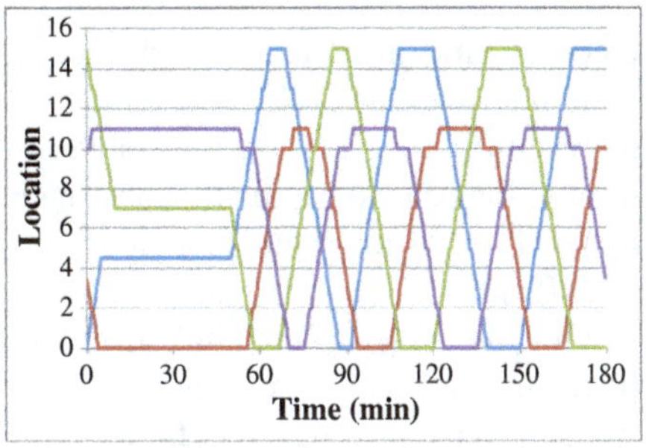

(c) Rescheduling without vehicle change. (Recovery time = 150(min))

(d) Rescheduling without vehicle change. (Recovery time = 95(min))

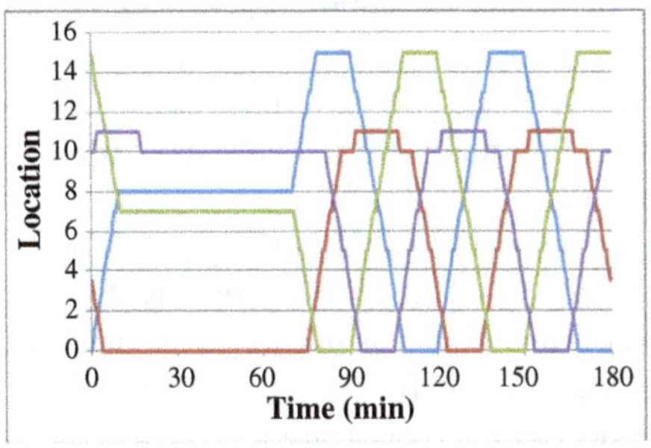

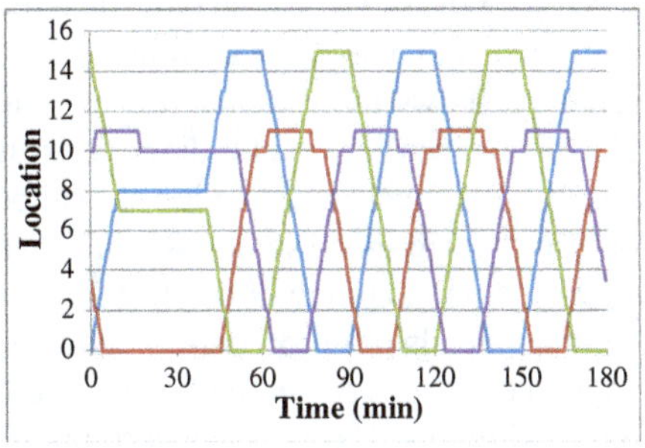

(e) Discontinuation without vehicle change until the cycle is repeated.

(f) Discontinuation with vehicle change until the cycle is repeated.

Figure 10: Generated diagrams and conditions.

4.3 Discussion and considering

4.3.1 Discussion

Figures 10(a) and 10(b) show rescheduling without changing vehicle operation. In the diagram, any trains turn at station no. 3, which is generally called the 'mountain cut', and the diagrams give a tricky impression. Figures 10(b) and 10(d) have the smallest evaluation value based on optimum recovery time. The solution of Fig. 10(d) is obtained by only shortening the turning time at the terminal stations. When comparing Figs. 10(a) and 10(b) with Figs. 10(c) and 10(d) in terms of flexible operation effect for the same vehicle type, Figs. 10(c) and 10(d) have smaller evaluation values. For operation with change vehicle,

Table 3: Parameter conditions for train rescheduling.

	Method	Resumption time (min)	Recovery time (min)	Connection type	Evaluation value (min^2)
a	Rescheduling	40	150	No change	5,772
b	Rescheduling	40	178	No change	1,400
c	Rescheduling	40	150	Change	2,258
d	Rescheduling	40	95	Change	1,391
e	Waiting	30	78	No change	4,100
f	Waiting	30	48	Change	2,120

the recovery time on Table 3(d) is shortened. Free operation shortens the complete recovery time and improves the evaluation value.

Figures 10(a)–(d) show the generated rescheduling diagrams; however, sometimes there are no solutions due to parameter combinations.

Figure 11 shows the evaluation values versus recovery time. Resumption time is set constant at 40 min.

The blue line indicates that the operation is back to the originally scheduled diagram.

The red line indicates connections to other operations with the same vehicle type. It is generally said that a longer recovery time will result in a good rescheduling diagram. If the vehicle operation remains the same as the original one (Fig. 6(a)), then the longer recovery time leads to a smaller evaluation value. On the other hand, if the vehicle operation is changed while keeping the same vehicle type, then the evaluation value feature is hardly related to the recovery time and does not vary much either.

Figure 11 shows that with a recovery time shorter than 175 min, the vehicle operation should be changed if a cyclic diagram is used on this model line. If the recovery time is shorter than the time, the operation should be changed.

Sometimes there are no solutions available for any recovery time, depending on the timetable cycle and the running time of one turning. The reason is that the recovery time is shorter than one turning time. In this procedure, waiting time is not distributed on turning time but used after resumption time only. Accordingly, the generated rescheduling timetable is a tight one.

Figures 10(e) and 10(f) indicate a solution that only consists of waiting using the repeating cycle timetable. Especially Fig. 10(f) results in an earlier recovery time, thanks to a flexible combination of same vehicle types. However, because both Figs. 10(e) and 10(f) are made to wait after the possible resumption time, the evaluation values are not so favourable.

The computation time is about 20 s on a general personal computer, because the modelled line size and vehicle numbers are small.

4.3.2 The difference with the procedure by an expert operator

The proposed rescheduling procedure is similar for the way of thinking by an expert operator. In other words, the flow to make minor adjustments to decide the number of possible turnings for a given vehicle in a given area till the complete recovery time and to avoid crossing hindrance is similar. However, because waiting time is not included in the turning time, only rigid solutions are obtained, and sometimes there are no solutions, depending on the way the complete recovery time is set. In addition, this procedure does not have typical rule-based knowledge. Therefore, solution-searching efficiency is not so high.

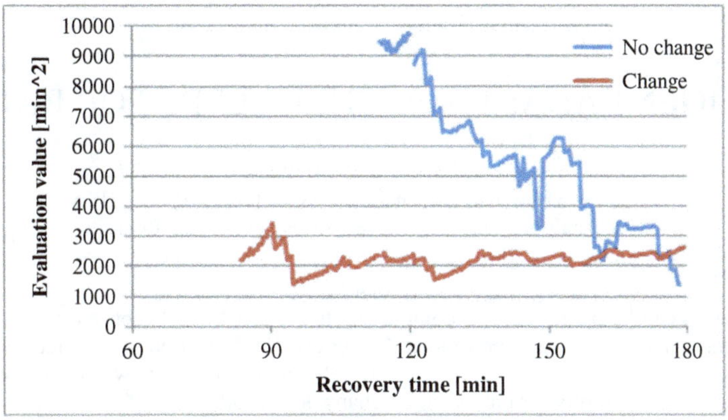

Figure 11: Evaluation values versus recovery time (with or without changing vehicles operation). Resumption time: 40 min.

5 CONCLUSION

We suggested how to generate an automatic rescheduling timetable when traffic accidents occur. We considered vehicle types in particular. Possible routes per vehicle type were registered in order to deal with different private companies running through on each other's tracks. The rescheduling timetable is a combination of route combinations including information on vehicle types that can be used.

Important elements for the rescheduling procedure are train locations at the resumption time and how to deal with vehicle operation, i.e. changing or not changing the vehicle operation. Both conditions can still be improved.

In future research, we will focus on automatic determination of vehicle connection and the addition of rapid train operation. In addition, shorter division unit time (track length) can lead to higher accuracy for complex timetables with a larger size. We will also evaluate the method for spider operation as is usual in Europe.

ACKNOWLEDGEMENT

We thank Associate Professor Ruth Vanbaelen, Nihon University, for English language assistance.

REFERENCES

[1] Kunimatsu, T., Hirai, C. & Tomii, N., Train rescheduling evaluation from the viewpoint of passengers by microsimulation of train operation and passenger flow. *IEEJ Transactions on Industry Applications*, **133(7)**, pp. 756–764, 2013. DOI: 10.1541/ieejias.133.756. (In Japanese)

[2] Takagi, R., Newly developed simple railway timetable evaluation program Sujic with the new model to deal with re-scheduling. *Computers in Railways XIII*, **(13)**, pp. 513–520, 2012. DOI: 10.2495/CR120431.

[3] 'Railway diagram recover technology', Ohm-sya, 2010. (In Japanese)

[4] Katori, T., Yamazaki, H. & Izumi, T., Generating a train rescheduling timetable considering the train routes on tracks and car types. *Computers in Railways XIV*, **(14)**, pp. 675–685, 2014. DOI: 10.2495/CR140561.

INTELLIGENT ADAPTIVE SCHEDULERS FOR RAILWAYS

G. RZEVSKI[1] & P. SKOBELEV[2]
[1]The Open University, UK and Multi-Agent Technology Ltd, UK.
[2]Samara University and Smart Solutions Ltd, Samara, Russia.

ABSTRACT
Railway operation can be perceived as a complex system consisting of thousands of constituent elements, including demands for transportation and a variety of transportation resources, such as tracks, track blocks, stations, sidings, trains, crews, etc. Overall behaviour of a railway operation is characterized by unpredictable disruptive events such as changes in availability of resources due to failures, weather conditions or human errors. In addition transportation demands tend to change over time whilst changing transportation resources, such as tracks, is not always possible or practical. The key task of railway management is the allocation of transportation resources to transportation demands with a goal of achieving a complete match, ensuring a smooth operation. The difficulty of this task primarily depends on the variability of demand and reliability of resources, in particular, tracks, trains and human resources. The paper describes how to design complex adaptive railway schedulers, which allocate resources to demands in real time and ensure rapid rescheduling in reaction to unpredictable disruptive events.
Keywords: adaptability, complexity, railway schedules, real-time schedulers, self-organization.

1 INTRODUCTION

This paper describes how concepts, principles and methods of complexity science have been used to revolutionize timetabling and scheduling of railways.

1.1 Complexity science basics

Complexity is a property of open systems consisting of a large number of diverse, partially autonomous components, called agents, engaged in interaction by exchanging information, without having central control. Global behaviour of complex systems emerges from the interaction of agents and is unpredictable but not random; it follows discernible patterns.

There is a well-developed theory of complex systems [1–4], which provides rich variety of methods and tools for modelling and managing complexity [5], and documented experience in applying these methods and tools to variety of business and engineering problems, including real-time space logistics [6], real-time aircraft life cycle management [7], real-time supply chain management [8], real-time taxi scheduling [9], real-time car rental scheduling [10] and many others.

1.2 Railway operation as a complex system

Railway operation can be perceived as a complex system. It consists of thousands of interacting, diverse constituent elements, including demands for transportation and transportation resources such as tracks, track blocks, stations, station crews, trains, train crews, resource

DOI: 10.2495/TDI-V1-N3-414-420

maintenance crews, maintenance plants and tools, energy supplies (normally, electricity or diesel), energy supply equipment, cargos, cargo loading equipment, passengers, passenger services (e.g. ticketing, inspection, safety and security) and passenger supplies (e.g. food, drinks, waste disposal equipment). Overall behaviour of a railway operation is unpredictable in detail. All railways experience, to a greater or lesser degree, unexpected delays and disruptions of service and occasional disasters due to unpredictable disruptive events such as changes in availability of resources due to resource failures, weather conditions or human errors. In addition transportation demands tend to change over time whilst changing transportation resources such as trucks is not always possible or practical.

The key task of railway management is the allocation of available transportation resources to transportation demands, aiming at achieving a complete match. The difficulty of this task primarily depends on the variability of demand and availability of resources, in particular, tracks, trains and human resources. As a rule, the variability is such that scheduling of resources must be done in real time.

2 THE RAILWAY SCHEDULING PROBLEM

The railway-scheduling problem can be formulated as the allocation of trains to track blocks with the aim of ensuring movement of trains according to specified constraints and priorities, where a block is a section of a railway line, which can be occupied only by one train. Standard operational conditions include the following:

- Different trains travel with different speeds.
- Different trains have different priorities.
- A train can overtake another train at a station or, if necessary, use a free track block on the parallel line (even on a line normally reserved for trains travelling in the opposite direction).
- The capacity of a station to hold trains, which are being overtaken, is limited.

Train scheduling is subject to the occurrence of unpredictable disruptive events, including train failures, track failures, train delays and track repair delays.

Whenever a disruptive event occurs, the schedule must be changed to accommodate the disruption with the minimal consequences to trains that are not directly affected.

3 CASE STUDY 1

Our task was to improve scheduling of trains on Moscow–St Petersburg main line, which is one of the busiest railway lines in Russia. The line has 700 km of tracks, 49 stations, 3,700 track blocks and 48 sections, including 2 four-track sections, 14 three-track sections, 31 double-track sections and 1 single-track section. The diversity of 810 trains is very high (passenger slow, passenger suburban, passenger express, passenger direct super express, freight, service, etc.) with 50 different priorities. An exceptional traffic density in suburban areas of Moscow and St Petersburg was a particular problem. Unpredictability of disruptions was very high: on average 100 unpredictable disruptive events are occurring during a single dispatcher shift.

A specific requirement was that the direct high-speed trains were not allowed to be late whatever the conditions. Track failures were frequent and repair works occasionally required up to 8 h of track closures. The official train timetable produced by railway headquarters was given to us as a reference and the starting point for real-time scheduling.

Table 1: Case study 2: train delay reasons

Delay Reasons	Passenger train delays p.a.	Freight train Delays p.a.
Locomotive & crew	432	15819
Passenger/freight carriages	155	6659
Track failures/bad conditions	174	10667
Signalling failures	174	672
Power supply failures	99	451
Wrong operational decisions	442	667

4 CASE STUDY 2

The second task was to develop a system of real-time schedulers for a Regional Siberian Railways with 5,000 km of tracks, carrying 5 millions of passengers and transporting 70 million tonnes of freight per annum (Table 1).

Trains often operate in this region under extreme weather conditions. The second case study is significant because of the volume of railway traffic in Russia. Russia is second only to China in railway traffic density in the world and third in railway traffic volume after the USA and China.

5 THE SOLUTION

5.1 Brief overview

The proposed solution for both case studies was a network of cooperating real-time schedulers. However, each network of schedulers required a considerable customization effort to accommodate specific client requirements.

The constituent schedulers of both networks are capable of entering the official train timetable as the initial state of the system, producing the schedule covering all constituent resources and then, most importantly, rapidly reacting to any unpredictable disruptive event, such as a failure or a delay, and rescheduling affected trains in real time ensuring a continuous strict correspondence between the schedule and the corresponding situation in the field.

The real-time scheduling of trains capable of rapidly eliminating consequences of unpredictable disruptions is our key achievement and represents, to the best of our knowledge, an original contribution to railway operation.

Our real-time schedulers enable to split railway timetabling into two parts:

- Producing master timetable applying evolutionary, heuristic or optimization methods, which can be used as a guide by railway authorities and users [11–13].
- Rapidly rescheduling master timetable whenever an unpredictable disruption occurs with the aim of eliminating consequences of disruptive events, as described in this paper.

5.2 Network architecture

The architecture of the event-driven, multi-agent, real-time scheduling network is illustrated in Fig. 1.

The network consists of a number of real-time schedulers, which share the scheduling load and cooperate or compete with each other.

5.3 Scheduler architecture

Each scheduler consists of three key components: a knowledge base, a virtual world of software agents and the links between the virtual world and the real world.

Knowledge base consists of ontology, containing conceptual knowledge and databases containing factual knowledge.

Ontology contained classes of objects such as train, station, sidetrack and block, linked by relations, such as 'Train occupies Block', 'Train travels to Station', 'Sidetrack belongs to Station', etc. Each object was described by appropriate attributes.

Databases contain values defining instances of object classes.

Virtual world is the world of software agents. Typical agent types for the railway operation are train agent, station agent, sidetrack agent and block agent.

Scheduling is done by negotiation between demand agents and resource agents, where a typical demand agent is a train agent and a typical resource agent is a block agent. The basis for the negotiation is the premise that for a train to travel from A to B all blocks between these two points must be free from other trains at the time when the train is scheduled to reach them. When a train seeks a sidetrack at a station to wait to be overtaken by a higher priority train, resource agents involved in negotiation will be station agents and sidetrack agents. At any time there are over 40,000 agents simultaneously active.

Decisions on how to reschedule trains to eliminate consequences of disruptions are made by agent negotiations and conveyed to appropriate resources, as illustrated in Fig. 2.

The schedule is constructed in several stages.

The first stage is the production of a rough schedule for all trains based on the official timetable. The schedule is built for all trains in parallel, exposing points of conflicts between trains (more than one train on a block), if any.

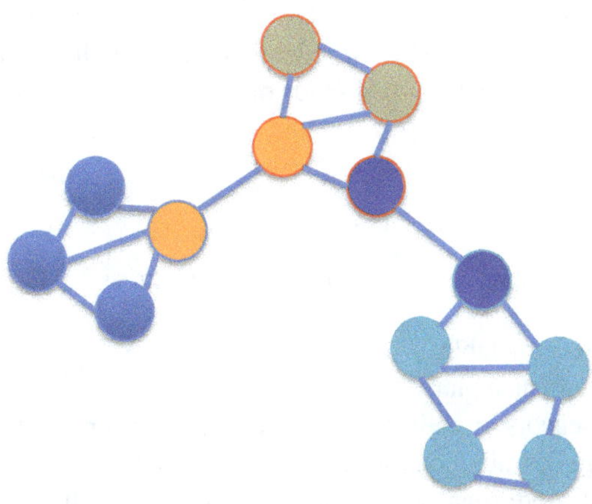

Figure 1: Scheduling of trains by a network of cooperating/competing schedulers.

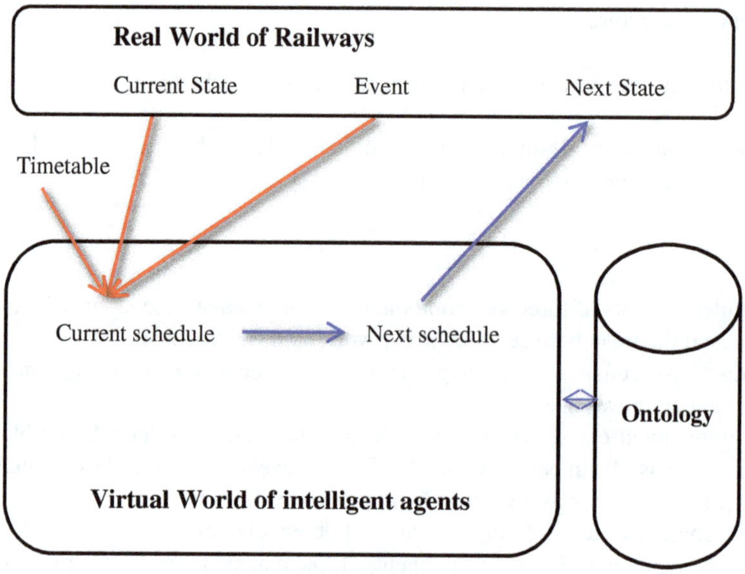

Figure 2: Real-time scheduler architecture.

The second stage is concerned with the elimination of conflicts. Train agents and block agents negotiate which of the trains that are in conflict will enter the contested block first, respecting, whenever possible, specified train priorities. Whenever a train has to overtake another train, the affected station agents and sidetrack agents get involved in the negotiation.

The third stage irons out anomalies, such as excessive delays imposed on some of the low-priority trains, aiming to produce a well-balanced schedule.

All negotiations are conducted with the aim of maximizing enterprise value, which is a complex function that balances different, often conflicting, goals, i.e. respecting specified train priorities, minimizing train delays, minimizing the use of blocks of the line reserved for trains travelling in opposite direction, special train constrains, etc. In the third stage attempts to increase enterprise value are carried out until the process reaches the point of diminishing returns.

If the selected scheduling strategy does not lead to the satisfactory enterprise value, agents may autonomously decide to destroy the inadequate schedule and start the process from scratch, following alternative strategies (constructive destruction). The same may happen if the user of the system (a dispatcher) decides that the schedule produced by the agents is inadequate.

The scheduler receives inputs from and sends outputs to the railways information processing system.

5.4 Salient features

The constituent schedulers have the following salient features:

- Loading of the master timetable
- Loading of the maintenance plans
- Loading data from sensors on railway infrastructure and train positions
- Real-time processing of train locations and track occupancy
- Taking into account more than 80 criteria when scheduling trains (train length, weight, running/accelerating/braking times, track layouts, security, etc.)

- Reacting in real time to unpredictable disruptive events (track or train failures, delays) and rescheduling affected trains
- Rapidly resolving unforeseeable conflicts (two trains demanding access to the same siding at the overlapping times) and rescheduling affected trains
- Visualization of output data to help dispatchers to make decisions
- Creation of and storing a variety of reports

5.5 The key advantage of multi-agent technology

Multi-agent schedulers are capable of rapidly (a) detecting a disruptive event, (b) searching for trains that will be affected and (c) rescheduling affected trains in real time before the next disruption occurs.

This unique feature is possible because computational search is replaced with search by interaction (communication) between agents.

6 RESULTS

The early version of the adaptive multi-agent scheduling system has substantially reduced the time required to reschedule the affected trains after the occurrence of disruptive events (failures or delays) and improved the utilization of railway resources. The following are some of the improvements achieved within operational and security constraints.

- Delays of high-speed trains practically eliminated
- Average train delays reduced to below 8%
- Trains regained time lost due to disruptive events one and a half times faster
- Productivity of dispatchers doubled
- Losses due to delays reduced by 50%
- Speed of rescheduling to eliminate consequences of failures increased dramatically (800 trains rescheduled within a minute)
- Customer satisfaction increased
- Opportunities opened for the increase of traffic density

7 CONCLUSIONS

Railway timetabling is a well-established art. However, recent developments in complexity science offer new superior methods for scheduling of trains in real time, which can increase profitability of railways and improve quality of service to railway clients. This paper describes how these new scheduling methods have been applied to two very different railway operations with a considerable success. To the best of authors' knowledge the real-time schedulers developed by the authors' teams are the first in the world actually working and delivering value to a railway authority under very difficult operating conditions.

The further work will include polishing agent negotiation and growing the network of competing/cooperating schedulers.

REFERENCES

[1] Prigogine, I., *Is Future Given?,* Singapore: World Scientific Publishing Co., 2003.
[2] Prigogine, I., *The End of Certainty: Time, Chaos and the New Laws of Nature,* New York: Free Press, 1997.
[3] Kaufman, S., *At Home in the Universe: The Search for the Laws of Self-Organization and Complexity*, Oxford, NY: Oxford Press, 1995.

[4] Holland, J., *Emergence: from Chaos to Order*, Oxford, NY: Oxford University Press, 1998.
[5] Rzevski, G. & Skobelev, P., *Managing Complexity*, WIT Press: Southampton, Boston, 2014.
[6] Rzevski, G., Soloviev, V., Skobelev, P., & Lakhin, O., Complex adaptive logistics for the international space station. *International Journal of Design & Nature and Ecodynamics*, 11(3), pp. 459-472, 2016. DOI: 10.2495/DNE-V11-N3-459-472.
[7] Rzevski, G., Knezevic, J., Skobelev, P., Borgest, N. & Lakhin, O., Managing aircraft lifecycle complexity. *International Journal of Design & Nature and Ecodynamics*, **11(2)**, pp. 77–87, 2016. DOI: 10.2495/DNE-V11-N2-77-87.
[8] Madsen, B., Rzevski, G., Skobelev, P., & Tsarev, A., A strategy for managing complexity of the global market and prototype real-time scheduler for LEGO supply chain, *International Journal of Software Innovation*, **1(2)**, pp. 28–39, 2013. DOI: 10.4018/ijsi.2013040103.
[9] Glaschenko, A., Ivaschenko, A., Rzevski, G. & Skobelev, P., Multi-agent real time scheduling system for taxi companies. *Proceedings of 8th International Conference on Autonomous Agents and Multiagent Systems (AAMAS 2009)*, eds. Decker, Sichman, Sierra, and Castelfranchi, Budapest, Hungary, May, 10–15, 2009,
[10] Andreev, S., Rzevski, G., Shveykin, P., Skobelev, P. & Yankov, I., Multi-agent scheduler for rent-a-car companies**.** *Lecture Notes in Computer Science, Volume 5696, Holonic and Multi-Agent Systems for Manufacturing: Forth International Conference on Industrial Applications of Holonic and Multi-Agent Systems, HoloMAS 2009,* Springer: Linz, Austria, pp. 305–314.
[11] Tormos, P., Lova, A., Barber, F., Ingolotti, L., Abril, M. & Salido, M. A., A genetic algorithm for railway scheduling problems. *Studies in Computational Intelligence (SCI)*, **128**, pp. 255–276, 2008. DOI: 10.1007/978-3-540-78985-7_10.
[12] Cai, X. & Goh, C. J., A fast heuristics for the train scheduling problem. *Computers and Operations Research*, **50**, pp. 738–753, 2002.
[13] Caprara, A., Monaci, M., Toth, P. & Gvida, P., A Lagrangian heuristic algorithm for real-world train timetabling problem. *Discrete Applied Mathematics,* 154, pp. 738–753, 2006. DOI: 10.1016/j.dam.2005.05.026.

DELAYS FOR PASSENGER TRAINS ON A REGIONAL RAILWAY LINE IN SOUTHERN SWEDEN

C.W. PALMQVIST[1], N.O.E. OLSSON[2], L. HISELIUS[1]
[1] Lund University, Department of Traffic and Roads, Sweden.
[2] NTNU Norwegian University of Science and Technology, Norway.

ABSTRACT

The aim of this article is to study how selected variables influence delays in train traffic. Data has been collected on train movements, timetables, weather and capacity utilization on a highly utilized single-track railway line in southern Sweden during 2014. Based on this dataset, we have analysed how different factors affect delays in passenger traffic. We measure delays in a novel way, as deviations from the scheduled duration for each line section and station stop, not as deviations from a published or operational timetable, and this allows us to identify when and where the delays first occur. Average delays were much larger at station stops. The most significant factor affecting delays was the scheduled duration time at station stops and the existence of margins on line sections. If trains arrive to a line section or station stop slightly delayed they speed up the activity, otherwise they are typically delayed. The influence of weather was less significant and somewhat contradictory: snow and cold temperatures increase delays on line sections but reduce them at station stops, while precipitation made no difference. Capacity utilization seems to have a negative correlation with delays, but we have too little variation in the levels to be confident. All studied variables, except for precipitation, have impacts that are statistically significant to a very high degree of confidence, using both t-tests and regression analysis. The results of this study have important practical implications for timetable construction; for instance we estimate that a reallocation of scheduled time at stations could reduce delays by as much as 80%.

Keywords: allowances, delays, margins, punctuality, railroad, railway, station stop, timetable, trains, travel time variation.

1 INTRODUCTION

Punctuality is a key performance indicator and important success factor for the railway systems, and a failure in this key factor affects the competitiveness of railway transports relative to other transport modes [1].

A number of factors can influence the precision of the railway traffic [2]. Harris [3] found that distance covered and train length were statistically significant in determining punctuality.

Running time margins are generally added to the nominal running time when setting up a timetable, as described by Goverde [4]. Nominal running times are based on a maximal speed profile in normal conditions. To make a timetable realistic, margins are added to allow for variations in rolling stock performance, individual driving patterns for drivers, variations in weather conditions and other factors that may influence the running time.

Running time margins are typically expressed as a percentage of the nominal running time. International guidelines for running time margins are given in [5]. According to Pachl [6], common running time margins in Europe are in the range of 3% to 7% and 6% to 8% in North America. According to Goverde [4], the Dutch Railways use a margin of 7% of the nominal

DOI: 10.2495/TDI-V1-N3-421-431

running time. Palmqvist [7] found that the actual running time margins for a selected line in Sweden were about 10%. Punctuality, disturbances and the optimal distribution of supplements are discussed at length in Vromans [8]. Kroon et al. [9] developed a measure to assess at which part of the line time supplements were added. Parbo et al. [10] discuss the use of percentiles of a sample of actual times instead of defining theoretical running times and then add a percentage as margin.

Ceder and Hassold [11] found that one of the main causes for delays in New Zealand was heavy passenger load, which increases dwell times. Bender et al. [12] state that time required for passenger boarding and alighting at stations is a critical element of overall train service performance. Harris et al. [13] studied delays at stations in the Oslo area. They claim that these delays are often small in nature, poorly recorded and not well understood.

Weather conditions can influence the railway traffic. Xia et al. [14] estimate the effects of weather conditions on railway operator performance of passenger train services. They found that wind gust, snow, precipitation, temperature and leaves contributed to infrastructure disruptions. Wei et al. [15] studied the cascade dynamics of delay propagation during severe weather. Ludvigsen and Klaboe [16] and UIC [17] present how weather conditions affect train operator and railway infrastructure performance.

Railway capacity is recognized as an important factor regarding train punctuality [18, 19].

A growing data volume and data availability has enabled researches to study delays and delay causes on a relatively detailed level. Markovic et al. [20] proposed machine learning models for analysing the relation between train arrival delays and selected characteristics of the railway system. Gorman [21] used econometric methods on US freight rail data to predict congestion delays. Total train running time was predicted based on free running time predictors (horsepower per ton, track topography and slow orders) and congestion-related factors (meets, passes, overtakes, number of trains, total train hours, train spacing variability and train departure headway). Wallander and Mäkitalo [22] used a data-mining approach for analysing rail transport delays with the aim of developing a more robust timetable structure and provide tools for rail network planning.

This article has a similar objective as Wallander and Mäkitalo [22] and shows an example of how this type of analysis could be carried out. We relate delays to deviations from the scheduled running times between stations and to scheduled station times in order to distinguish between not only the places of occurrence of the delay, but also the actual use of the infrastructure.

There are well-established models based on physics for calculating nominal train running times, and such calculations are implemented in timetabling support tools. Time at stations has traditionally been set more empirically or based on tradition. Our aim is to map deviations from these scheduled reference times, for each section along a selected line, and to investigate influencing factors. By focusing on how long an activity should take and how long it actually took, this approach also facilitates discussions related to the most efficient use of the infrastructure. We perform the analyses for both line sections and station stops.

The main aim of this article is consequently to contribute to the methods by which deviations can be understood through this novel approach on defining deviations. The article furthermore contributes to the much-needed knowledgebase on deviations and punctuality in the railway system in order to reach a robust system, yet keeping an efficient use of the infrastructure. The analysis is based on data from a single railway track in Sweden and various variables possibly influencing the occurrence and size of the delays.

2 METHOD

2.1 Key terminology and calculations

Punctuality is measured as percentage of late trains [1], while delays are measured by a time unit, such as minutes. Punctuality and delays provide an overview of the performance of a railway system and can be summarized to easily communicated numbers, such as punctuality of all trains on a line for a period, average delay or aggregated delay hours. However, from an analytical point of view, a lot of information is hidden in such aggregation.

When analysing railway traffic, nuanced presentations with higher resolution are desired. Rietveld et al. [23] measure unreliability as deviations from a timetable. Noland and Polak [24] discuss travel time variability using a distribution of arrival times, without referencing a timetable. Nicholson et al. [25] instead propose three measures for the term ‘resilience’.

The analyses in this article are based on differences between the actual and scheduled durations of activities, be they movements along line sections or station stops. If a stop at a station is scheduled to take two minutes but actually takes three, we would call that a delay of one minute. In the rare case that the stop only took one minute instead of two, that would register as a *negative* delay of one minute (-1). This article thus differs from what is typically meant by ‘delay’ in that, beyond the scheduled duration and any margins in it, we are not concerned with the timetable per se. If a train is scheduled to arrive at a station at 12:00 and depart at 12:02, thus with a scheduled duration of two minutes, but it actually arrives at 12:01 and departs at 12:03, with an actual duration of two minutes, we would not register that as a delay. Instead, we would have registered the delay, with relation to the arrival and departure times, where and when the delay first occurred.

The article furthermore differs from what is typically meant by ‘travel time variability’ in that we use the timetable as the baseline for how long activities ‘should’ take, instead of the average duration as measured over some time span. If a stop at a station is scheduled to take two minutes, but actually takes three minutes, we would consider that a delay of one minute even if the average dwell time is three minutes. From the perspective of travel time variability, however, there would be no deviation.

In the article, we use the term ‘average delay’. What we mean is an average *conditional* on the value of an explanatory variable. We also discuss *relative risk*, which indicates how much a factor increases or decreases the average delay in *relative* terms. The relative risk of delays is calculated as the average delay conditional on some value of an explanatory value, divided by the average delay across all station stops or line sections, as the case may be. This makes it easier to discern the effect of factors where the values or differences are small in absolute terms.

The analysis of this article also makes use of the *delay contribution.* Here we multiply the average delay with the number of observations and compare this to the total amount of delays. This allows us to identify which levels lead to the most delays, overall.

To ease comparison between the differing lengths of line sections, between the speeds of trains, and so on, we have chosen to use ratios instead of absolute time units. Throughout we have chosen to use the scheduled duration as the denominator. If the scheduled duration is 100 seconds and the total margin for that activity is 10 seconds, we would register that as an allowance of 0.10 or 10%.

For durations we normalize with the average duration for the respective activity. Across our sample those averages are approximately 100 seconds for station stops and 197 seconds for line sections. A scheduled stop of two minutes is thus translated to a duration of 1.20 while a scheduled movement along a line section of two minutes would translate to 0.61. We have then rounded these ratios to one or two decimal points to limit the number of distinct values we work with and to ensure that there are enough observations in each bin for their average delays to be reasonably stable.

2.2 Data

The railway line analysed in this study is 113 km long, electrified, single track, with dense and heterogeneous regional traffic. This makes it a fairly typical case in a Scandinavian context. It is located in the far south of Sweden and goes almost straight to the east from Helsingborg on the western coast to Kristianstad near the eastern shore, connecting the West Coast Line and the Southern Main Line.

The article is based on four main datasets: train movements on the railway line for the year of 2015, detailed timetables for these trains, weather observations from around the region and the calculated capacity utilization along the railway. These datasets have been linked together through a series of Access databases, resulting in data for over 198,000 stops at 13 stations and 363,000 movements along 20 line sections. Details about the component datasets follow below.

The core of our data is that of registered train movements. We do not have data from every signal, but at certain points, typically where trains can meet. Arrival and departure of the trains are logged for purposes of traffic control, punctuality measurement, etc. From these logs we can calculate the actual duration of the activity, whether it was a stop at a station or a movement between them.

Timetables for trains in Sweden are created and stored in the tool TrainPlan. The geographical resolution is almost the same as above. We have received exports from this programme, in which we can see every planned activity and its allotted time. This covers both stops and movements. From this we extract the scheduled duration and any margins that may have been allotted.

We have meteorological observations from measuring stations in connection to the studied railway line, over the variables – snow depth, precipitation and temperature. The two former variables are measured once per day, the latter every hour. Snow depth is presented as metres, with a precision of two decimal points. Temperature is in degrees Celsius, precipitation in millimetres.

The infrastructure manager has adopted an interpretation of the UIC [18, 19] guidelines for how to calculate the capacity utilization on line sections and provided us with the results. However, this framework does not include a method to calculate the utilization of capacity at stations. In order to look at this, we define our own measure. First we calculated, for every day and hour, how many trains passed through the station. Then we took the maximum value and set it as the maximum capacity for every given station. Finally, we compared the utilization over every day and hour with the maximum for that station to get two ratios, one for the daily utilization and one for the peak utilization.

A summary of studied variables is found in Table 1.

Table 1: Description of studied variables.

Variable	Description
Earlier delay	The train's delay at the beginning of the activity, as a fraction of the scheduled duration of the activity. Early arrivals are counted as negative delays.
Scheduled duration	The amount of time scheduled for an activity, be it line section or station stop, before margins are applied. Normalized by the average scheduled duration for the respective activity and presented as a fraction.
Margins	The amount of margins in the timetable for the current activity. Normalized by the scheduled duration and presented as a percentage.
Temperature	Degree Celsius, measured hourly.
Precipitation	Rain or snowfall [mm], measured daily.
Snow	Depth of snow [m], measured daily.
Day of the week	Monday (1) through Sunday (7)
Month of the year	January (1) through December (12)
Capacity utilization	Percentage of maximum theoretical capacity on line section or at station. Calculated and analysed both for an entire day and for peak load.

2.3 Analysis

In order to analyse the impact of these variables on delays we make use of three basic steps.

The first is to determine if the studied variables had a statistically significant impact on the delays. We used Welch's t-test to do this, which is a more general version of Student's t-test, which permits the studied samples to have different sizes and variances. Here the data is segmented based on the value of the explanatory variable. For example, we look at temperatures below zero and temperatures that are above zero and find out whether the average delays for these samples are significantly different from each other.

A second step is to perform linear regressions for delays at station stops and line sections respectively in order to give an overview of the trend of each studied variable. These are not intended to be the focus of the article or to provide good predictive models. This regression analysis assumes that, instead of there being two categories within each variable, there is a linear relationship between the studied variable and delays occurring.

The third step of the analysis includes three different plots for each variable across line sections and station stops, each emphasizing different aspects of the data. Each explanatory variable is plotted against the average delay, the relative risk and the delay *contribution* as defined in Section 2.1 for line sections and station stops respectively. These plots together give a more complete and nuanced picture than either the t-tests or the regressions, and visual inspection can reveal if the studied variables vary along with delays in any recognizable pattern or if there is no pattern at all. Due to limitation of space this article presents only two diagrams. The result is thus mainly presented in text.

3 RESULTS AND DISCUSSION

3.1 Line sections and station stops

The presentation of the results begins with an overview over the average delays and probabilities of being delayed on line sections and at station stops. As Table 2 shows, the average delay for a train on a line section is very close to zero, while significant delays are to be expected at the scheduled station stops. The values in Table 2 are normalized with regard to the scheduled duration.

Table 2 also illustrates that for movements on line sections, the probability of being exactly on time is higher than at stations, and the probability of gaining time is higher than that of losing it. At stations almost half of all stops take longer than scheduled. The delays that do occur also tend to be larger at station stops than on line sections.

3.2 T-test and regression analysis

To test the significance of the studied variables we first use Welch's t-test. The results of these are summarized in Table 3. Most of the variables were found to have a highly significant impact on delays.

We then perform regression analyses on the line section and station stop data respectively. Both of these yield results similar to the t-test, and the key results are also summarized in Table 3.

A number of coefficients differ in sign between the regression for line sections and station stops: the three weather variables increase delays on line sections while reducing them at stations, while month by month the delays decrease on line sections and increase at stations.

3.3 About the explanatory variables

In the following section we discuss the results of visual analysis of more than 60 plots over the explanatory variables impacts on delays. This analysis allows us to see patterns that are not visible in Table 3. We do not have the space to show more than a couple of these diagrams but briefly discuss the others in text.

3.3.1 Timetable factors

Visual analysis indicates that small earlier deviations from the timetable affect the occurrence delays on line sections. One possible interpretation is that if trains arrive a little before the

Table 2: Average delays and probabilities of delay.

	Summary statistics		Probability of finishing activity			
	Average delay	Standard deviation	Early	On time	Delayed	Delayed by more than 100%
Line section	0.00	0.01	24%	58%	18%	0%
Station stop	0.44	1.21	19%	34%	47%	41%

Table 3: Results of t-tests and regression analysis

Input		Welch's t-test							Regression	
Variable	Source	Value	Average	St. Dev.	Value	Average	St. Dev.	p-value	Coefficient	p-value
Earlier delay of train	Line section	>0	0.00	0.00	<=0	0.00	0.01	0.00	0.00	0.00
	Station stop		0.14	1.11		0.78	1.22	0.00	−0.05	0.00
Scheduled duration	Line section	>1	0.00	0.01	<=1	0.00	0.01	0.27	0.00	0.00
relative to average	Station stop		0.08	0.73		0.54	1.30	0.00	−0.22	0.00
Margins in timetable	Line section	>0	0.00	0.01	<=0	0.00	0.01	0.00	−0.01	0.00
	Station stop		−0.43	0.92		0.44	1.21	0.00	0.00	0.00
Temperature	Line section	<0	0.00	0.01	>=0	0.00	0.01	0.00	0.00	0.00
	Station stop		0.35	1.20		0.44	1.21	0.00	0.00	0.01
Precipitation	Line section	>0	0.00	0.01	0	0.00	0.01	0.05	0.00	0.00
	Station stop		0.44	1.16		0.44	1.24	1.00	0.00	0.39
Snow depth	Line section	>0	0.00	0.01	0	0.00	0.01	0.00	0.02	0.00
	Station stop		0.29	1.14		0.45	1.21	0.00	−1.16	0.00
Day of week	Line section	6-7	0.00	0.01	1-5	0.00	0.01	0.00	0.00	0.00
	Station stop		0.42	1.17		0.44	1.22	0.01	−0.01	0.00
Month of year	Line section								0.00	0.00
	Station stop								0.02	0.00
Capacity utilization	Line section								0.01	0.00
calculated daily	Station stop								0.06	0.00
Capacity utilization	Line section								−0.01	0.00
calculated at peak	Station stop								−0.03	0.10

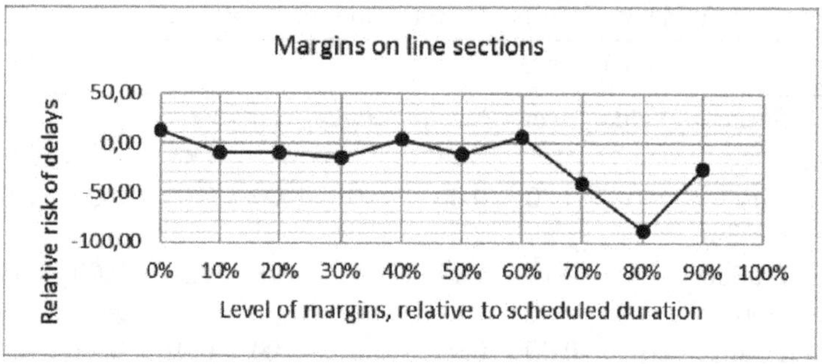

Figure 1: Relative risk of delays at different margin levels on line sections.

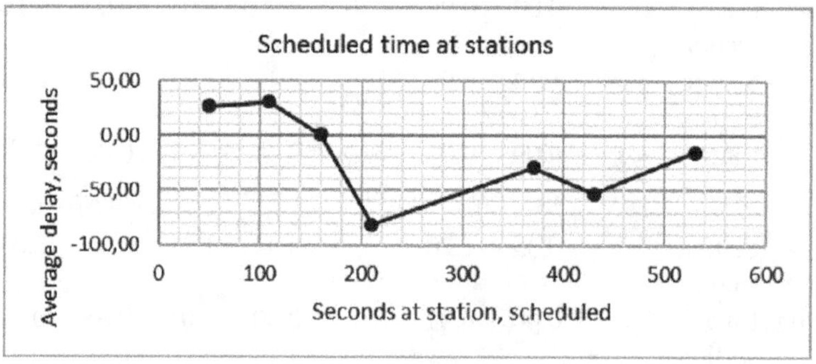

Figure 2: Average delays at stations, scheduled duration.

scheduled arrival time the drivers seem to slow down, and if they are a little bit late, they try to catch up as best they can. This effect disappears as the deviations become larger, both in positive and negative values.

Station stops were shortest when trains arrived a few minutes after the scheduled arrival time. Trains that arrived either on time or early to the station were on average delayed at the stations so that they depart with a delay of around half a minute, regardless of how early they were. Trains that arrived more than a few minutes late were on average delayed even more at the station.

On line sections margins have greater impact on delays than the scheduled duration. While longer line sections on average had slightly lower delays than shorter ones, the effect is not statistically significant, as we see in Table 3. Most delays on line sections in our sample are caused when trains have no margins, although negative margins do occur and contribute to a small degree. The relationship is illustrated in Fig. 1. The most effective level for delay reduction is around 10%, while the greatest delay reduction is achieved with margins around 80%. On line sections we find this to be the single most important variable for explaining delays.

Margins are less well defined at station stops compared to margins on line sections. In the sample we study only a few hundred trains have clearly labelled margins, and those are on the order of hours per station stop, clearly intended for other purposes. Instead, the scheduled duration of the stop is more important and is manually set by the timetable constructor. Figure 2 illustrates that the average delay is significant if the scheduled duration is lower than 160 seconds. The greatest reduction in delays occurs at a duration of 210 seconds, and thereafter additional station time increases the average delays. This variable, scheduled time at station stops, explains more of the delays than any other, both at stations and in total.

Knowing this, we might seek to reallocate the time at stations so as to reduce the delays. To do this we chose a strategy where the total scheduled duration across all stations stops is not allowed to increase, while trying to minimize the total delay time. As a result, we set 21% of the station stops to 210 seconds and the rest at 50 seconds. Using the estimated effect on delays from Fig. 2, we found that this reallocation could reduce delays by as much as 80% without increasing the scheduled travel times.

3.3.2 Weather variables

None of the weather variables we study are found to have a large impact on delays; the coefficients and differences in average delays are relatively small in Table 3, even if the impact is statistically significant for both snow and temperature.

Visual inspection reveals that an increasing snow depth increases the delays on line sections and *reduces* delays at stations, while precipitation does not seem to have a significant effect on delays at all. Temperatures below zero contribute to raising the average delays on line sections, while milder temperatures pull the average towards trains arriving early. At stations we find quite the opposite: the risk of delays *decreases* around the freezing point, while more normal temperatures are associated with delays.

3.3.3 Time: days of the week, months of the year

Visual inspection reveals a noticeable and statistically significant difference between the average delays across the seven days of the week, with delays being the smallest at the weekend and largest on Mondays. This is almost entirely explained by how many trains run each day, as the correlation between the daily averages and the number of trains running on the same days is over 92% on line sections and 70% at stations.

The risk of being delayed also varies from month to month. On line sections we do not discern any pattern or seasonal effect, while at stations the risk increased slowly as the months went by. The risk was around 25% lower than average during the early part of the year and about as much higher towards the end.

3.3.4 Capacity utilization

Capacity utilization is the factor that we are the least confident about, both for line sections and station stops, and especially using the peak measure.

The railway line we study is not very long and only has three different levels of capacity utilization for line sections, calculated by the infrastructure manager. The regression analysis reported in Table 3 showed a small positive correlation, but our visual inspection instead shows a *negative* correlation between capacity utilization and delays. We hypothesize that this mismatch is due to a too small number of distinct capacity utilization levels in our sample and does not draw any strong conclusions about this variable.

Capacity utilization is not defined and calculated in the same way for stations as for line sections, so we define and use a measure of our own, described in the method chapter. When we use the daily measure for stations, the risk is somewhat elevated when the capacity utilization is 70%. Using the peak measure, which we are less confident about methodologically, we instead see a reduction at the same level. Again, we do not think any strong conclusions should be drawn from this.

4 CONCLUSIONS

In this article we have used a novel approach of looking at where delays occur, a method which makes it easy to distinguish between delays occurring on line sections and those occurring at station stops. By focusing on how long an activity should take and how long it actually took, this approach also facilitates discussions about efficient use of the infrastructure.

The results indicate that variables influencing delays do so to different degrees, sometimes even with different signs, on line sections and at station stops. We also showed that, at least on this railway line, delays at station stops were by far the biggest problem.

We found that the timetable was the most important factor for explaining delays. Trains with slight delays make better time on average, both on line sections and at station stops. Those that instead are early or with larger delays typically require more time, increasing delays. When margins of around 10% exist on line sections the risk of delays falls, while larger margins are less efficient. On station stops a duration of 210 seconds is optimal for reducing delays, and we propose that these should be mixed with stops of 50 seconds to obtain a good balance between minimizing delays and minimizing the expected travel time. We estimate that by reallocating existing station time in this manner, it is possible to reduce delays by as much as 80%.

Weather had some contradictory impacts. On line sections delays increase as temperatures fall, or if there is snow. At stations the same conditions actually improve performance, and in our sample the effect at stations was more important. Precipitation had no significant impact on delays. The average delays are slightly smaller on weekends, especially on line sections, and this is mostly because there are fewer trains running then than during the week. The number of trains also explains part of the variation of delays between months when we look at stations, but not on line sections. We do not detect a clear pattern regarding capacity utilization and its effect on delays.

The results of this study have important practical implications for railway traffic planning. Some factors, such as weather, cannot be influenced, but their impact on traffic is still important to know when establishing plans. Other factors, such as the distribution of margins, and length of station stops are factors that planners can influence.

From a research perspective we have shown that it is both possible and useful to gather and analyse large volumes of data about train movements in this manner: analysing even small deviations from the timetable, identifying when and where the deviations occur, considering the differences between line sections and station stops, and combining several datasets and explanatory variables.

We shall continue our research expanding the network size, time period and number of variables studied, and increasing the depth of analysis.

REFERENCES

[1] Nyström, B., *Aspects of Improving Punctuality*, Luleå University of Technology Doctoral Thesis, 2008.

[2] Olsson, N.O.E. & Haugland, H., Influencing factors on train punctuality – results from some Norwegian studies. *Transport Policy*, **11(4)**, pp. 387–397, 2004.

[3] Harris, N.G., *Planning Passenger Railways: A Handbook*. A & N Harris, London, 1992.

[4] Goverde, R.M.P. *Punctuality of Railway Operations and Timetable Stability Analysis*. The Netherlands TRAIL Research School Thesis Series no. T2005/10, 2005.

[5] International Union of Railways, UIC Code 451 Timetable recovery margins to guarantee timekeeping – Recovery margins, 2000.

[6] Pachl, J., *Railway Operation and Control.* VTD Rail Publishing: Mountlake Terrace, 2002.
[7] Palmqvist, C.W., *Running Time Supplements for High Speed Trains.* Lund University Master Thesis 257, Lund, 2014.
[8] Vromans, M.J.C.M., *Reliability of Railway Systems.* ERIM PhD series Research in Management 62, 2005.
[9] Kroon, L., Dekker, R. & Vromans, M., Cyclic railway timetabling: A stochastic optimization approach. *Algorithmic Methods for Railway Optimization*, **16**, pp. 41–66, 2007. DOI: 10.1007/978-3-540-74247-0_2
[10] Parbo, J., Nielsen, O.A. & Prato, C.G., Passenger perspectives in railway timetabling: A literature review. *Transport Reviews*, pp. 1–27, 2015.
[11] Ceder, A. & Hassold, S., Applied analysis for improving rail-network operations. *Journal of Rail Transport Planning & Management*, **5**, pp. 50–63, 2015.
[12] Bender, M., Büttner, S. & Krumke, S.O., Online delay management on a single train line: beyond competitive analysis. *Public Transport,* **5**, pp. 243–266, 2013. DOI: 10.1007/s12469-013-0070-z
[13] Harris, N.G., Mjøsund, C.S. & Haugland, H., Improving railway performance in Norway. *Journal of Rail Transport Planning & Management*, **3**(4), pp. 172–180, 2013.
[14] Xia, Y., Van Ommeren, J.N., Rietveld, P. & Verhagen, W., Railway infrastructure disturbances and train operator performance: The role of weather. *Transportation Research Part D*, **18**, pp. 97–102, 2013. DOI:10.1016/j.trd.2012.09.008.
[15] Wei, D., Liu, H. & Qin, Y., Modeling cascade dynamics of railway networks under inclement weather. *Transportation Research Part E*, **80**, pp. 95–122, 2015.
[16] Ludvigsen, J. & Klæboe, R., Extreme weather impacts on freight railways in Europe. *Natural Hazards*, **70**, pp. 707–787, 2014.
[17] International Union of Railways, Winter and Railways, UIC/SIAFI Report, 2011.
[18] International Union of Railways, UIC CODE 406 Capacity, 2004.
[19] International Union of Railways, UIC CODE 406 Capacity, 2013.
[20] Markovic, N., Milinkovic, S., Tikhonov, K.S. & Schonfeld, P., Analyzing passenger train arrival delays with support vector regression. *Transportation Research Part C*, **56**, pp. 251–262, 2015.
[21] Gorman, M.F., Statistical estimation of railroad congestion delay. *Transportation Research Part E*, **45**, pp. 446–456, 2009.
[22] Wallander, J. & Mäkitalo, M., Data mining in rail transport delay chain analysis. *International Journal of Shipping and Transport Logistics*, **4**(**3**), pp. 269–285, 2012.
[23] Rietveld, P., Bruinsma, F. R. & van Vuuren D.J., Coping with unreliability in public transportation chains: A case study for Netherlands. *Transportation Research Part A*, **35**, pp. 539–559, 2001.
[24] Noland, R. B. & Polak, J. W., Travel time variability: a review of theoretical and empirical issues. *Transport Reviews*, **22**(**1**), pp. 39–45, 2002.
[25] Nicholson, G., Kirkwood, D., Roberts, C. & Schmid, F., Benchmarking and evaluation of railway operations performance. *Journal of Rail Transport Planning & Management*, **5**(4), pp. 274–293, 2015. DOI: doi:10.1016/j.jrtpm.2015.11.004

INCREASING ROBUSTNESS OF TIMETABLES BY DELIBERATE OPERATION OF TRAINS TO SHORTEN HEADWAYS

H. YABUKI[1,2], Y. TAKATORI[1], M. KORESAWA[1], & N. TOMII[2]
[1]Tokyo Metro Co. ltd., Japan.
[2]Department of Computer Science, Chiba Institute of Technology, Japan.

ABSTRACT
In railway lines where trains are running densely, a small increase of a dwell time causes a delay and the delay tends to expand and propagate to many other trains. One idea to avoid such delays is to increase dwell times of trains in timetables. Increase of dwell times, however, may cause an expansion of headways and we may have to decrease the frequency, which is never allowed in a congested railway line. In this article, we propose a procedure to get a timetable in which each train is given an enough amount of dwell time without reducing the frequency of trains. Our key idea is to find a train performance curve which minimizes the departure–arrival interval of trains and to make use of the generated time to increase the dwell times. In this process, we do not stick to an idea that travel times of trains have to be minimized. Following this procedure, we can increase the dwell times without reducing the frequency. We have applied our procedure to actual timetables and confirmed that our procedure works very well.
Keywords: headway, robustness, timetable, train performance curve.

1 INTRODUCTION

In Tokyo, the capital of Japan, a great number of people commute using railways. In order to fulfil the big demand for railways, trains are operated very densely. As a matter of fact, during morning rush hours, 25 to 30 trains are running per hour per direction on a double track in many railway lines in Tokyo. Trains typically consist of 8 to 10 cars, which are 160 m to 200 m long. Nevertheless, congestion rates of trains are very high. It is not unusual to find trains during rush hours of which the congestion rates are more than 150%, which means more than 2,000 passengers are on board.

One of the recent problems in the railway network of Tokyo area is that short delays very often happen during morning rush hours. Small, say several minutes, delays happen almost every weekday in the morning.

The main cause of such delays is an excess of dwell times. If more passengers than expected get on and/or off, dwell times increase. It is often observed that a passenger's bag or an umbrella is caught by doors of a train due to congestion inside the train. In such cases, the doors must be opened again and the dwell time increases. Another case is rescue of a sick passenger. If a passenger gets sick inside a train, he/she is rescued by an ambulance and the dwell time increases.

One of the characteristics of railway systems in Tokyo area is that so-called direct operation is widely realized. This is an operation which allows trains from suburban areas go directly into subway lines and vice versa. Passengers can arrive at downtown without transferring,

DOI: 10.2495/TDI-V1-N3-432-441

which is very convenient for them. On the other hand, the direct operation has a drawback that a delay tends to propagate to other railway lines. Thus, railway companies are keen to prevent such delays from happening and making every effort to make their timetable more robust [1–3].

In order to make timetables more robust, it is effective to increase dwell times in the timetables. But the problem is that because we are required to set up a certain number of trains to fulfil the demands, it is impossible to increase the dwell times any more, which we will discuss later.

In this article, we introduce a procedure to increase dwell times of trains for a given timetable without decreasing the frequency of trains (we define 'a frequency' as the number of trains during rush hours in this article). If we can increase the dwell times, delays due to an increase of dwell times do not occur anymore and the timetable becomes more robust. Our key idea is to find a train performance curve which minimizes the arrival–departure interval of trains and to make use of the generated time to increase the dwell times. In this process, we do not stick to an idea that travel times of trains have to be minimized.

The process to seek for such a train performance curve is not so simple. One reason is that we have to find a practical performance curve. We have to be careful about the ride comfort and should avoid losses. It is not desirable if it is difficult to drive. Another reason is that we have to assure that the performance curve has to be applicable for all the trains, because it is not reasonable to specify different performance curves for each train. One more reason is that we have to consider an influence from the preceding train because trains are running densely. These mean that in the procedure, we have to consider the preceding train, the current train and the succeeding train all at the same time. To solve the problems, we introduce a procedure which consists of iteration and feedback.

We have applied our procedure to an actual timetable and confirmed that our procedure successfully increases the dwell times of a given timetable.

2 MAKING A TIMETABLE

2.1 Constraints in making a timetable

To make a timetable means fixing the arrival and departure times of trains at stations together with finding out whether trains pass or stop at stations and the track which the trains are going to use. In Tokyo Metro, arrival and departure times are specified with a unit of 5 seconds.

There exist three major constraints we have to take into account when we make a timetable.

1. Minimum running time
   A minimum running time of a train means the minimum time required when a train runs from a station to the next station. It is common that we consider only one train and we do not care about an interaction between two trains. This is because what we want to get here is the minimum running time when a train runs without any interference from other trains.
   Minimum running times are nowadays obtained by calculation using a computer software. The output of the software is a so-called train performance curve (hereafter, we may call it just 'a performance curve' if there is no ambiguity), which consists of the speed curve and the time curve. We show an example of a train performance curve from one station (Station A) to the next station (Station B) in Fig. 1. The x-axis is the distance and the y-axis is the speed and the time. The blue line expresses the speed limit imposed by ATP. The speed

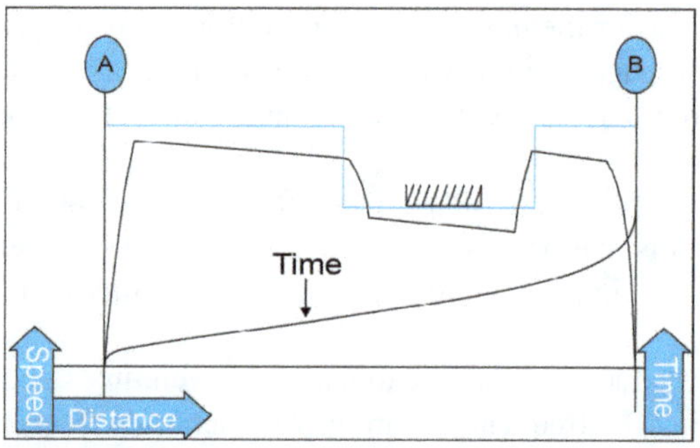

Figure 1: Train performance curve (example).

curve depicts the running speed of the train at each location together with how the train should be driven, namely, where the train should accelerate and where the train should decelerate. From the time curve, we can know how much time is needed for the train to run between the stations. The unit of time for the results of calculation is usually 1 second and is rounded to a time with a unit of 5 seconds in Tokyo Metro (and in other railway companies in Japan). Hereafter, we call the rounded time a technically minimum running time and if there is no ambiguity we just call it a *minimum running time*.

2. Dwell times
We have to set the dwell time of trains at each station. We estimate or observe how many passengers get on/off at the station and fix an appropriate dwell time. Thus, the dwell times vary according to the time, the direction and the station. Dwell times during rush hours are set to be longer than those in less congested time. In Tokyo Metro, dwell times are typically from 30 seconds to 1 minute.

3. Minimum headway
When two trains consecutively use the same track at a station, we have to keep a certain time interval between these two trains (it is true that we also have to consider the time interval outside a station. But because the necessary interval outside stations is usually larger than that of inside a station, we focus on the necessary interval inside a station). In this article, we call the minimum time interval between two trains which consecutively run 'a technically minimum headway' or '*a minimum headway*' if there is no ambiguity. More exactly speaking, the minimum headway in this article is defined as the minimum necessary time from the departure of a train to the arrival of the next train (Sometimes, a headway is thought to be an interval of trains in the timetable. But in this article, we use 'headway' as the minimum interval as described above. We do not want to change the frequency of trains. This is the most important prerequisite of this article.).

The minimum headway is influenced by the signalling system (ATP) to a large extent. There are no wayside signals in Tokyo Metro and the allowable speed at that moment is shown in the cab of the train. The signal continuously changes as the situation such as the location of the preceding train changes.

Minimum headways are also calculated by computers nowadays. The computer software calculates the minimum headways using the train performance curves and outputs a picture called a minimum headway chart as shown in Fig. 2. Unlike the train

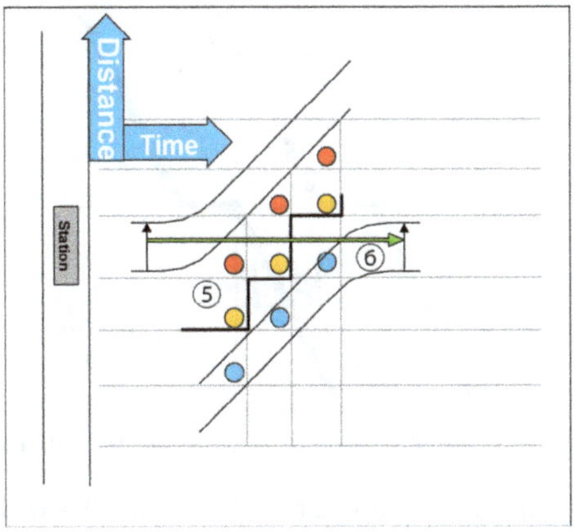

Figure 2: A minimum headway chart (example).

performance curve, the x-axis is the time and the y-axis is the distance. The minimum headways are also rounded with a time unit of 5 seconds in Tokyo Metro.

We would like to explain how Tokyo Metro is dealing with the margins in running times when we make a timetable. We understand it is common to include some amount of margin in the running times of trains in timetables. Namely, in a timetable, the following formula holds:

Arrival time of a train at a station – departure time of a train at the prior station
= technically minimum running time + margin. (1)

In Tokyo Metro (and many other railway companies in Japan which operate trains in urban area), however, we do not give a margin explicitly. Instead, we impose rather strict conditions when we calculate the minimum running times (such as the voltage of the catenary, the diameter of the wheels and so on) and we expect some margin is included in the minimum running time eventually. This is because in Tokyo Metro, distances between stations are rather short and the types of the trains are almost the same and it is not necessary to give different amount of margins to trains. Thus, it is more reasonable that a certain amount of margin is implicitly included in the minimum running times.

2.2 Conventional Process to make a timetable

When a railway line is newly constructed, we have to make a timetable from scratch. The basic process in this case is as follows:

1. From the forecast of demands, we fix the necessary frequency of trains during the busiest hours.
2. We draw a train performance curve to get the minimum running times.
3. We fix necessary dwell times for each station based on an estimation of the number of passengers. Dwell times may be different depending on times of the day, direction, etc.
4. We design the facilities including signalling systems so that we can realize the minimum running times and the frequency during the busiest hours.
5. We fix arrival and departure times of each train.

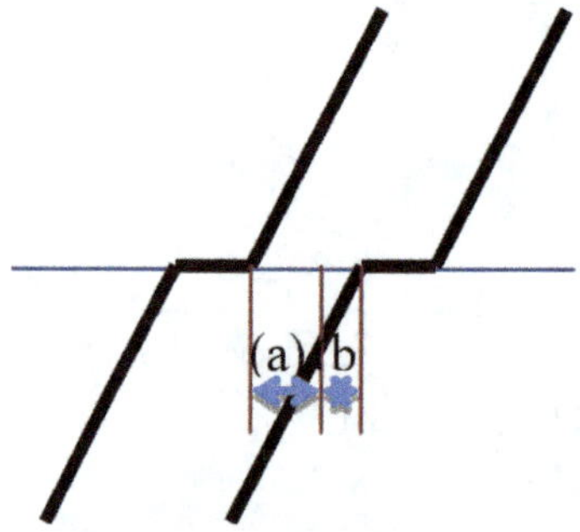

Figure 3: Dwell time, headway and buffer time.

When we want to revise an existing timetable, minimum headways and minimum running times become constraints. If necessary, we might change the facilities so that we can relax these constraints. But this is usually very difficult or almost impossible because the facilities such as signalling systems are already optimized and further improvement is not possible any more.

We have to note that this process is aiming at reducing the travel time and so much attention was not paid to increasing the dwell times.

2.3 Current issues concerning timetables

As we have discussed before, the major cause of short delays is an increase of dwell times. In order to prevent such delays, dwell times should be long enough in the timetable. But we cannot make the dwell time longer unless we decrease the frequency. This is because we have to consider the constraint of the minimum headway and if we increase dwell times, it becomes impossible to keep the frequency of trains. Figure 3 expresses the reason. In Fig. 3, (a) means the minimum headway and (b) is a buffer time. The dwell time could be increased by (b) (the buffer time), but if we make the dwell time longer than this, we have to change the arrival time of the next train due to the constraint of the minimum headway. This means the frequency decreases. Trains are very congested; so it is never allowed to reduce the frequency. In addition, in the timetables of Tokyo Metro, dwell times are already prolonged almost to the limit and it is almost impossible to increase the dwell times any more. Hence, we need to introduce a brand new idea to increase the dwell times in the timetable.

3 INCREASING ROBUSTNESS OF TIMETABLES BY DELIBERATE OPERATION OF TRAINS TO SHORTEN HEADWAYS

3.1 Basic ideas

As discussed above, it is difficult or almost impossible to increase dwell times in the timetables. But it might be possible to increase dwell times if we can reduce minimum headways. If a minimum headway becomes smaller, we can increase the dwell time without reducing the frequency by making use of the generated time. Figure 4 explains the idea. If we can reduce the headway to (a'), we can increase the dwell time by (c) at the largest, which is calculated as

$$(c) = (b) + ((a) - (a')). \tag{2}$$

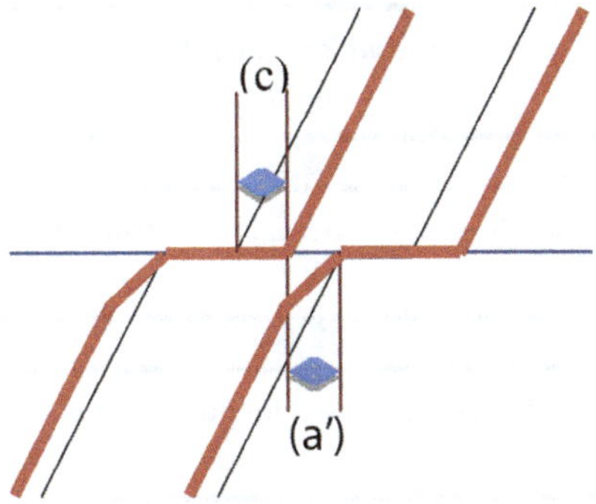

Figure 4: To increase dwell time by reducing headway.

In order to reduce minimum headways, we have to allow trains enter a station with a lower speed. Conventionally, we assume a train runs seeing a green signal (this is a kind of metaphor. As we have explained, in Tokyo Metro, there are no wayside signals) when we make a train performance curve. This is because we want to get the 'minimum' running time. We change this assumption so that we admit a train sees an aspect other than green. Thus, trains may run with a lower speed and the running time may increase, which might be regarded that the level of service is degraded. But we believe punctuality is more important and it is all right if we can increase robustness by slightly making the travel times longer.

3.2 Issues

In order to implement this idea, it is indispensable to establish a procedure to find a train performance curve which realizes the minimum interval. In establishing this procedure, there exist three issues to be settled.

1. The train performance curve must be feasible from practical point of view.
   This means that we have to consider the ride comfort for one thing. In addition, it is not preferable if it is difficult to drive. For example, a sudden change of driving mode is not desirable. We also should consider avoiding losses. For example, if there exists a speed limit ahead, the train should not accelerate and hit the brakes but should coast.
2. The same train performance curve has to be applicable to all the trains for the segment (in this article, we define a *segment* as a track between a station to the next station).
   Trains are driven by drivers. We have to indicate them how to drive a train in advance. It is not reasonable to prepare different train performance curves for each train but we have to prepare only one train performance curve which all trains can follow.
3. We have to consider the preceding train at the same time.
   In a railway line where trains are operated densely, when a train is arriving at a station, the signal aspect for the train is influenced by the preceding train which is just departing from the station. Namely, we have to make the train performance curve considering the current train and the preceding train at the same time.

Although there exist a lot of papers concerning algorithms to make train performance curves, none of them deal with these three issues [4].

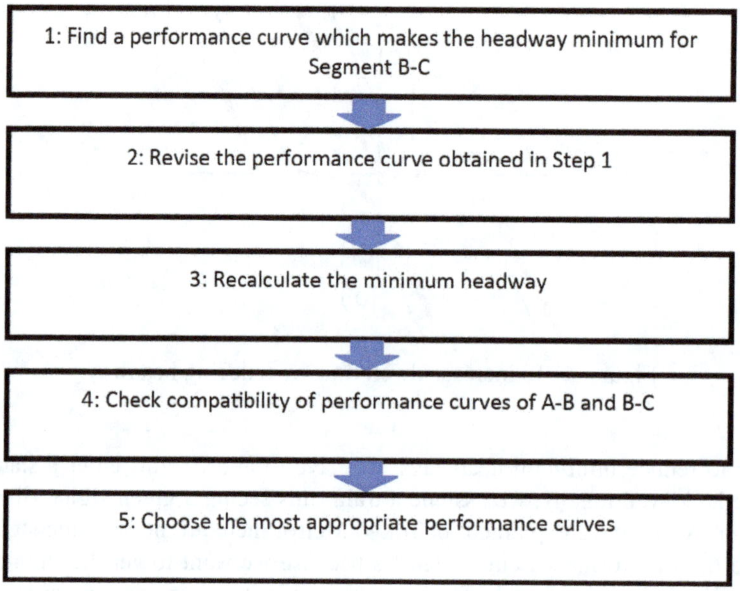

Figure 5: Overall structure of the procedure.

3.3 Procedure

We show our procedure to get a timetable in which dwell times are made longer. The overall structure of the procedure is depicted in Fig. 5. The procedure is basically an iteration of making and confirming the train performance curves, in which a feedback to check the mutual influence is included. More exactly speaking, we try to find a performance curve for one segment and we iterate this process to the following segments. There exist a couple of problems inherent in this process. The first one is that the performance curve obtained in a segment is not applicable to the succeeding train in the same segment. The succeeding train cannot always run as specified by the performance curve obtained. The second problem is that it may reveal that a performance curve obtained for the next segment is not compatible with the performance curve obtained for the current segment. So, during the procedure, we occasionally confirm these compatibilities and if some incompatibility is found, we modify the performance curves so that we can assure the compatibility.

For simplicity, we show a procedure in which only three stations appear: Station A, Station B and Station C (trains proceed from Station A to Station C) and we want to increase the dwell times at Station C. By iterating this procedure, however, we can deal with more complicated situation such as more stations are involved and dwell times of more than one station are increased.

Step 1: Find a performance curve which makes the headway minimum for Segment B-C
We first try to find a train performance curve from Station B to Station C (Segment B-C) which realizes the minimum headway when trains arrive at Station C. Such performance curve must exist between a performance curve for the minimum running time and a performance curve in which the train is almost compelled to stop before it arrives at Station C. We enumerate all train performance curves by fixing the preceding train and changing the departure time of a train at Station B by 1 second and choose one in which

the interval between the departure of the preceding train at Station C and the arrival of the train at Station C is the smallest.

Step2: Revise the performance curve obtained in Step 1

The train performance curve generated in Step 1 is not always appropriate from practical point of view. So we modify it manually so that the performance curve is practically appropriate as discussed earlier.

We also confirm if the obtained performance curve can be applied by the succeeding train(s) as well. We simulate the movement of one train (the first train) and the succeeding train simultaneously and confirm if the succeeding train can run following the same performance curve. If not, we modify the performance curve manually so that the trains can run following the same performance curve.

Step 3: Recalculate the minimum headway

If we have modified the performance curve in Step 2, this means the modified performance curve does not always realize the minimum headway. Hence, we calculate the interval between the train (the first train) and the succeeding train based on the modified performance curve. In this process, we confirm how closely the succeeding train can run considering the signals which are influenced by the first train.

Step 4: Check compatibility of performance curves of A-B and B-C

Now, we may have a new performance curve for Segment B-C. This means that trains may not be able to run in Segment A-B according to the current performance curve. This is because when a train is arriving at Station B, the preceding train is departing from Station B based on the modified performance curve and the signal aspects for the train are influenced by the preceding train. Thus, we have to check the compatibility of performance curve for Segment A-B and that for Segment B-C. In order to check the compatibility of the performance curves, we simulate two trains for these two segments simultaneously.

Step 5: Choose the most appropriate performance curves

Here, we may have the following four options.

If the two performance curves are proved to be compatible in Step 4,

(1) we can just adopt the current performance curve for Segment A-B and the obtained performance curve for Segment B-C.

If the performance curves are proved to be incompatible,

(2) we modify the performance curve for Segment A-B (we do not change the current performance curve for Segment B-C), which we can obtain by applying a procedure similar to Steps 1–3 to Segment A-B.

(3) we modify the performance curve for Segment B-C (we do not change the current performance curve for Segment A-B). We can accomplish this by paying attention to the change of the signal aspects and by modifying the performance curve for Segment B-C so that it does not give an influence to the succeeding train in Segment A-B.

(4) we change the performance curves for Segment A-B and Segment B-C. This could be done as a combination of the two procedures described above.

Then, we choose one of these options which we believe is the most appropriate for the situation.

4 NUMERICAL EXPERIMENTS

We have conducted numerical experiments using actual data. The data we used are ones of Hibiya line of Tokyo Metro. We intended to get a timetable in which dwell times at Ueno station (this is the busiest station in Hibiya line. Hereafter, we call the station 'Station U') are increased. The station prior to Ueno is Iriya (hereafter, 'Station I') and the station prior to Iriya is Minowa (hereafter, 'Station M'). Station U corresponds to Station C, Station I corresponds to Station B and Station M corresponds to Station A, respectively, in the procedure we showed above.

We show the results in Table 1. It was proved that we cannot realize option (1). Namely, if we change the performance curve for Segment B-C, trains cannot run according to the current performance curve for Segment A-B. Hence, we have three options (2), (3) and (4) as shown in Table 1.

Now, we examine and compare these options from various kinds of viewpoints such as whether the riding comfort is considerably good, whether it is not difficult to drive, how much the travel time increases, how much we can increase the dwell times and so on. If we examine the above three options, the discussions are as follows:

- If we adopt option (2), we can increase the dwell time of Station U by 10 seconds. But we are concerned that drivers have to rather suddenly change the speed in Segment M-I because the signalling system indicates a low speed limit for a short time.
- For option (3), it is preferable that the performance curve for Segment M-I is the same as the current one but the increase of the dwell time at Station U is the smallest. The increase of the travel time is the smallest.
- We can get the largest increase of the dwell time at Station U if we adopt option (4).

So one conclusion might be that if we do not care to introduce new performance curves, we should adopt option (4). But if we prefer not to change the performance curve for Segment M-I and if we do not like the running time increases so much, we should adopt option (3).

Table 1: Results of the numerical experiments

	Headway (sec.)		Increase dwell (sec.)	Performance curve, Running time			
	M-I	I-U	U	M-I		I-U	
current	71	60		90 sec.		95 sec.	
(1)	-	-		Not possible			
(2)	71	50	+10	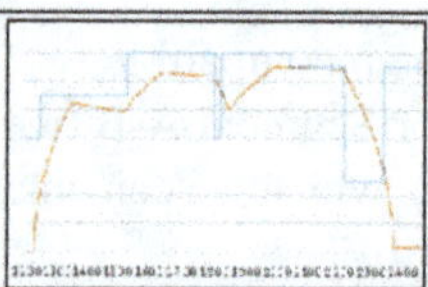	89 sec (-1 sec)	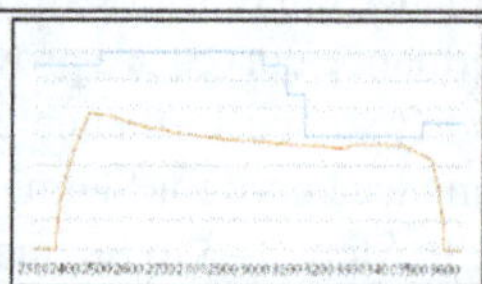	125 sec (+30 sec)
(3)	71	52	+8	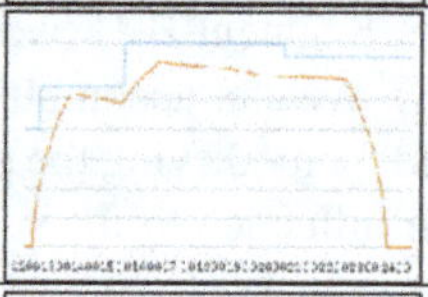	90 sec (±0 sec)	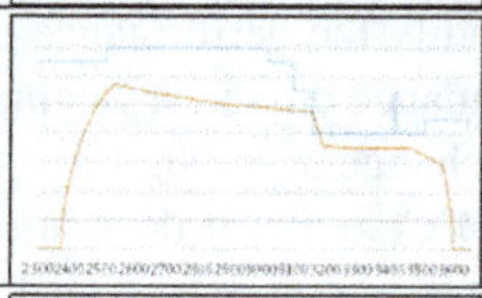	114 sec (+19 sec)
(4)	71	48	+12	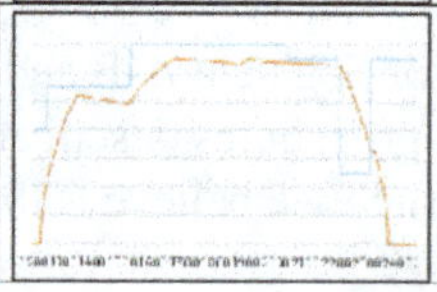	88 sec (-2 sec)	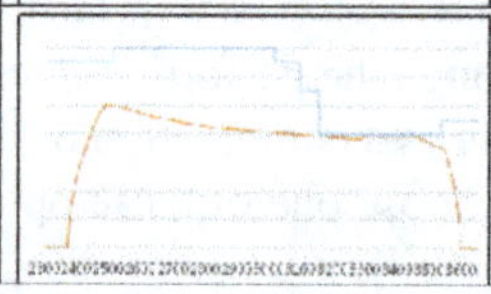	119 sec (+24 sec)

It is a great merit of our procedure that it can show several options and leaves a freedom of choice for the users to examine them from various kinds of viewpoints.

5 CONCLUSIONS

We have proposed a procedure to increase dwell times for a given timetable without reducing frequency. The key idea of the procedure is to find a train performance curve which realizes the minimum headway. In this process, we do not stick to an idea to make trains run as fast as possible but we rather decrease the running speed so that we can obtain a performance curve which makes the headway minimum. Thus, trains may enter with a lower speed but the headway becomes smaller. If the headway becomes smaller, we can consume the generated time to increase the dwell times.

Based on this idea, we have introduced a procedure with which we can know how we can increase the dwell times in a timetable. The procedure could show several alternatives. The outputs contain a performance curve and other figures such as how much we can increase the dwell times, how much the running times increase and so on, which are useful information for users. We have examined how the procedure works using actual data and confirmed that our procedure works very well.

Some of our future works are the following:

- Application of this idea to real-time environment.
  We have introduced a procedure to revise a timetable so that the train can have longer dwell times. It is useful to give an indication to a driver to run so that the train can arrive with a minimum headway when the dwell time of the preceding train has become longer and is delayed.

- Apply this idea when renovation of the signalling systems is allowed.
  In this article, we assumed that we do not change the signalling systems. But it is more effective if we can renovate the signalling system because signalling systems are very influential to headways.

ACKNOWLEDGEMENTS

This research is partly supported by the Ministry of Education, Science, Sports and Culture, Grant-in-Aid for Scientific Research (C) 15K01199.

REFERENCES

[1] Yamamura, A., Koresawa, M., Adachi, S. & Tomii, N., How we have succeeded in regaining Punctuality in Tokyo Metropolitan Railway Network? *WCRR2013-World Congress on Railway Research*, Sydney, Australia, 2013.

[2] Yamamura, A., Koresawa, M., Adachi, S. & Tomii, N., Taking effective delay reduction measures using delay elements of indices on the Tokyo metropolitan railways. *COMPRAIL 2014*, Rome, Italy, 2014.

[3] Ochiai, Y. & Tomii, N., Punctuality analysis using a microscopic simulation in which drivers' behaviour is considered. *Journal of Rail Transport, Planning and Management*, **5**(3), pp. 128–145, 2015.

[4] Goodman, C., Overview of electric railway systems and the calculation of train performance. *2008 IET Professional Development course on Traction Systems, Manchester,* Institution of Engineering and Technology, 2006.

FEASIBILITY EVALUATION AND CRITICAL FACTOR ANALYSIS FOR SUBWAY SCHEDULING

T. ZHU, J.M. MERA, E. CASTELLOTE & J. LÓPEZ
Railway Technology Research Centre, Polytechnic University of Madrid, Spain.

ABSTRACT

In strategic subway scheduling stage, the conflict sometimes comes from different requirements of the subway operator. This study aims to investigate the significant factors concerning strategic subway scheduling problem and to develop an automatic procedure of feasibility analysis in subway scheduling. To this end, accurate simulation of train movement (via a simulator, named HAMLET) is applied first by considering the line geography, train performances, actual speed restrictions, etc. The critical elements of subway scheduling and their correlations are then studied and a bound structure of the critical factors is established. The feasibility of primary plan requirements is analysed with the restrictions of the bound structure. Infeasible aspects and possible adjustments are shortly discussed. Finally, the subsequent applications including schedule generation and optimization according to various objectives are indicated as well.

Keywords: factor analysis, feasibility evaluation, periodic scheduling, subway, train scheduling.

1 INTRODUCTION

There are several distinctive features of subway schedule:

- Subway schedules are remarkably periodic. And the period of a subway schedule is actually short, typically ranging from tens of minutes to several hours. Thus, for both a train and a station the daily schedule they follow consists of multiple periods. Each period of the schedule for a train is called a cycle of the train.
- The service of subway is required to be highly frequent. In other words, the intervals between trains are quite short.
- Within a certain travel zone, the satisfaction of passengers does not critically rely on the punctualities of all trains in the timetable (except the first and last trains on schedule), but the average waiting time and total travel time.
- Only a few stations are equipped with siding tracks (for trains in failure); i.e. it is difficult to prevent the impact of some local failure to propagate into a global scope.
- Possible conflicts come from multi-objective requirements of the subway operator.

There are a wide variety of elements involved in railway scheduling. As in other kinds of scheduling problems, over a time, periodic event scheduling problem (PESP) has played a dominant role in railway scheduling problem by considering the basic aspects, including the arrival and departure times of trains at each station and the assignment of train routes based on the infrastructure resource. For example, as an early study to exploit the PESP to model and construct railway timetables, Odijk [1] first built a visual graph to express the timetable structure where a vertex represents an event and an arc represents a constraint.

DOI: 10.2495/TDI-V1-N3-442-451

The scheduling problem is solved by searching a feasible and good timetable in this constraint graph. Similarly Liebchen et al. [2] also modelled a periodic timetable optimization task as a PESP. Besides, it takes the amount of rolling stock required, average passenger changing time (connection time), average speed of lines, etc., into account.

Besides the reasonable travel times and waiting times, other interests have also been concerned. In the scheduling and rescheduling model proposed by [3], the total final delay and the cost associated with delays are considered. Miyatake et al. [4] investigated the relation between running time and the energy consumption in the whole section inter-stations and further proposed an adjustment law to minimize the total energy consumption by optimizing all running times. In this study the dwell time at each station is unchanged. In contrast, in order to optimize the utilization of regenerated energy in train braking stage, Yang et al. [5] proposed a cooperative scheduling approach, which allows the dwell times to vary. It particularly considers the successive trains and allows the regenerated energy from the braking train to be used by the accelerating train simultaneously. In the earlier model, the energy is the primary factor being taken into consideration. Still in their subsequent work, an extension version using two-objective model was proposed where the energy saving is maximized and the passenger waiting time is minimized [6].

To sum up, in subway train scheduling, the main concerns are arrival times, stop and waiting time at each station, passenger travel times, delays and energy consumption, etc. However, the minimum train intervals, train turning time at terminals and required number of trains are not widely discussed. Instead, these elements are supposed to be a set of static and known parameters in most subway scheduling problems. Thus, this paper studies these elements first at a more fundamental layer, integrating the physical properties of line and the comprehensive behaviours of train. Then it aims to develop an automatic procedure of feasibility analysis in subway scheduling based on the investigation of the key factors in periodic subway scheduling problem. Furthermore, some experience rules to solve the infeasibility are shortly discussed in a demonstrative instance.

2 CRITICAL FACTORS OF SUBWAY SCHEDULING

2.1 Accurate simulation of train movement

First, basic technical factors for accurate simulation of train movement are analysed to study the actual information required in real world.

A simulation tool named HAMLET (Herramienta Avanzada Multidisciplinar para Líneas Eléctricas de Trenes) is applied to reconstruct the infrastructure and signalling system and simulate the train behaviours. As described in the previous work of HAMLET [7], aspects of infrastructure, train, and electrical and signalling systems are all considered (Fig. 1).

There are two classes of infrastructure data of line. The static configurations include positions of key point, lengths, slopes, curves, tunnels and bridges, civil speed restrictions, segment operational speed limitation, feasible route table, etc. And the dynamic information here refers to the track description, which mainly depends on the signalling system and train movement on the lines.

Signalling and electrical systems represent the most important trackside support systems. They are integrally designed to satisfy the applications of train control system.

Train data mainly depend on the train type (locomotive and car specifications, the composition of train, weight, length, rated number of passengers, electrical and mechanical

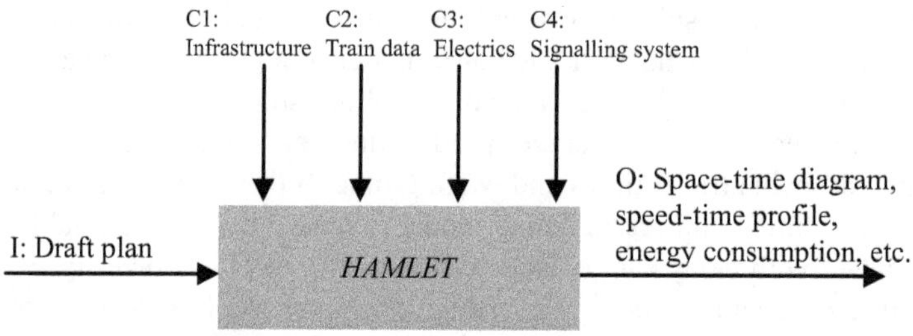

Figure 1: System chart of HAMLET.

parameters, etc.). The traction and braking curves are important references in the accurate simulation of a train. It is noted, however, that each train is considered to be as a whole and its specific performances are represented in the actual speed profile and energy consumption profile.

Actually, the input of HAMLET can be a route description for a single train as well as a draft plan of multi-trains according to the user requirements. When it is a route description for a particular train type, the HAMLET outputs the simulation results of this type of train based on the given routes, such as the running time on each piece of section, actual speed curves, driving modes, energy estimations, etc. Besides, actions such as changing driving mode or using the temperate speed limitations can still be added to a draft plan, and correspondingly it leads to a distinct set of results of the train. Multi-train simulation is based on each single train performance, required plan (input), and track description and the control levels. Its outcomes may indicate the feasibility or even potential conflicts of the draft plan.

2.2 Varied requirements

In many cases the uniform interval I_0 between given trains is required by the subway operator once the information of line facility, signalling system and train types are known. The total number of needed trains N is the variable to be determined.

On the contrary, in some other cases, the feasible interval I is questioned whilst the total number of available trains N_0 (and the composition of each train) is provided as a precondition.

Besides, the average stop and waiting time at stations W_0 and the average time for changing train direction at terminals C_0 are usually given. Nonetheless, some of the waiting times W and direction turning around times C in actual scheduling can be determined in any situation.

2.3 Fundamental factors

Other critical factors considered to analyse and optimize the operating plans are expressed as follows:

- In case there is more than one type of trains, synthetic train speed index $\overline{vp}$ of train type i is defined. It is a relative value that is calculated by

$$\overline{vp}_i = \frac{V_i}{V_0}, \tag{1}$$

where V_0 is the mean velocity of the referred train type and V_i is the average speed of train type i. As the trains for a line are known, $\overline{vp_i}$ of each train type is determined. The index $\overline{vp_i}$ values 1 if there is only a unique train type used in a line.

- Minimum operation times: waiting time at station $W_{\min}$, time to switch on the locomotive of the train $C_{\min}$, i.e. direction turning time for a train.
- Minimum intervals $I_{\min}$ depend on the most restrictive conditions of key points, such as the limited speed and the additional operation time of a switch, etc. Thus, it should be a key point to obtain its minimum interval. As there usually are a set of key points on a line, $I_{\min}$ is a vector. The finally adopted interval I of the general schedule is not smaller than the maximum value of $I_{\min}$.
- Route decision r: r is the selected route set from the route table. If a route is selected, correspondingly the running time set B_r of all segments of the route (related to complement time and required interval and switch positions) is also referred.

2.4 Integral effect

There are some variables that can be only integers. A frequent one is the total number of trains planned to use in a particular line, i.e. N is an integer.

As the periodic characteristic of subway schedule, the second cycle of the train that departs first has to be scheduled with the same interval I from the train that departs last; commonly an additional time has to be introduced to complement the difference between the schedule period and the single cycling time of train. This complement time t_{comp} is expressed by

$$t_{\text{comp}} = T - t_{cyc}, \tag{2}$$

where t_{cyc} is the total travel time of a particular train type to complete a cycle in a condition that the train path (a set of routes) and the basic time for waiting and for changing train direction are given. T is the schedule period.

$$T = NI. \tag{3}$$

The complement time sometimes can be a flexible buffer for slight delays. Nevertheless, it might also cause potential conflicts at terminals since in many cases the complement time is distributed to the train dwelling time or turning time at terminals and it increases the time to complete the usage of the platforms for a train (the trains usually are not planned to enter the depot).

3 FACTOR CORRELATIONS

The critical factors can be further categorized into technical and decidable groups. As the train data are given, the minimum operation times $W_{\min}$ and $C_{\min}$ and the train speed index are determined. Taking the line data and signalling system into account, the interval restrictions $I_{\min}$ are technically definite, yet they are to be measured. By contrast, the waiting times, direction turning around times, intervals in service, total number of trains and even the train path are decidable, either by operator's expectations or other objectives. Figure 2 shows the logical associations of these factors.

To determine the variables in the latter group, the correlations of factors are further studied by statistical analysis. Correlation matrix of factors is shown in Table 1 where each row

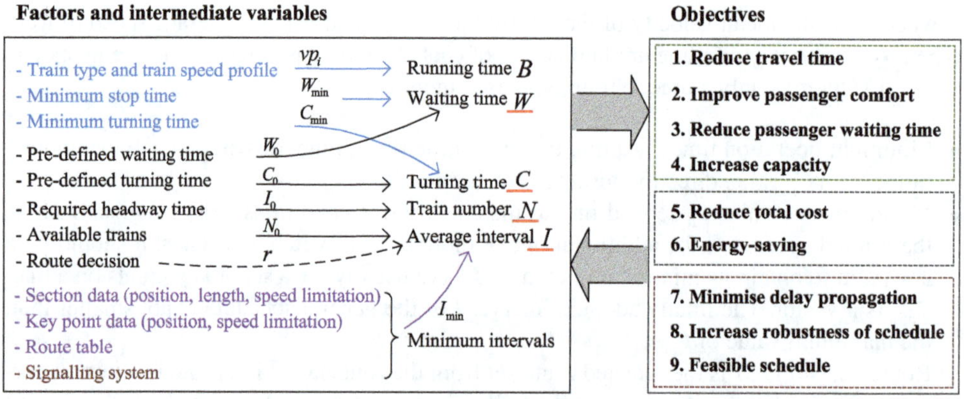

Figure 2: The principal associations of factors.

implies how a decision variable depends on other factors and the bigger a correlation value indicates the more related that pair of factors. Some important correlations (bold values) are coincident with the common rules we obtain. First we note that between two order stops of train only the strictest key point is denoted by the index K. The frontal dwelling time (in the train movement direction) of key point K is denoted by C_k or W_k. Thus, some important correlations showed in Table 1 can be partly explained as follows:

- The minimum interval of a key point depends on the position and the type of train stop (to keep or to change the direction of train); besides, it is related to the average speed to travel through the key point. They are expressed, respectively, by

$$I_{\min}(k) = aW_k + b\overline{vp}_i + c\,, \tag{4}$$

$$I_{\min}(k) = dC_k + e\overline{vp}_i + f\,, \tag{5}$$

  where $a,b,\ c,d,\ e,\ f$ are the parameters of a given system.
- The final determined interval should not be smaller than the minimum allowed interval of any key point; that is,

$$I \geq \max\{I_{\min}\}\,. \tag{6}$$

  If the number of trains and the train performances are given, then

$$I \geq t_{cyc} / N_0\,, \tag{7a}$$

or

$$I = (t_{cyc} + t_{comp}) / N_0\,. \tag{7b}$$

- In a contrary case the interval in service is known, and the number of trains satisfies

$$N = (t_{cyc} + t_{\text{comp}}) / I_0\,. \tag{8}$$

Table 1: Correlation matrix of factors in statistical analysis.

	Pos*	$I_{\min}$	C_{k}.	W_k	V_i	I_0	N_0	t_{cyc}	t_{comp}	r.
Pos	1.000	–0.387	–0.322	0.294	0.320	0.021	–0.028	–0.029	–0.004	–0.003
$I_{\min}$	***–0.387***	1.000	***0.401***	***–0.530***	***–0.670***	0.213	–0.184	–0.043	–0.028	–0.079
C	–0.322	0.401	1.000	–0.8			–0.049	–0.021	–0.072	–0.074
W	0.294	–0.530	–0.831	1.000	0.654	–0.006	0.014	0.019	0.006	0.038
V_i	0.320	–0.670	–0.499	0.654	1.000	0.052	–0.132	–0.209	–0.084	–0.139
I	0.021	***0.213***	0.033	–0.006	0.052	1.000	***–0.890***	***–0.226***	–0.127	***–0.487***
N	–0.028	–0.184	–0.049	0.014	–0.132	***–0.890***	1.000	***0.605***	***0.365***	***0.701***
t_{cyc}	–0.029	–0.043	–0.021	0.019	–0.209	***–0.226***	***0.605***	1.000	***0.325***	***0.779***
t_{comp}	–0.004	–0.028	–0.072	0.006	–0.084	***–0.127***	***0.365***	***0.325***	1.000	0.092
r.	–0.003	–0.079	–0.074	0.038	***–0.139***	***–0.487***	***0.701***	***0.779***	0.092	1.000

*Position of the key point on the line.

- When there are many types of trains to run on the line, the train type and its speed index are one of the influencing factors of the single cycling time of the train.

$$t_{cyc} = h\ \overline{vp}_i + l\,. \tag{9}$$

- Route decision also has an effect on the train cycling time and the variable pair: the adopted number of trains N and interval I.
- Last but not least, the complement time has to be distributed to the dwelling times and it satisfies eqns (10) and (11).

$$t_{\text{comp}} = \sum(\ W_k - W_0) + \sum(\ C_k - C_0)\,, \tag{10}$$

$$t_{\text{comp}} < \sum(\ I - I_{\min}(k))\,, \tag{11}$$

W_0 and C_0 are pre-defined and $W_0,\ W_k \geq W_{\min}\,,\ C_0, C_k \geq C_{\min}$.

4 PLANNING FEASIBILITY

The difficulty to estimate the feasibility of the user requirements lies on the correlative variation of the general restriction structure. In other words, all the restrictions are not constant. Fortunately, most factors can be related to the key factor, minimum interval vector $I_{\min}$. To sum up, the procedure of feasibility evaluation may include the following steps:

(s1) Calculate $I_{\min}$.
(s2) Obtain t_{cyc} (on the basis of $W_k = W_0$ and $C_k = C_0$).

(s3) Obtain the coefficients in eqns (4), (5) and (9).
(s4) Check the feasibility by inequalities in eqns (6), (7a), (7b), (8) and (11).

After step (s3), linear bounds of each desired variable may be visualized in a bound diagram; e.g. Fig. 3 shows a bound diagram of acceptable interval of a line where the train performance and train path are definitive.

Step (s4) deduces a conclusion by verifying all bounds of the variables. It results in a positive outcome when all requirements can be satisfied within the bounded domain. Otherwise, it identifies that the user requirements of plan are infeasible. The user may obtain inspired information to adjust the requirements.

Take the line given in Fig. 4 as an example. 21 key points of the line are taken into account in I_{min} investigation, among which 1, 3, 5 and 14 are the key points that can be related to the times to change train direction, i.e. C_k; others are related to the waiting times W_k. The available number of trains of certain type is known, $N_0 = 22$. It is queried whether it is feasible to schedule the trains at an interval of 100 s.

When the average speeds of train types are not obviously different ($\overline{vp_i} \in \{1, 0.881\}$ in this example), some parameters appear as Table 2 shows. In detecting $I_{\min}$ the initial setting

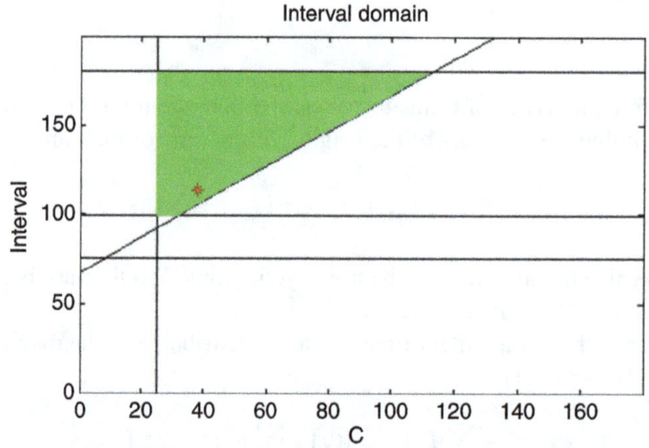

Figure 3: An instance of linear bounds of the interval (along a given train path).

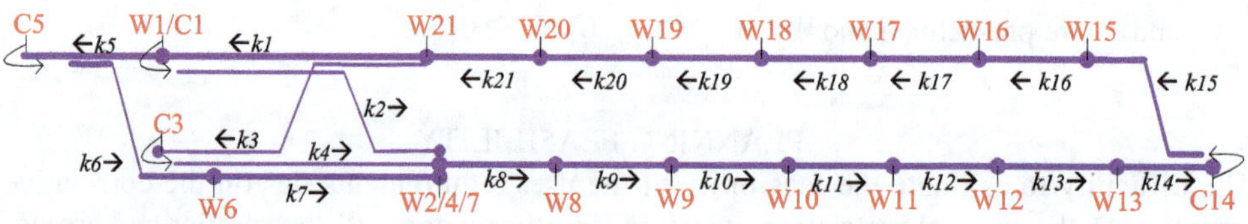

Key points and times in Route set 1: *k1*-C1-*k2*-W2-*k8*-W8-*k9*-W9-...-*k14*-C14-*k15*-W15-*k16*-W16-...-*k21*-W21-*k1*-...

Key points and times in Route set 2: *k3*-C3-*k4*-W4-*k8*-W8-*k9*-W9-...-*k14*-C14-*k15*-W15-*k16*-W16-...-*k21*-W21-*k1*-...

Key points and times in Route set 3: *k1*-W1-*k5*-C5-*k6*-W6-*k7*-W7-*k8*-W8-...-*k14*-C14-*k15*-W15-*k16*-W16-...-*k21*-W21-*k1*-...

Figure 4: A subway line with 7 stations and trains can only change the direction and track at terminal stations; 3 train paths with corresponding key points and routes to be selected.

includes $W_0 = 30, C_0 = 30$, and $\overline{vp_i} = 1$. Table 3 shows the results of feasibility estimation varying from train path decision.

From the results we can conclude that the requirement of running trains at an interval of 100 s in the given case is possible. It is infeasible only when there is no route set satisfying the requirement and restrictions.

Table 2: Parameters of correlations and measured minimum intervals of partial key points.

a, b, c, d, e, f, h, l	$I_{\min}\_C_1$	$I_{\min}\_C_3$	$I_{\min}\_C_5$	$I_{\min}\_C_{14}$	$\max\{I_{\min}\_W\}$
1,0,45,1,–42,112,–866,2939	115	106	100	99	75

Table 3: Feasibility estimation by route set.

Route set r	t_{cyc}	I domain	I	$N.$		$\sum(I - I_{min}(k))$	Feasibility
1	1911	$[115, \hat{I}]^*$	100	–	–	–	No
2	1910	$[106, \hat{I}]$	100	–	–	–	No
3	2053	$[100, \hat{I}]$	100	21	47	401	Yes
1&2**	1971	$[75, C_1 + 15]$	100	20		$321 + C_1$	Yes

*$\hat{I}$ is the maximum value of the acceptable interval, given by the user.

** 1&2 means successive trains alternately take the route set 1 and route set 2.

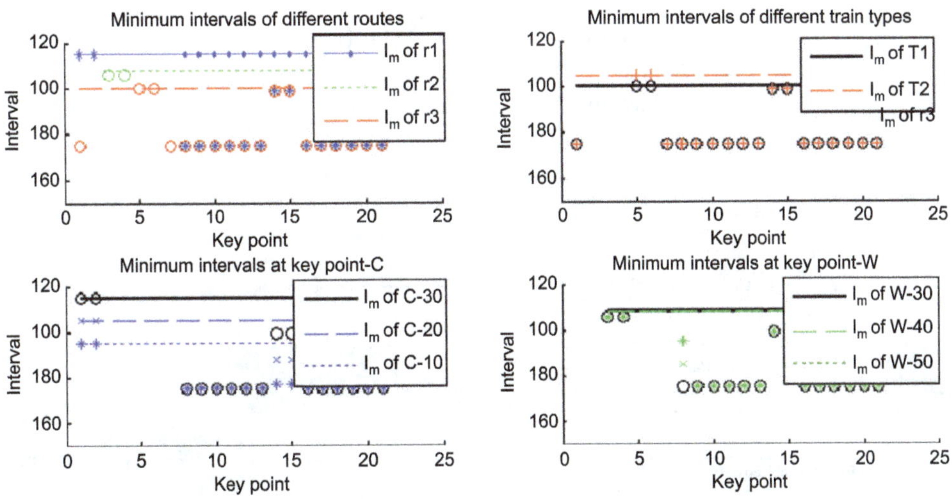

Figure 5: Distribution information of the intermediate variable I_{min}.

Besides further information of the minimum interval values we obtained by setting different routes, train types or distributing different dwelling times are illustrated in Fig. 5, from which two experience rules are summed up.

- #1: Only the key point whose $I_{\min}(k)$ equals $\max(I_{\min})$ of certain route set directly affects the acceptable interval I of the route set;
- #2: The acceptable interval I of a regular periodic schedule depends on I_{min} of the lower-speed train type.

Thus, the user may decide whether to adjust the dwelling time related to this kind of key point in order to obtain an appropriate interval I. But certainly, other variables and restrictions may be re-analysed as well.

5 APPLICATIONS

As shown in Fig. 6, further applications of feasibility evaluation include plan generation and optimization. Actually, only when the general feasibility is verified, it comes to have practical significance to apply algorithms and methods to generate and optimize the plan.

To generate a feasible plan, we only have to follow two more steps after the feasibility evaluation steps (s1 to s4):

(s5) Determine the favourable route set r and corresponding t_{comp}.
(s6) Determine the feasible C and W by distributing t_{comp}.

Here we consider the data of the train type(s) are pre-defined.

To optimize the plan, objectives are to be clarified first. Some of the objectives concluded in Fig. 2 can only be achieved by combining with the train control (e.g. passenger comfort), and some can be approached by adding objective functions to the scheduling model. For example, functions (12) and (13) are, respectively, defined to minimize the total travel time and to maximize the robustness of the plan.

$$\text{Minimise:} \quad B_r + \sum (W(k)) \tag{12}$$

$$\text{Maximise:} \quad \min(I - I_{\min}(k)) \tag{13}$$

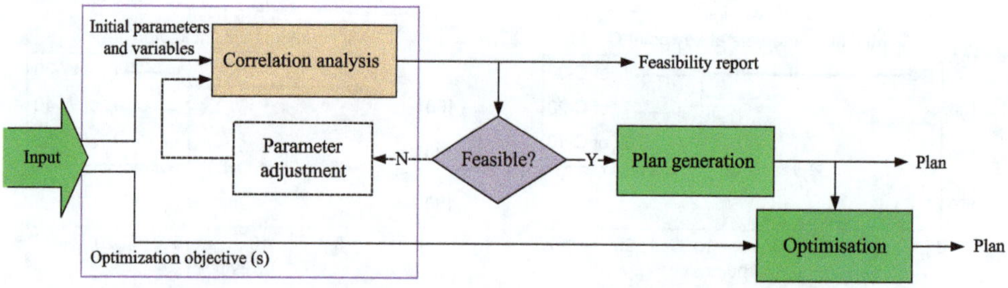

Figure 6: Possible applications based on feasibility analysis.

6 CONCLUSIONS

Actually, there are a number of factors to be concerned in practical subway scheduling process. These factors are quite basic and detailed but sometimes difficult to obtain for their variability and correlations. In many subway scheduling models, detailed factors are not considered and constant parameters are commonly used to simplify the problem. In this research, we investigated the main factors which might be considered in periodic subway scheduling problem from a more fundamental level.

Based on the correlation study of the critical factors, we established a set of internal restrictions among the factors, namely bound structure. Via this dynamic bound structure we realized the feasibility evaluation of the original requirements proposed by the subway operator (the user). Subsequently, the subway schedule (plan) can be generated if the user requirements are practicable and further optimized according to particular objectives. Otherwise, infeasible aspects are analysed, and possible adjustments which might produce a feasible plan can be proposed according to the available resources of the system.

Feasibility analysis is an indispensable part of work in practical subway scheduling, and more complex situations (such as multi-train type, more shared conjunctions on different routes, etc.) are to be included in the research.

REFERENCES

[1] Odijk, M.A., A constraint generation algorithm for the construction of periodic railway timetable. *Transportation Research Part B: Methodological,* **30**, pp. 455–464, 1996. DOI: 10.1016/0191-2615(96)00005-7.

[2] Liebchen, C. & Möhring, R. H., A case study in periodic timetabling. *Electronic Notes in Theoretical Computer Science,* **66**, pp. 18–31, 2002. DOI: 10.1016/S1571-0661(04)80526-7.

[3] Törnquist, J. & Persson, J.A., N-tracked railway traffic re-scheduling during disturbances. *Transportation Research Part B: Methodological,* **41**, pp. 342–362, 2007. DOI: 10.1016/j.trb.2006.06.001.

[4] Miyatake, M., Kuwahara, R. & Nakasa, S., A simple adjustment of runtimes between stations for saving traction energy by means of mathematical programming, presented at the *Computers in Railways XIII*, New Forest, UK, 2012.

[5] Yang, X., Li, X., Gao, Z., Wang, H. & Tang, T., A cooperative scheduling model for timetable optimization in subway systems. *IEEE Transactions on Intelligent Transportation Systems,* **14**, pp. 438–447, 2013. DOI: 10.1109/TITS.2012.2219620.

[6] Yang, X., Ning, B., Li, X. & Tang, T., A Two-objective timetable optimization model in sybway systems. *IEEE Transactions on Intelligent Transportation Systems,* **15**, pp. 1913–1921, 2014. DOI: 10.1109/TITS.2014.2303146.

[7] Soler, M., López, J. & Mera, J.M., Simulation system for the optimization of a block distribution under the ERTMS-1 signalling system, presented at the *Computers in Railways XIII,* New Forest, UK, 2012.

AN OPTIMIZATION APPROACH TO RAILWAY TRACK ALLOCATION CONSIDERING THE TRAIN-SET ROUTINGS

BUM HWAN PARK
Dept. of Railroad Management and Logistics, Korea National University of Transportation, Korea

ABSTRACT

This paper deals with an optimization approach to railway track allocation, which is one of the most important decision problems in the presence of multiple train operating companies (TOCs). In fact there has been deep discussion about how to boost the competition environment in Korean railway since the functional reform in 2004, which at last resulted in introducing a new entrant to high-speed railway passenger transportation market. Finally, in August 2016, two operating companies will compete on the major routes in the Korean high-speed railway network. The infra manager, KR Network, who is responsible for allocating the slots, has been developing their own allocation procedure which partly uses an optimization model for adjusting the times of requested train-paths. But one of the TOCs' concerns with respect to the adjustment is that their train-set routing plan could be in disorder by the adjustment of the arrival/departure times. Assuming TOCs submit their routing plan as well as their desired train-paths, we present an optimization model and algorithm for track allocation problem, considering the routing plan requested by TOCs. The model is developed on a time-space network, where a train-path can be described as the sequence of the arcs. Based on the network, we developed an column-generation approach to dynamically generate the promising train-paths for each requested one so as to maximize the total profit while preventing the routing plans from disrupting by means of setting up the arcs only among the two successive train-paths in the routing plan. Also we present the experimental results applied to the Korean high-speed railway network.

Keywords : track allocation, column generation, train-set routing.

1 INTRODUCTION

After the functional reform of the Korean railway to separate the train operating company (TOC) from the infrastructure manager in 2004, the Korean railway has been trying to foster the competitive environment in the railway passenger transportation market. At last, the efforts resulted in introducing a new entrant to high-speed railway passenger transportation market. Finally, in August 2016, two TOCs, KORAIL and SR, will compete on the major routes in the Korean high-speed railway network.

The infrastructure manager, the KR Network, has been developing its own procedure to allocate the tracks to the operating companies. The procedure is based on an optimization model to find the conflict-free timetable by adjusting arrival/departure times of each train submitted by the operating companies. But one of the operating companies' biggest concerns with respect to the adjustment is that their train-set routing plan could be in disorder by the adjustment of the arrival/departure times. In fact, the operating companies try to convince the infra manager to respect the train-set routing plan. However, the consideration of the routing plans could make the adjustment procedure more difficult.

Our study deals with an optimization approach to railway track allocation, respecting the routing plans considered by the TOCs.

DOI: 10.2495/TDI-V1-N3-452-459

In fact, the track allocation problems, which is in essence a combination of timetabling problem and bidding procedure, have received attention in a few literature: see, for example, Borndörfer et al. [1], Nilsson [2], Brewer and Plott [3] and Parkes and Ungar [4]. These works need a variant of timetabling problem which is called a *winner determination problem* in the context of *auction* mechanism.

The remainder of this paper is organized as follows. Section 2 explains our problem in detail, and the time-space network to mathematically formulate our problem. Section 3 presents an integer programme model by using path variables. And then we suggest a column-generation approach to find promising paths in the time-space network to improve the master problem at each iteration. In Section 4, we show the experimental results applied to the Korean high-speed railway network.

2 PROBLEM DEFINITION

Each operating company submits their favourite schedule of trains together with the routing plan. The routing plan means the sequence of the trains covered by each train set, so we can assume each train has its previous and next connection trains. And the Korean high-speed trains, so-called KTX, are being run with various halting patterns.

Our model can be easily explained on a time-space network where each node represents discretized time and each arc train movements. The approaches to use time-space network are found in other works [5–7].

In Fig. 1, each node on a horizontal line represents a discretized time while all the nodes together on any line belong to a single station. An arc between two nodes belonging to different stations represents the travelling times between those two stations. Where nodes corresponding to an intermediate station are split, the arc between them represents the dwell time. Each train can be expressed as a sequence of arcs. We assume each train has its own profit but the profit is decreased, depending on the deviation from the departure time at the original station and from requested dwell time at each stopping station.

And the trains are combined with each other by the train-set routing plan. For example, train 1 should be connected to train 2 and train 2 connected to train 4. If the departure time of train 1 would be delayed and there could not be enough turnaround time between the arrival time of train 1 and the departure time of train 2, the shift of departure time of train 1 should not be admitted. In this case, we say there are two routings, each of which has two trains. In the next section, we suggest an optimization model to consider the routing plan.

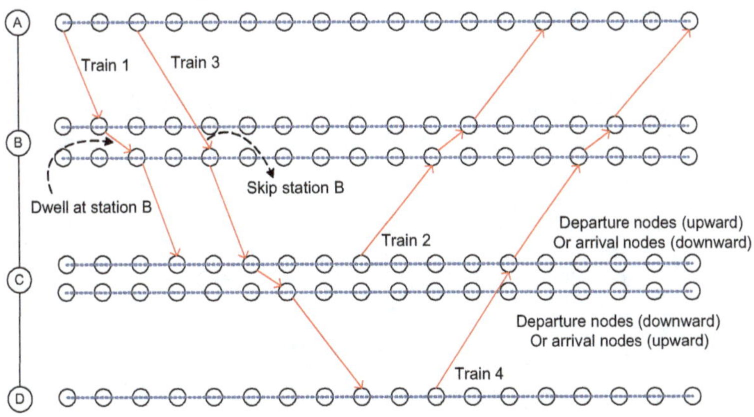

Figure 1: Representing the trains in the time-space network.

3 MODEL AND ALGORITHM

3.1 Optimization model

To develop our optimization model, the time-space network should be modified as follows. For the first train in a routing, we include an artificial source node, which is denoted as s_k, to express the time windows of departure at the original station. The size of the window is assumed to be a priori given by the infra manager.

And for the last train in a routing, we include an artificial terminal node, which is denoted as t_k for the routing k, and the associated arcs to express an alternative routing as a path to the node. As a result, a path from s_k to t_k corresponds to an alternative routing, which should be optimally adjusted in the model.

Finally, the arcs representing the dwelling time are added within a certain acceptable deviation. Figure 2 shows the time-space network representation of our model.

Operational constraints in the timetabling problem can be easily stated as *set packing* constraints [1], which means there is a restriction on the number of arcs in a set defining a *conflict*.

These constraints can be categorized as one of two types: those for guaranteeing the minimum headway at departure and arrival, and those for restricting the number of trains stopping simultaneously at any station. In fact we ignored the constraints preventing the trains from overtaking each other because we consider the case of the homogeneous traffic; in other words, all trains considered have the same running times between two stations. Figure 3 shows the way to express the operational constraints.

Based on the network representation as the above, we use the following notations to define our mathematical model.

Notation

- $G = (V, A)$: the time-space network. The nodes are indexed by i and j, and the arcs are a.
- c_a: the penalized profit of the arc a.

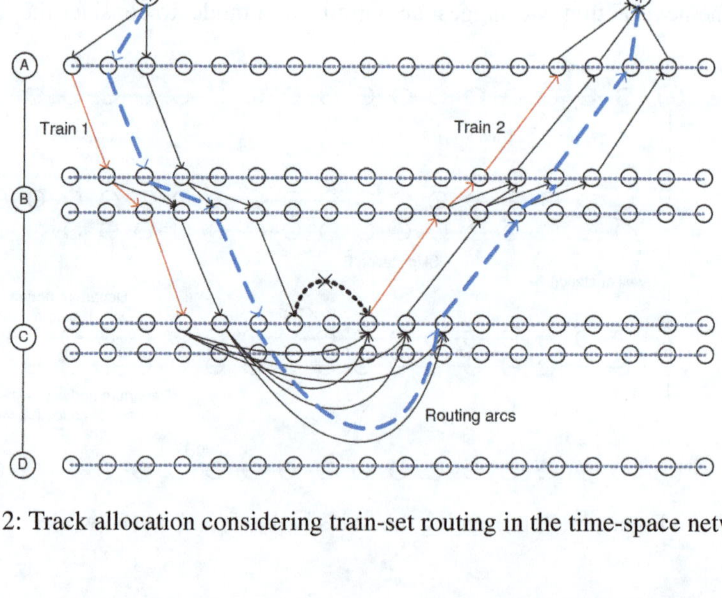

Figure 2: Track allocation considering train-set routing in the time-space network.

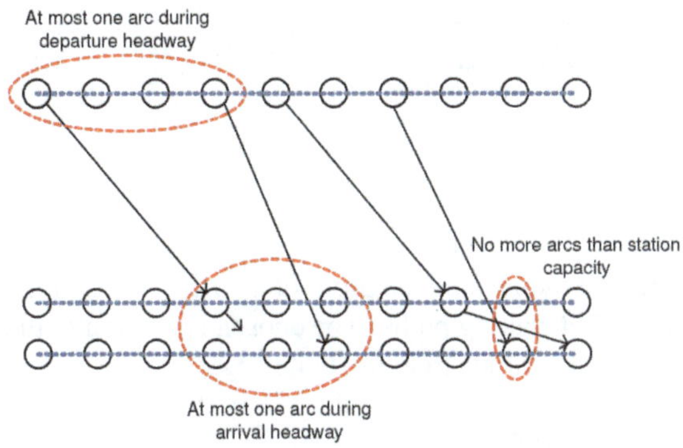

Figure 3: Establishing the conflict-free conditions in the time-space network.

- $A = A_1 \cup A_2 \cup A_3 \cup A_4 \cup A_5$ where $A_1, A_2 \ldots$, and A_5 is the set of arcs which represent the train movement between two stations, dwelling at a station, connection to the next train in the routing plan, the arcs from the artificial source nodes and the arcs to the artificial terminal nodes, respectively.
- K: the set of request routing, the elements of which are indexed by k.
- T_k: the set of request trains belonging to the routing k, the elements of which are indexed by t.
- $P^k(t)$: the profit of the train t belonging to the routing k.
- P_k: the set of paths in G from S_k to t_k for the routing k, the elements of which are denoted by P.
- $c^k(P) = \sum_{a \in p} c_a$: the profit of the path P for the routing k. The path will be called *routing path*.
- C: the set of arcs that make conflicts. The set of C denoted by Ω.
- lc: the allowed number of arcs in the conflict set C.

We assume for simplicity that penalties are assigned only to the delayed departure time of the first train belonging to a routing and the delayed dwelling time at stations. If we assign the penalties to the associated arcs, we can easily compute the penalized profit of a request train. For the train connected from the previous train in a routing like train 2 in Fig. 2, the penalties are assigned to the *routing arcs* which means ones connecting two adjacent trains belonging to the same routing with sufficient turnaround time. In conclusion, we can compute the penalized profit of the routing path P by $\sum_{a \in p} c_a$.

In our study, we present a mathematical formulation based on path flow variables and a column-generation approach. The formulation, which is called *master problem* in the context of Dantzig-Wolfe decomposition [9], then becomes the following. If the path P is used for routing k, the decision variable $f^k(P)$ equals to 1, and $f^k(P)$ equals to 0 otherwise.

Master problem

$$\max \sum_{k \in K} \sum_{P \in P_k} c^k(P) f^k(P), \tag{1}$$

$$\text{s.t.} \sum_{P \in P_k} f^k(P) \le 1, \forall k \in K, \tag{2}$$

$$\sum_{k\in K}\ \sum_{P\in P_k|P\cap C\neq\phi} f^k(P)\leq lc, \forall C\in\Omega, \tag{3}$$

$$f^k(P)\in\{0,1\}, \tag{4}$$

where $c^k(P) = \sum_{a\in p(\in pk)} c_a$.

The model maximizes the total penalized profit of all the routing paths according to eqn (1). By the eqn (2), the number of routing paths to be chosen is no greater than 1 for each routing. Finally, eqn (3) guarantees a conflict-free timetable.

3.2 Column-generation approach

If $f^k(P)\in\{0,1\}$ is replaced with $f^k(P)\geq 0$ and the non-negative dual variables are denoted as π_k, μ_C corresponding to eqn (2) and (3), respectively, the optimality condition of the relaxed master problem, which can be stated as dual feasibility condition, is as follows. For all $P\in P_k$,

$$\bar{c}^k(P) = c^k(P) - \sum_{\{C|P\cap C\neq\phi\}} l_C\mu_C = \sum_{a\in P} ca - \sum_{a\in P}\sum_{\{C|a\in C\}} l_C\mu_C - \pi_k$$

$$\sum_{a\in P}[c_a - \sum_{\{C|a\in C\}} l_C\mu_C] - \pi_k \leq 0. \tag{5}$$

We can find new columns to potentially improve the master problem by finding new paths $P\in P_k$ such that $c^{-k}(P) > 0$. To find such a path, we impose a modified cost on the arcs in the associated network with the routing k. It is then sufficient to find a longest path in the network, which can be easily solved because the network is acyclic [8]. Figure 4 shows the overall algorithm.

First, the initial relaxed master problem is established using the request trains and routing paths submitted by the operating companies. And then the above optimality conditions in eqn. 5 are checked, and if there is at least a routing path violating the optimality conditions,

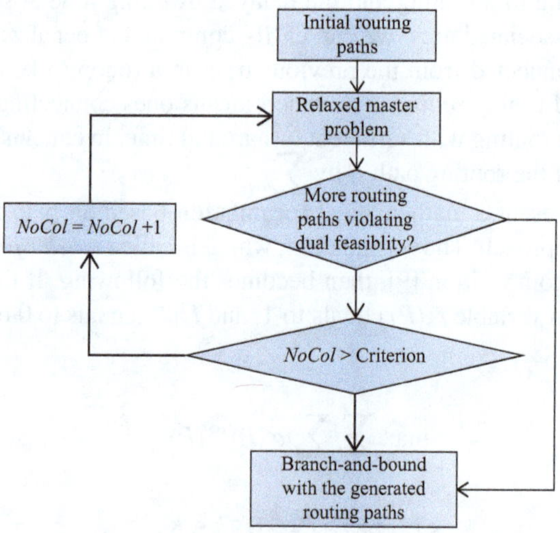

Figure 4: Column-generation heuristic.

new columns are added until the number of added columns, which is denoted as *NoCol* in Fig. 4, is no greater than 'Criterion'. If no more routing paths are found or *NoCol* is greater than 'Criterion', the algorithm goes in the branch-and-bound algorithm with the generated routing paths so far.

4 EXPERIMENTAL RESULTS

Figure 5 shows the Korean high-speed railway network used in our experiments. It consists of 40 stations and the bold lines (including dashed lines) indicate the dedicated tracks for the high-speed trains. The dashed lines show the dedicated tracks used by a new operating company SR. The incumbent operating company is running their high-speed trains (KTX) on the routes Seoul-Mokpo, Seoul-Yeosoo, Seoul-Jinjoo, Seoul-Busan and Seoul-Pohang. But the new entrant SR is supposed to run their high-speed trains (SRT) on the routes Suseo-Mokpo and Suseo-Busan.

For the experiments, we used the KORAIL's timetable in 2013 which consists of 200 trains a day. And to make the timetable of trains departing Suseo station, which means the new operating company's timetable, KORAIL's timetable was copied and then shifted to some minutes keeping the running times between new stations in minds. Totally, we generated 392 trains. And we assume that the driving time between two stations is fixed for both KTX and SRT. Therefore there is no need to include the conflict-sets preventing overtaking in Ω.

And we assume that there are 140 routings each of which contains one to six trains submitted by the companies. We simply generated 140 routings by keeping connecting trains if possible. So there are many routings with just one train. The following table (Table 1) shows the information about our routings.

Figure 5: The Korean high-speed railway network.

Table 1: Routing information.

# of trains in routing	1	2	3	4	5	6
# of routings	52	13	19	30	19	2

For the dwell times, tolerances [0,5] are imposed on the dwell time so that the network includes five additional dwelling arcs for each request dwelling arc. And for departure times of the first train belonging to a routing, we set the tolerance interval as [0,10] . As mentioned before, the profit is penalized, depending on the deviation from the request times. We set the penalty 1 for each minute delay. The profits of trains are all set to be a constant. Finally, the algorithm terminates if 'Criterion' in Fig. 4 becomes no smaller than 7,000. The algorithm was implemented using CPLEX 12.5 on a PC with an Intel Core 3.20 GHz CPU.

We implemented our model by two steps: the first step is to construct the time-space network and the set of conflict sets Ω given the list of trains and routing plans by the operating companies. In case of Fig. 5, it took about 19 min to construct the overall time-space network and Ω. The second step is to find an optimal routing paths by Fig. 4. Table 2 shows the performances of our algorithm in Fig. 4.

As shown in the table, it was assumed that two operating companies submitted 392 trains and 140 routings, and all the trains have the profit of 100.

First, we investigated the difference between the solutions of two cases $d = 0$ and $d = 10$. $d = 0$ means there is no room to adjust the request times of trains so that the number of chosen trains as well as routings are much less than the case of $d = 10$.

Table 3 shows how many trains or routings were adjusted and how much time was shifted. The first row indicates the total amount of shifted times that resulted in our algorithm and the second and third rows show the number of routings and trains with no more than the associated delay, respectively.

However, it takes a lot of time to find a solution even though the number of generated columns is restricted to 7,000. In fact, most of computation times was taken to find a good integer solution by branch-and-bound procedure even though the MIP gap was set to be 15% in CPLEX.

Table 2: Experimental results.

Departure [0, d]	# of routings: $\|K\|$	# of trains: $\sum_{k \in K} \|T_k\|$	Total profit	Computation time (min.)
given	140	392	39,200	n/a
$d = 0$	101	259	25,543*	4
$d = 0$	108	285	27,640*	31

* terminated if MIP gap is within 15%.

Table 3: Total delay.

Delay (min.)	= 0	≤ 5	≤ 10	≤ 20	≤ 60
# of routings	33	63	82	94	108
# of trains	124	234	261	285	285

5 CONCLUDING REMARKS

We suggested a new optimization model and algorithm for the track allocation model to consider the train-set routing plan. The formulation was based on path-flow variable defined on the time-space network, and then a column-generation approach was presented. But the algorithm should be more elaborated since there still needs a lot of time to find an integer solution by branch-and-bound procedure.

ACKNOWLEDGEMENT

This article was supported by the Korea National University of Transportation in 2016.

REFERENCES

[1] Borndörfer, R., Grotschel, M., Lukac, S., Mitusch, K., Schlechte, T., Schultz, S. & Tanner, A., An auctioning approach to railway slot allocation. *ZIB-Report*, 05-45, 2005.

[2] J.-E. Nilsson., Allocation of track capacity: Experimental evidence on the use of priority auctioning in the railway industry. *International Journal of Industrial Organization*, **17**, pp. 1139–1162, 1999 DOI: 10.1016/S0167-7187(99)00016-8.

[3] Brewer, P.J. & Plott, C.R., A binary conflict ascending price (bicap) mechanism for the decentralized allocation of the right to use railroads tracks. *International Journal of Industrial Organization*, **14**, pp. 857–886, 1996 DOI: 10.1016/0167-7187(96)01014-4.

[4] Parkes, D.C. & Ungar, L.H., An auction-based method for decentralized train scheduling. *Proceedings of the Fifth International Conference on Autonomous Agents*, 2001.

[5] Caprara, A., Fischetti, M. & Toth, P., Modeling and solving the train timetabling problem. *Operations Research*, **50(5)**, pp. 851–861, 2002.

[6] Caprara, A., Cacchiani, V. & Toth, P., A column generation approach to train timetabling on a corridor. *4OR*, **6**, pp. 125–142, 2008.

[7] Caprara, A., Cacchiani, V. & Toth, P., Scheduling extra freight trains on railway networks. *Transportation Research Part B*, **44**, pp. 215–231, 2010 DOI: 10.1016/j.trb.2009.07.007.

[8] Ahuja, R.K., Magnanti, T.L. & Orlin, J.B., *Network Flows : Theory, Algorithms, and Applications*. Prentice Hall: Upper Saddle River, NJ, 1993.

[9] Wosley, L.A., *Integer Programming*. John Wiley and Sons: Hoboken, NJ, 1998.

EVALUATION OF TRAIN SCHEDULE BY USING INFORMATION THEORY

K. MORO & Y. SUGAI
Graduate School of Engineering, Chiba University, Japan.

ABSTRACT

Convenience is an important requirement of a scheduled transit system. The authors proposed a new method to evaluate a train schedule by using entropy of the information theory. By the aspect of statistical mechanics, the randomness of a schedule is associated with the convenience of the schedule. However, the way of using entropy requires special conditions. In this paper, we show that we can relax the conditions by interpreting the entropy as information.

First, we explain the representation of a schedule by a tree structure. Introducing the tree structure, we recognize the more composite construction of schedules. The tree structure leads to the concept of conditional entropy. Using the concept of conditional entropy, we can evaluate schedules of the states of different places or times. Secondly, we can understand what the information of the schedule is. We can grasp that the information is the benefit of knowing the schedule. The increase of the entropy corresponds to the increase of the benefit. Moreover, it becomes possible to express an unexpected disbenefit by the entropy with the loss of information. In other words, the benefit is a positive entropy and the disbenefit is negative entropy. Finally, we show that we are able to deal with an influence of wasteful time in a schedule, such as an unnecessary waiting time and an extra duration by using the entropy. It is very important that we can evaluate negative effects, namely the discrepancy of supply and demand from the point of view of time.

Keywords: entropy, evaluate timetable, loss of information, tree structure.

1 INTRODUCTION

The schedule of a transit system, such as a railway network, has a great influence on the convenience of the transit system. We sense the transit system is convenient in that a great number of trains run and the distribution of departures is even. In other words, we can get a train in a short time at anytime we come in a station. Moro and Sugai [1] presented the method of evaluating schedules from the viewpoint of the number of trains and the distribution of departures. This method uses Shannon's entropy [2] of the information theory. Nevertheless it assumes that each passenger chooses his/her train. It does not cope with a case that a passenger cannot select any train.

In this paper, we introduce a method that deals with such cases, in which some passengers fail to select their trains, not only in cases in which passengers are able to select their trains. First, we show a way of expressing a schedule by a tree structure. Employing this structure we will be able to evaluate various states of schedules together by using entropy. Secondly, we explain what the information of the schedule is. By using the information of the schedule, we will be able to evaluate the adverse influence of the schedule according to its entropy. Finally, we show that we are able to deal with the influence of wasteful time in a schedule, such as unnecessary waiting time and an extra duration by the entropy.

DOI: 10.2495/TDI-V1-N3-460-470

2 ENTROPY AND INFORMATION OF THE INFORMATION THEORY

In the beginning, we will explain Shannon's entropy and information of the information theory [2].

2.1 Definition

Let X be a discrete random variable on a finite set $\tilde{X} = \{x_1, \ldots, x_n\}$, with probability distribution function $p(x) := Pr(X = x)$.

2.1.1 Information

We introduce information as a means to measure a rarity of an event. Information of the event $x \in \tilde{X}$ with probability $p(x)$ is defined as

$$I(x) := -\log p(x). \tag{1}$$

2.1.2 Shannon's entropy

The entropy $H(X)$ of X is defined as

$$H(X) := -\sum_{x \in X} p(x) \cdot \log p(x). \tag{2}$$

The convention 0 log 0 = 0 is adopted in the definition. The logarithm is usually taken to the base 2, in which case the entropy is measured in 'bits'. This definition shows that we can calculate entropy only if we define the probability.

2.2 Implications of entropy

Shannon's entropy of X corresponds to the expected value of the information of X.

3 EXPRESSION OF SCHEDULE BY THE TREE STRUCTURE

In this section we introduce the expression of a departure schedule at a station by a tree structure.

3.1 Definition

Let S be a schedule consisting of trains as

$$S = \{tr_1, tr_2, \cdots, tr_n\}. \tag{3}$$

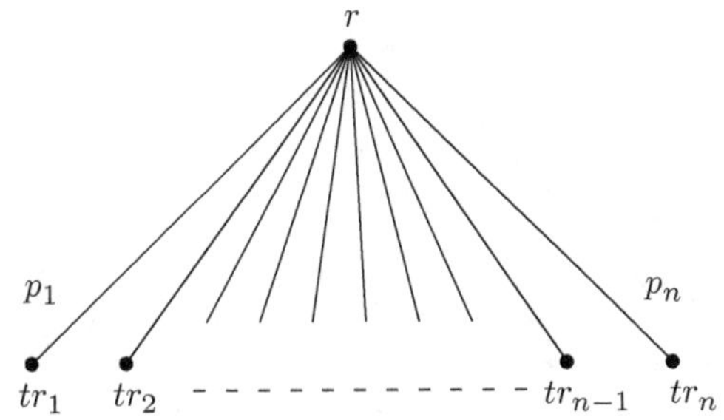

Figure 1: Tree structure of a schedule.

S leads a graph $T_S = \{N_S, A_S\}$. N_S is the set of trains and root node r. A_S is the set of arcs. For every $a_i = (r, tr_i) \in A_S$ is a pair of nodes of N_S with probability p_i. The tree structure of the schedule is represented in Fig. 1. In addition, we call the node described at the upper *parent* and at the lower *child*.

3.2 Nested schedule

Now, we think about integrated plural schedules. Consider the integrated schedules of the days of week for example. Suppose that the trains $\{tr_k\}^i$ are scheduled in the ith day of the week d_i. Moreover, the $p_j^{(i)}$ is the probability of train $tr_j \in \{tr_k\}^{(i)}$ in the day d_i. Furthermore, we assume that π_i is the ratio of ridership on the day d_i to the ridership of whole term. Namely, the set of trains $\{tr_k\}^i$ nests under the day d_i, and the set of days makes an intermediate class.

Specifically, we can say that D is a discrete random variable on the finite set $\tilde{D} = \{d_1, \cdots, d_n\}$ with probability distribution function $\pi_i = Pr(D = d_i)$. Besides, we can think of d_i as the set of the trains $\{tr_k\}^i$. Additionally, $p_j^{(i)}$ is the conditional probability under d_i, and what is more, if tr_j is not a member of d_i, then $p_j^{(i)} = 0$.

Figure 2 indicates a nested schedule. The intermediate class E contains two elements: η_1 and η_2. The elements η_1 and η_2 are formed as follows:

$$\eta_1 = \{e_1, e_2, e_3\}, \qquad \eta_2 = \{e_4, e_5\}. \tag{4}$$

The entropies of E, η_1 and η_2 are

$$H(E) = -\sum_{i \in \{1,2\}} \pi_i \cdot \log \pi_i, \tag{5}$$

$$H(\eta_1) = -\sum_{j \in \{1,2,3\}} p_j^{(1)} \cdot \log p_j^{(1)}, \tag{6}$$

$$H(\eta_2) = -\sum_{j \in \{4,5\}} p_j^{(2)} \cdot \log p_j^{(2)}. \tag{7}$$

Then the entropy of the entire system S becomes as follows:

$$H(S) = H(E) + \sum_{i \in \{1,2\}} \pi_i \cdot H(\eta_i). \tag{8}$$

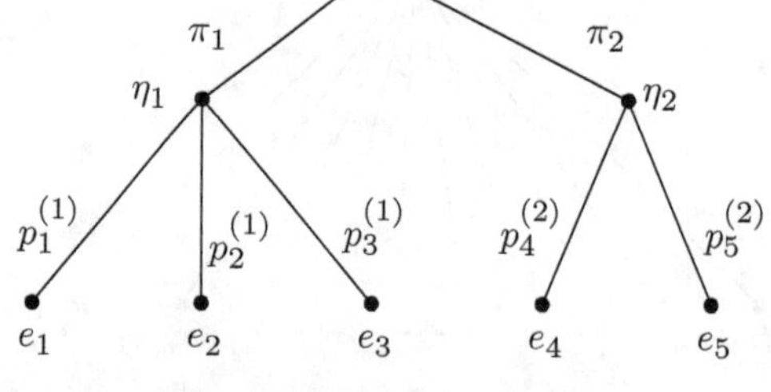

Figure 2: Nested schedule.

4 INTERPRETATION OF THE INFORMATION OF A SCHEDULE

We will discuss the relation between the convenience of a schedule and the information.

4.1 Relation between convenience and information

Moro and Sugai asserted that the entropy increases when the convenience of schedule increases [1]. That is, the entropy of the useful schedule is larger than that of the less-useful schedule. Moreover, the entropy means the expected value of the information in information theory. From this viewpoint, one may say that the more convenient the schedule of transit system is, the more its information grows.

4.2 Missing information

In a previous study [1], only the case in which all passengers can select their trains was explained. However, some passengers fail to catch a train occasionally, e.g. the passenger arrives at the station after the last train has departed.

We take it for granted that the fact that some passengers miss their trains reduces the benefit provided by the transportation system. In other words, when some passengers miss their trains, the entropy must decrease compared to the case when all passengers catch trains successfully. Therefore, we propose an interpretation that the event which is disbenefit has negative information, i.e. the system loses information by worthless events.

4.2.1 General definition

How shall we calculate the loss of information? We consider the schedule as in Fig. 3. There are n trains planned in this schedule. The probability of train tr_i is p_i. And a child node labelled *failure* exists. This node corresponds to the passengers who catch no train. Its probability is p_f. When we suppose that the all-child nodes are useful choices, the entropy is as follows:

$$H_F^{\dagger} = -\sum_{i=0}^{n} p_i \cdot I_i + p_f \cdot I_f. \tag{9}$$

The value I_i denotes information of an event with probability p_i as

$$I_i = -\log p_i. \tag{10}$$

However, the information of *failure* is added in eqn (9), i.e. the disbenefit choice improves the convenience. Hence, we define that we do not add the information of unusable choices when we calculate the entropy of a schedule; namely, we dismiss the information of worthless choices. For instance, the entropy of the schedule as in Fig. 3 is as follows:

$$H_F^{*} = \sum_{i=1}^{n} p_i \cdot I_i. \tag{11}$$

4.2.2 Representation by nested tree

We will represent a schedule which is the same as Fig. 2 by a nested tree structure as in Fig. 4.

There is an intermediate class consisting of *success* and *failure*. The probability of *success* is π_s and the probability of *failure* is π_f. The conditional probability under *success* of the train tr_i is $\hat{p}_i$.

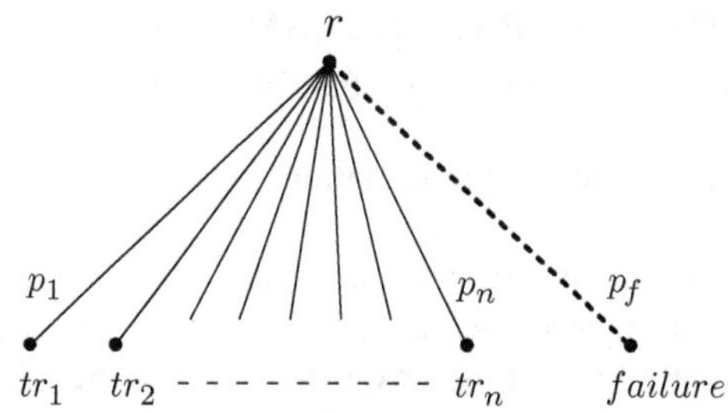

Figure 3: Tree structure with failure.

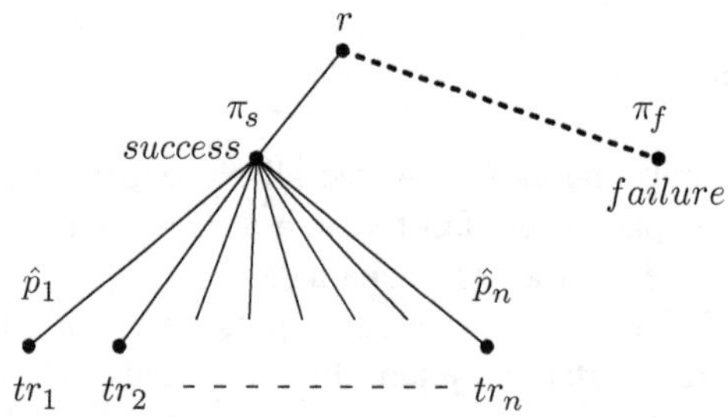

Figure 4: Nested tree structure with failure.

4.2.3 Example of calculation

We will calculate the entropy as follows in Table 1. Probabilities of the event of *success* and *failure* are as follows:

Table 1: Number of passengers: with failure.

	tr_1	tr_2	tr_3	tr_4	tr_5	tr_6	tr_7	tr_8	*Failure*	Total
Number of passengers	250	255	235	246	242	253	257	234	17	1,989
Subtotal of *success* 1,972										

$$\pi_s = 0.9915, \tag{12}$$

$$\pi_f = 0.0085. \tag{13}$$

Then the conditional probability $\hat{p}_i$ of tr_i under *success* and information I_i of tr_i calculated by $\hat{p}_i$ are as in Table 2.

The entropy of *success* is

$$H_s = \sum_{i=1}^{8} \hat{p}_i \cdot I_i = 2.9992. \tag{14}$$

The entropy of the whole is obtained as

$$\begin{aligned} H &= -\pi_s \cdot \log \pi_s + \pi_s \cdot H_s \\ &= 0.0123 + 0.9915 \cdot 2.9992 = 2.9860. \end{aligned} \tag{15}$$

Table 2: Probabilities and information: *success*.

Probability							
$\hat{p}_1$	$\hat{p}_2$	$\hat{p}_3$	$\hat{p}_4$	$\hat{p}_5$	$\hat{p}_6$	$\hat{p}_7$	$\hat{p}_8$
0.1268	0.1293	0.1192	0.1247	0.1227	0.1283	0.1303	0.1187
Information							
I_1	I_2	I_3	I_4	I_5	I_6	I_7	I_8
2.9797	2.9511	3.0689	3.0029	3.0266	2.9625	2.9398	3.0751
Probability × Information							
$\hat{p}_1 * I_1$	$\hat{p}_2 * I_2$	$\hat{p}_3 * I_3$	$\hat{p}_4 * I_4$	$\hat{p}_5 * I_5$	$\hat{p}_6 * I_6$	$\hat{p}_7 * I_7$	$\hat{p}_8 * I_8$
0.3777	0.3816	0.3657	0.3746	0.3714	0.3801	0.3831	0.3649

Including disbenefit phenomena reduces entropy.

5 EVALUATION OF WASTEFUL TIME

Nobody expects to waste time. Unexpected unuseful occurrences must reduce the entropy of a transit system. In this section, we explain the influence of wasteful time on the entropy.

5.1 Assumption

Previous studies of the evaluation of a transit system by its entropy have focussed only on the case in which the capacity of each train is infinity. Now, we would like to focus on a case in which the limited capacity of a train creates unexpected wasted time.

We consider that all passengers arrive at the station according to uniform distribution. We suppose some trains are closely packed before their departure.

5.2 Amount of waiting time

Under the conditions as stated above, we can calculate the amount of waiting time as in Fig. 5. We denote c is the capacity of the train. An empty train fills up with term $l^* > 0$. This train leaves station with term l later after filling up. The total waiting time of each passenger before the train fills up becomes w^* and after filling up becomes w. The value w is just equivalent to an idle time.

Now we can express w, w^* and $\exists\beta > 0$ as follows:

$$w = l \cdot c = \beta \cdot l^* \cdot c, \tag{16}$$

which denotes $l = \beta \cdot l^*$. Similarly, there exists $\alpha > 0$ such that

$$w^* = \alpha \cdot l^* \cdot c. \tag{17}$$

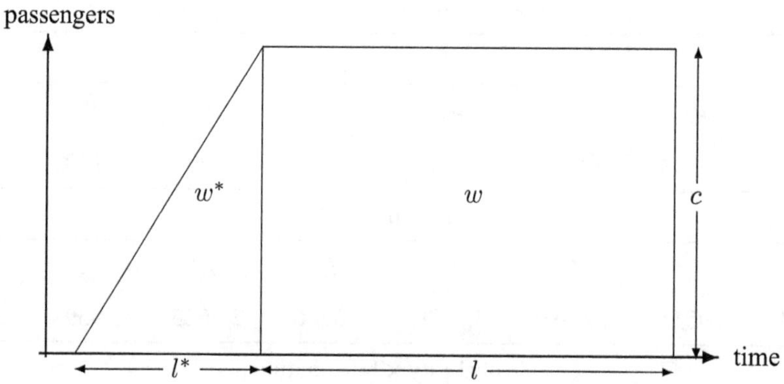

Figure 5: Amount of waiting time.

5.3 Ratio of effective time

We define the effective time rate r as

$$r := \frac{w*}{w*+w}. \tag{18}$$

If the train left before filling up, then we define $r := 1$. For the following discussions, we consider the condition to be limited to the prior condition.

Due to eqns (16) and (17), we can modify the effective time rate as follows:

$$r = \frac{\alpha \cdot l* \cdot c}{\alpha \cdot l* \cdot c + \beta \cdot l* \cdot c} = \frac{\alpha}{\alpha + \beta}. \tag{19}$$

5.4 Information of train with unexpected wasted time

We can express the part of train tr_i in a tree structure of schedule with probability p_i as in Fig. 6. The information of tr_i without useless time is

$$I_i = -\log p_i. \tag{20}$$

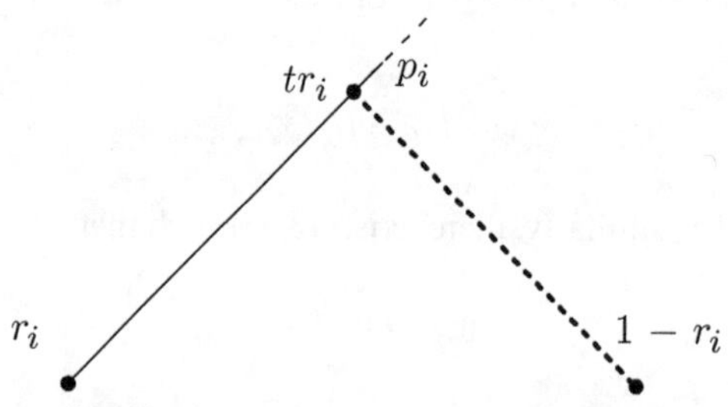

Figure 6: Example: with wasteful time.

We define that the information-reflected wasted time is calculated as

$$I_i^{\dagger} := -\log p_i^{\dagger}, \tag{21}$$

which denotes $p_i^{\dagger} := r_i \cdot p_i$. Furthermore, we define the entropy-reflected wasted time as

$$H^{\dagger} := \sum_i p_i^{\dagger} \cdot I_i^{\dagger}. \tag{22}$$

5.5 Influence of wasted waiting time

We calculated entropy of schedules with some variations of wasted waiting time.

5.5.1 Preconditions

We assume some preconditions to calculate entropy.

- Three trains are planned. The capacity of each train is the same.
- All passengers arrive at the origin station for first 15 min.
- The number of passengers equals 3 times the capacity of a train.
- Distribution of passengers arriving at the origin station is uniform, i.e. $\alpha = \frac{1}{2}$.

We will calculate the entropy about the schedules (a) to (f) as in Fig. 7.

5.5.2 Case (a): optimal schedule

The times of departure are {5, 10, 15} and the rates are {1, 1, 1}. All trains planned in (a) depart immediately when each train fills up. That is the optimal schedule.

The entropy is as follows:

$$H_{(a)} = \left(-\frac{1}{3} \cdot \log\frac{1}{3}\right) \cdot 3 = \log 3 = 1.585. \tag{23}$$

5.5.3 Case (b)

The times of departure are {5, 10, 20} and the rates are $\left\{1, 1, \frac{1}{2}\right\}$.

The entropy is as follows:

$$\begin{aligned} H_{(b)} &= \left(-\frac{1}{3} \cdot \log\frac{1}{3}\right) \cdot 2 - \left(\frac{1}{3} \cdot \frac{1}{3}\right) \cdot \log\left(\frac{1}{3} \cdot \frac{1}{3}\right) \\ &= \frac{8}{9} \cdot \log 3 = 1.409. \end{aligned} \tag{24}$$

5.5.4 Cases (c) to (f)

The entropies of case (c) to (f) are as follows:

$$\begin{aligned} H_{(c)} &= 1.371, \quad H_{(d)} = 1.266, \\ H_{(e)} &= 1.141, \quad H_{(f)} = 1.049. \end{aligned} \tag{25}$$

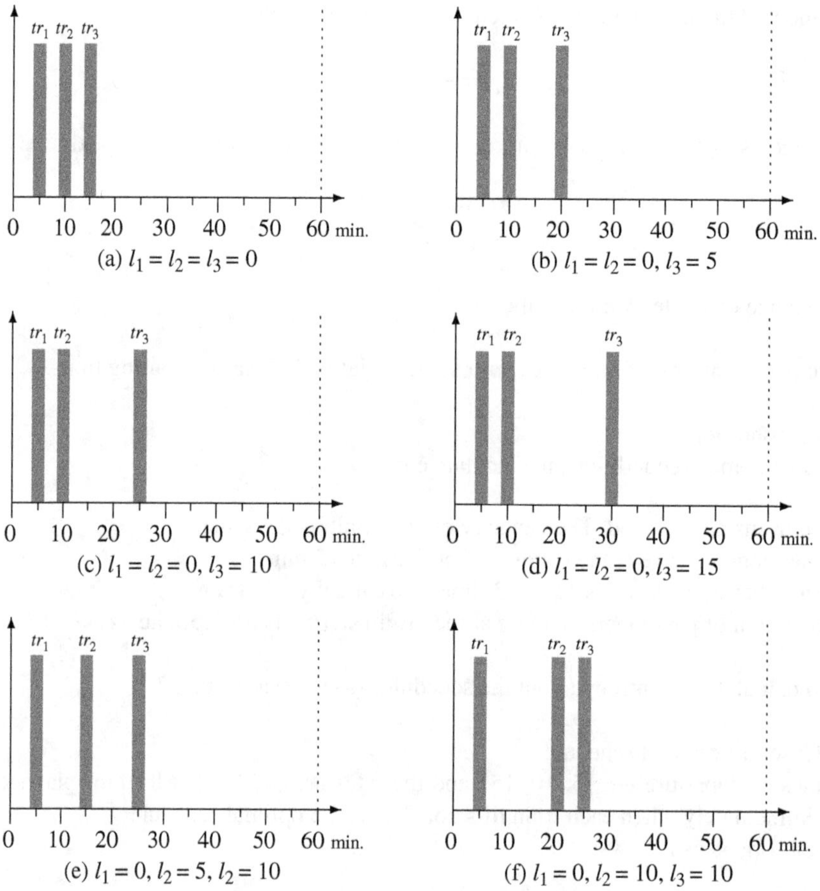

Figure 7: Timetables, for example.

5.6 Conclusions concerning the influence of wasted waiting time

Comparing with the results of (a), (b), (c) and (d), the entropy diminishes as the waiting time increases. Furthermore, comparing with the results of (d) and (e), when more passengers have to wait, the entropy decreases when the total waiting time equals each other. Thus we conclude that the proposed method is reasonable to evaluate the unexpected wasted time of a schedule.

5.7 Nested schedule with loss of information

We show a calculation example of a nested schedule with loss of information. We will consider the root with transfer at a station. Three trains are planned on each section. All passengers boarding tr_i of the predecessor section transfer to tr_i of the successor section as in Fig. 8.

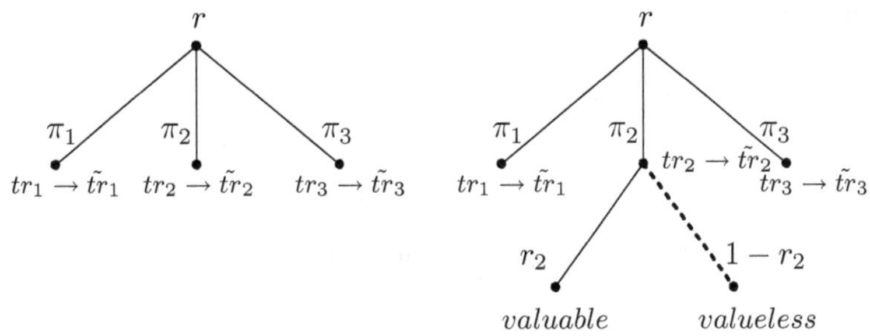

(A) No wasted time (B) With wasted time

Figure 8: Nested tree structure with wasteful time.

5.7.1 Assumptions

We assume some conditions of the schedule.

- The minimum transfer time at the station is 1 min, i.e. $l^* = 1$.
- The value α in eqn (17) is $\frac{1}{2}$.
- The number of passengers on each train is equal, i.e. $\pi_i = \frac{1}{3}$.

5.7.2 Optimal transfer

We consider the schedule (A) in Fig. 8 as the optimal connection. That is, all passengers retain at the station for only the minimum transfer time. The entropy is

$$H_{(A)} = \left(-\frac{1}{3}\cdot\log\frac{1}{3}\right)\cdot 3 = \log 3 = 1.585. \tag{26}$$

5.7.3 With wasted waiting time

See the schedule (B) in Fig. 8. Suppose the transfer time from tr_2 to tr_2 has a valueless time that equals the minimum transfer time, i.e. $l = l^*$. The value β in eqn (16) becomes 1 and the effective time rate r_2 becomes $\frac{1}{3}$.

We calculate the entropy of wasted time as

$$\begin{aligned} H_w &= -\pi_2 \cdot r_w \cdot \log(\pi_2 \cdot r_w) \\ &= r_w \cdot (-\pi_2 \cdot \log \pi_2) + \pi_2 \cdot (-r_w \cdot \log r_w) \\ &= r_w \cdot (\pi_2 \cdot I_{\pi_2}) + \pi_2 \cdot (r_w \cdot I_{r_w}), \end{aligned} \tag{27}$$

which denotes $r_w := 1 - r_2$. The entropy is

$$\begin{aligned} H_{(B)} &= H_{(A)} - H_w \\ &= \log 3 - \frac{2}{3}\cdot\frac{1}{3}\cdot\left(-\log\frac{1}{3}\right) - \frac{1}{3}\cdot\frac{2}{3}\cdot\left(-\log\frac{2}{3}\right) \\ &= \log 3 - \frac{2}{9}\cdot\log 3 - \frac{2}{9}\cdot(\log 3 - 1) \\ &= \frac{5}{9}\cdot\log 3 + \frac{2}{9} = 1.103. \end{aligned} \tag{28}$$

This result shows that we must not forget reduction in the entropy of the predecessor trains.

6 CONCLUSIONS

In this paper, we propose the interpretation that the entropy of a schedule is a metric of the usefulness of the schedule. Moreover, we can evaluate the disutility of schedule by using the proposed concept. In other words, we can express the benefit of a schedule as positive entropy and the disbenefit as negative entropy. In addition, the interpretation of a schedule as information raises the possibility that the techniques of the information theory are applicable to the evaluation of schedules and timetable planning. For instance, we will be able to identify the disruption of a schedule with erasure rate in a channel in information theory. And this identification leads to the explanation that we can consider the recovery of a schedule as correcting errors of received message.

REFERENCES

[1] Moro, K. & Sugai, Y., A relation between a robustness of the disruption of the train schedule and the entropy of information theory. *6th International Conference on Railway Operations Modelling and Analysis,* International Association of Railway Operation Research, pp. 074–1–19, 2015.

[2] Shannon, C.E., A mathematical theory of communication. *ACM SIGMOBILE Mobile Computing and Communications Review,* **5(1)**, pp. 3–55, 2001. DOI: 10.1145/584091.584093.

INFLUENCE OF HIGH-SPEED TRAIN POWER CONSUMPTION ON A NOVEL GROUND FAULT IDENTIFICATION METHOD FOR 2 × 25 KV RAILWAY POWER SUPPLY SYSTEMS

J. SERRANO, C.A. PLATERO, M. LÓPEZ-TOLEDO & R. GRANIZO
Department of Electric Engineering, E.T.S.I. Industriales, Politechnical University of Madrid, Spain.

ABSTRACT

Nowadays, the most commonly used configuration to supply high-speed trains is 2 × 25 kV power supply system. The location of ground faults in 2 × 25 kV power supply systems is a difficult task, since the use of distance protection relays to localize ground faults positions doesn't work properly as the relation between the distance and the impedance seen by the distance protection relays is not linear and therefore the location is not accurate enough. A new simple and economical method to identify the subsection between autotransformers (ATS) and the conductor (positive or negative) where the ground fault is happening, based on the comparison of the angle between the current and the voltage of the positive terminal in each autotransformer, was developed recently. Consequently, after the identification of the subsection and the conductor with the ground fault, only this subsection where the ground fault is present, will be quickly removed from service, with the minimum effect on rail traffic. The high-speed trains demand a power about 12–16 MW, hence a significant current flow through the conductors of the 2 × 25 kV power system. This paper presents a study about the influence of the current consumed by the high-speed trains on this novel ground fault identification method. The operation of the method is correct even with the high-speed train currents in the section with a ground fault. This fact has been validated through numerous computer simulations, obtaining excellent results.

Keywords: fault location, ground faults, protection, 2 × 25 kV, railways.

1 INTRODUCTION

At present, the usual method of providing high levels of power to feed high-speed trains is a traction system called 2 × 25 kV. This power supply system has a positive conductor (usually called catenary) at 25 kV AC voltage with a positive polarity with respect to ground, and a negative voltage conductor (usually called feeder) at 25 kV AC voltage with a negative polarity with respect to ground. The supply to the trains employs the catenary and the grounded rail.

In these power systems, the complete traction line is supplied by several traction substations (TS). Each TS supplies two sections. Moreover, in each section, there are several subsections at regular intervals which are delimited by stations with power ATS. At the end of each section another autotransformer station (SATS) is installed. In Fig. 1 a traction line section with three subsections is shown, as well as the theoretical current distribution.

In 2 × 25 kV power systems, the traction power demanded is delivered at 50 kV while it is used at 25 kV. This fact reduces the current needed to supply the power required by high-speed trains [1]. In consequence, the length of the sections can be greater and the number of TS lower.

DOI: 10.2495/TDI-V1-N3-471-480

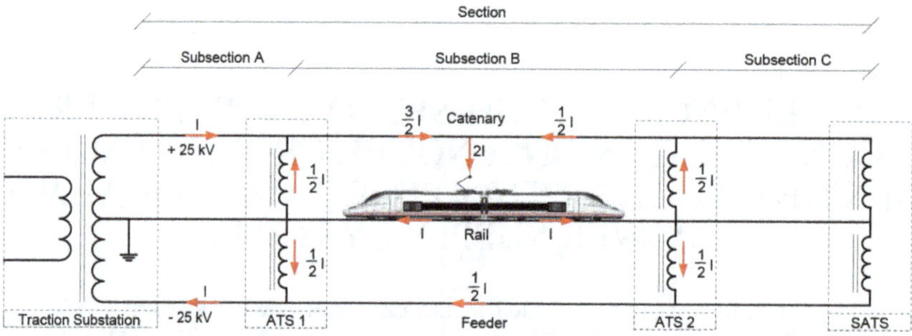

Figure 1: Simplified diagram of a 2 × 25 kV power system and currents distribution in a section comprising three subsections.

Another remarkable advantage of the 2 × 25 kV traction systems is the great reduction of electromagnetic interference on communication facilities and railway signalling circuits (signalling track circuits) as well as on nearby telecommunication lines [2, 3].

Railway traction systems have frequent external faults caused by accidental events. In the 2 × 25 kV power systems most of them are caused by outdoor short circuits between the catenary or the feeder to ground [4, 5]. But in this system, the ratio between the impedance (Z) seen from the TS and the distance to the ground fault is non-linear because of the use of ATS. This ratio is as shown in Fig. 2 [6, 7]. That is why it is so complex to use efficiently in the 2 × 25 kV lines the protection system based on distance protection relays [8].

This complexity of the detection and location of ground faults in the 2 × 25 kV lines means that the identification of the subsection and the conductor where the ground fault has happened [6, 9] is the main task of the protection systems in order to disconnect immediately the conductor with the ground fault in the corresponding subsection [10]. So only trains running in the subsection with the ground fault, in the case where the fault is in the catenary, will be disconnected from the power supply. Serrano et al. [11] describe a new method for identifying the subsection and the conductor (catenary or feeder) in which there is a ground fault in an easy and economical way. As the identification is done immediately, the subsection and conductor can be disconnected while keeping the rest of the power system in service. Besides, to implement this method, a small investment is required.

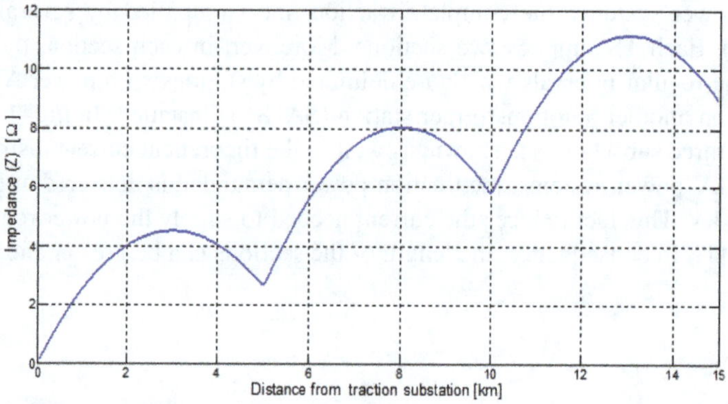

Figure 2: Modulus of the fault impedance seen from the traction substation in function of the distance in 2 × 25 kV traction power systems.

This paper presents a study about the influence of the current consumed by the high-speed trains on this novel ground fault identification method. In Section 2 an overview of this identification method is presented. In Section 3 the influence of high-speed train power consumption on this novel ground fault identification method is analysed. Finally, in Section 4 conclusions are shown.

2 GROUND FAULT IDENTIFICATION METHOD FOR 2 × 25 KV RAILWAY POWER SUPPLY SYSTEMS

2.1 Description of the method

In normal operation of 2 × 25 kV traction power supply system, the phase angle between voltage and current in the autotransformer is close to 180º (Fig. 3), according to the theoretical current distribution previously shown in Fig. 1 [11].

When there is a ground fault on a line between the catenary and rail or between the feeder and rail in a 2 × 25 kV traction power system, there is a substantial increase in the current circulating through the windings of the ATS closest to the defect location. This current increase can be detected easily by measuring the currents in the windings of the ATS. Furthermore, when the fault happens, the angles between the currents and voltages shift by 90º in the ATS closest to the fault location. In the case of a ground fault between the catenary and rail, the phase angle between the voltage and current in these ATS will result in 90º while in the case of ground fault between the feeder and rail, these angle will be 270º.

Figure 4 shows an example of a ground fault between catenary and rail in section B. In this particular case, the angles between I_{A1} and V_{C1} and between I_{A2} and V_{C2}, respectively, change their values from approximately 180º to 90º. Also, there is a great increase in the module of I_{A1} and I_{A2}.

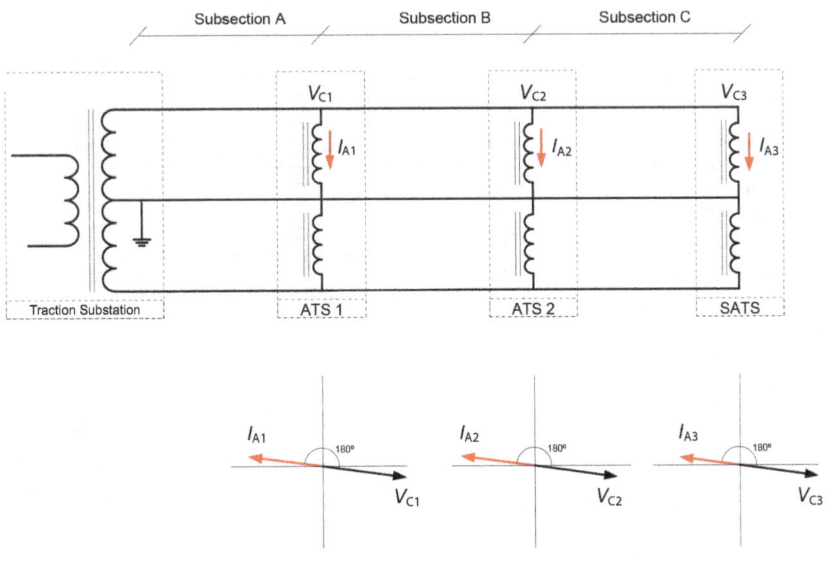

Figure 3: 2 × 25 kV power system currents (I_A) and voltages (V_C) distribution in normal operation.

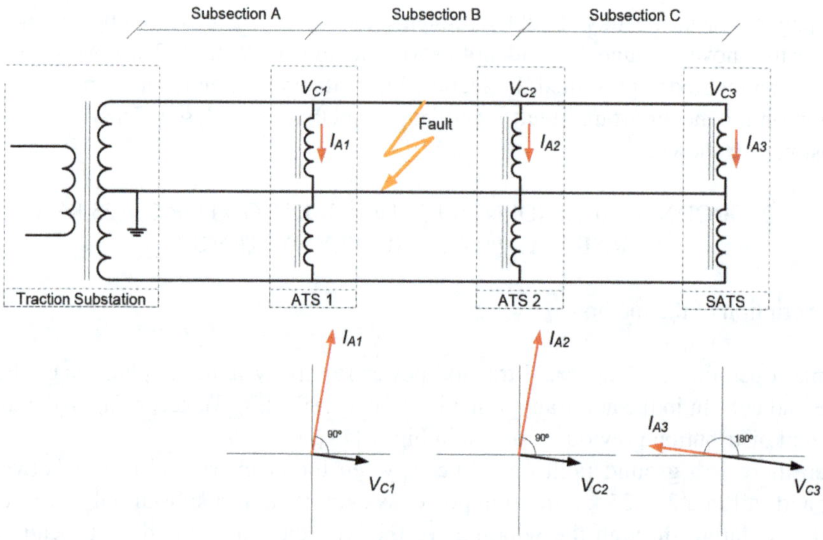

Figure 4: 2 × 25 kV power system currents (I_A) and voltages (V_C) distribution with a ground fault in the catenary in subsection B.

Another example can be seen in Fig. 5. It shows a ground fault between feeder and rail in section C. The current increase in this case will affect only currents I_{A2} and I_{A3} and there will be no remarkable change in current I_{A1}. The angles between I_{A2} and V_{C2}, and I_{A3} and V_{C3}, respectively, will change, but not the angle between I_{A1} and V_{C1}. These phase angle changes will be from 180° to 270°.

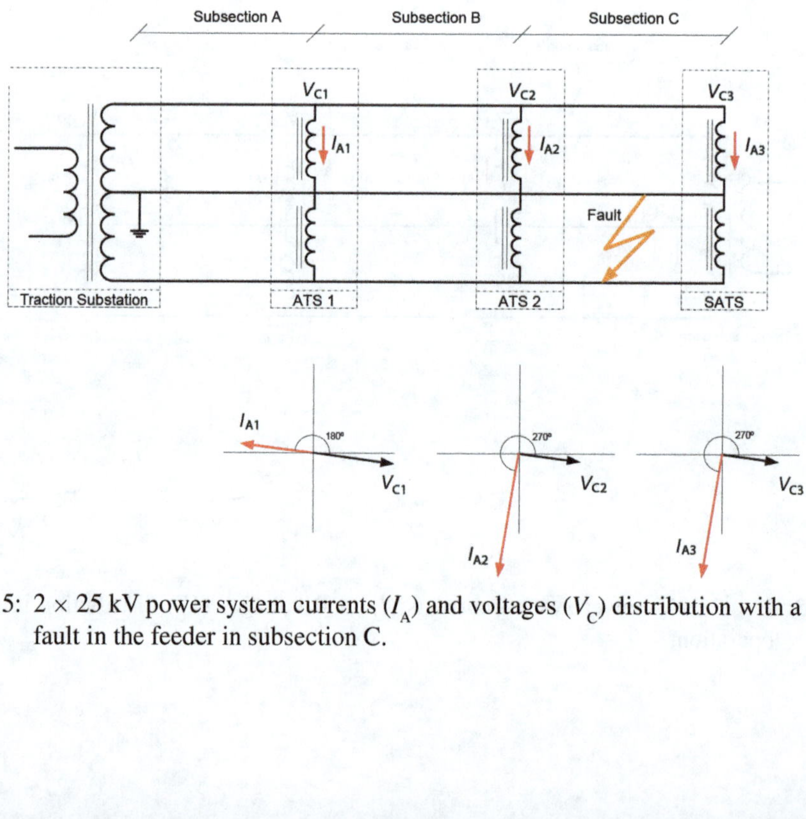

Figure 5: 2 × 25 kV power system currents (I_A) and voltages (V_C) distribution with a ground fault in the feeder in subsection C.

Table 1: Angle variation between I_A and V_C as a function of the subsection with the ground fault.

	ATS1	ATS2	ATS3
Fault at catenary in subsection 1	**90°**	180°	180°
Fault at feeder in subsection 1	**270°**	180°	180°
Fault at catenary in subsection 2	**90°**	**90°**	180°
Fault at feeder in subsection 2	**270°**	**270°**	180°
Fault at catenary in subsection 3	180°	**90°**	**90°**
Fault at feeder in subsection 3	180°	**270°**	**270°**

The changes of the angle between the currents and voltages when there is a ground fault in the system represented in Fig. 3 are listed in Table 1.

In this way, measuring the I_A module and the phase angle between V_C and I_A, it is possible to know the subsection and the conductor where the ground fault has happened.

2.2 Simulation model

In order to validate the new method [11] numerous computer simulations have been carried out using modified nodal circuit analysis method ran on MATLAB® software applied to the circuit shown in Fig. 6.

The experimental circuit is supplied from one end with two 25 kV AC sources with reverse polarity in each one. This circuit has three subsections A, B and C, each with a length of 10 km.

Self and mutual impedances values used in the simulations are listed in Table 2.

Table 2: Self and mutual impedances values of the railway conductors.

Conductor		Impedance[Ω/km]	Conductor		Impedance[Ω/km]
Catenary	Z_C	0.1197 + j0.6224	Catenary–feeder	Z_{CF}	0.0480 + j0.3401
Feeder	Z_F	0.1114 + j0.7389	Catenary–rail	Z_{CR}	0.0491 + j0.3222
Rail	Z_R	0.0637 + j0.5209	Feeder–rail	Z_{FR}	0.0488 + j0.2988

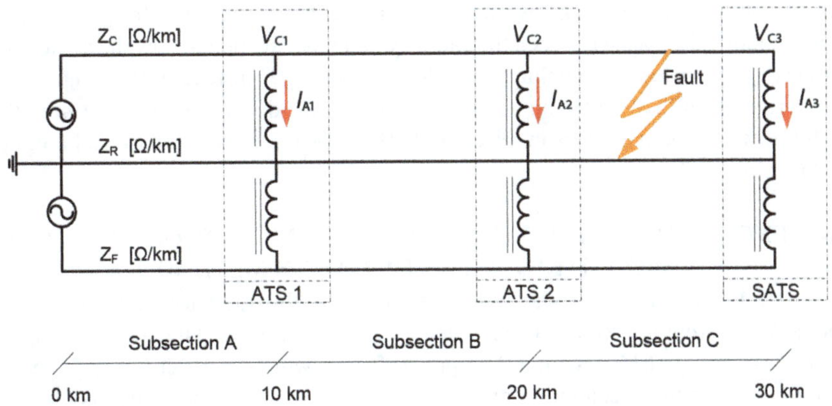

Figure 6: Simulated 2 × 25 kV power supply system.

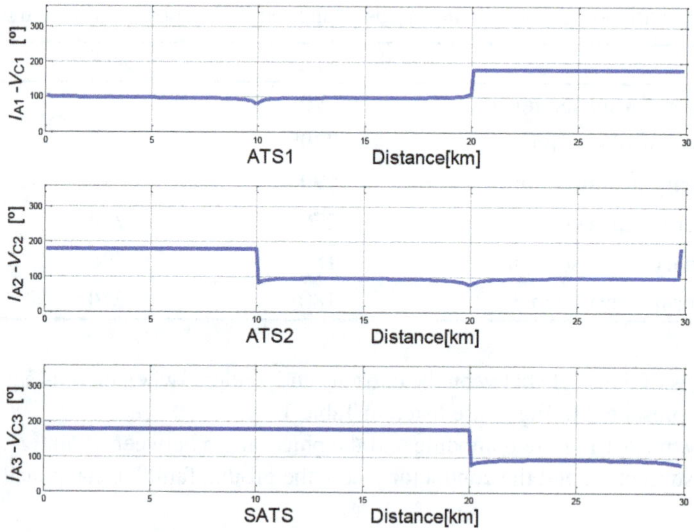

Figure 7: Phase angles between I_A and V_C at different ATS. Fault between catenary and rail.

Employing this circuit, different MATLAB® simulations were developed, performing short circuits between the catenary and rail, and between the feeder and rail. These short circuits were simulated at all the points along the three subsections A, B and C of the 30 km section. The phase angles between the currents I_{A1}–I_{A2}–I_{A3} and the voltages V_{C1}–V_{C2}–V_{C3} in the ATS ATS1, ATS2 and SATS, as a function of the distance from the TS where the fault has happened between the catenary and rail, are represented in Fig. 7. The axes used in Fig. 7 are scaled from 0 to 30 km and from 0° to 360°. It can be observed that if the fault happens in subsection A along the first 10 km, the angle between the current I_{A1} and voltage V_{C1} in the ATS1 is about 90°. However, such angles at ATS2 and SATS are 180°. Likewise, if the fault is in subsection B (between 10 and 20 km) the phase angles between the current I_{A1} and the voltage V_{C1}, and between the current I_{A2} and the voltage V_{C2}, will be about 90° at both ATS1 and ATS2, while at SATS this angle is close to 180°. It can also be seen that if the fault is in subsection C (between 20 and 30 km), the angle is about 90° at ATS2 and SATS but 180° at ATS1.

In the case where the fault happens between the feeder and rail, the results obtained are similar when the fault takes place between the catenary and rail, except that now the angles between the currents I_{A1}–I_{A2}–I_{A3} and the voltages V_{C1}–V_{C2}–V_{C3} are 270° instead of 90°. Figure 8 shows how the variations of these phase angles as a function of the distance to the fault from the substation follow the same pattern as in the case of a fault between the catenary and rail, but now with values close to 270°.

3 INFLUENCE OF HIGH-SPEED TRAIN POWER CONSUMPTION ON THIS NOVEL GROUND FAULT IDENTIFICATION METHOD

In order to study the influence of the train consumption in the operation of the new identification method, a current source has been added to the circuit previously shown in Fig. 6. The current I considered is 616 A with a 0.95 power factor, which represents a high-speed train consumption during the acceleration process. In this case, the current source is placed in the midpoint of the subsection B in Fig. 9, but different train locations have been studied.

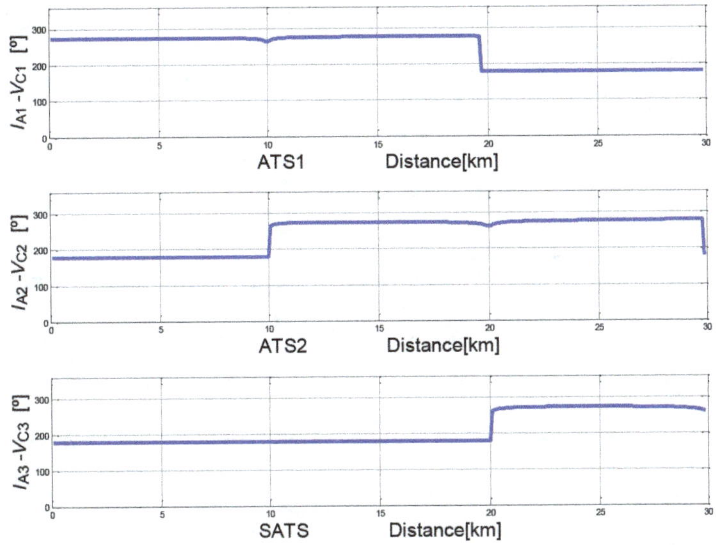

Figure 8: Phase angles between I_A and V_C at different ATS. Fault between feeder and rail.

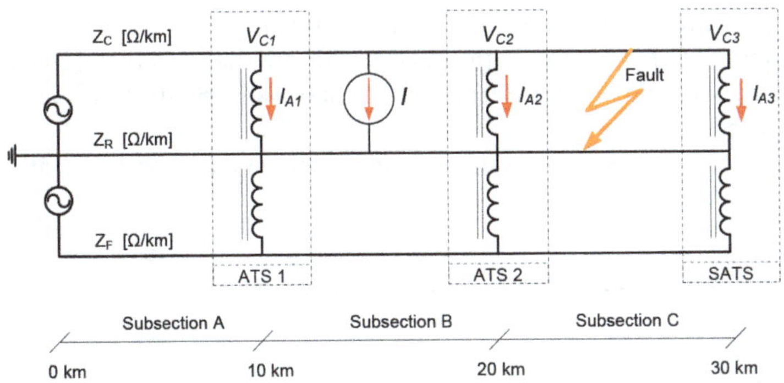

Figure 9: Simulated 2 × 25 kV power supply system including a train.

New MATLAB® simulations were developed on this modified circuit, performing short circuits between the catenary and rail, and between the feeder and rail. These short circuits were simulated at all points along the three subsections A, B and C of the 30 km section, including the presence of one or two trains in the middle of A, B and C subsections. In this way, the phase angles between the currents I_{A1}–I_{A2}–I_{A3} and the voltages V_{C1}–V_{C2}–V_{C3} in the ATS ATS1, ATS2 and SATS, as a function of the distance from the traction substation, are obtained when a ground fault has happened.

Among all the simulations performed, several results are presented in this section as example.

The phase angles in case of short circuit between catenary and rail are shown in Fig. 10, including a train in the midpoint of subsection B.

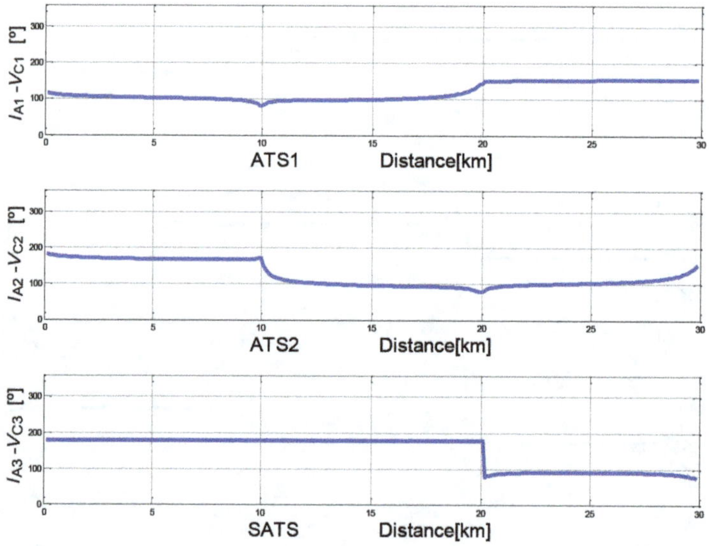

Figure 10: Phase angles between I_A and V_C at different ATS. Fault between catenary and rail. A train is simulated at the middle point of subsection B.

In case of short-circuit between feeder and rail, the results are represented in Fig. 11. In this case the train is placed at the midpoint of the subsection A.

The results in case of two trains in the same section are shown in Figs 12 and 13.

In Fig. 12 the short circuits considered are catenary–rail, and the trains are placed in the subsections A and B, respectively.

In Fig. 13 the case corresponds to short circuits between feeder and rail with two trains in the subsections A and C.

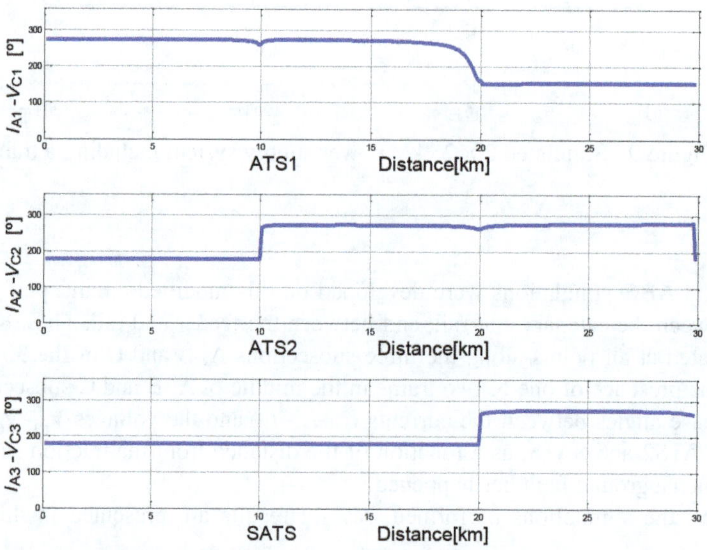

Figure 11: Phase angles between I_A and V_C at different ATS. Fault between feeder and rail. A train is simulated at the middle of subsection A.

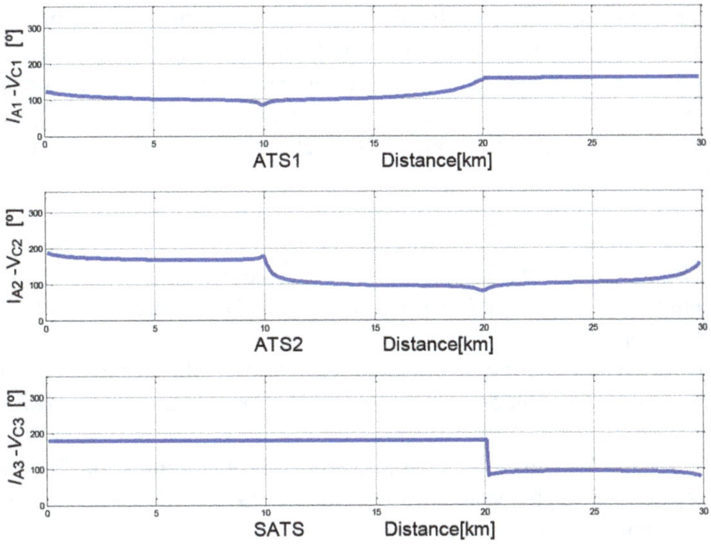

Figure 12: Phase angles between I_A and V_C at different ATS. Fault between catenary and rail. Two trains are placed at subsections A and B.

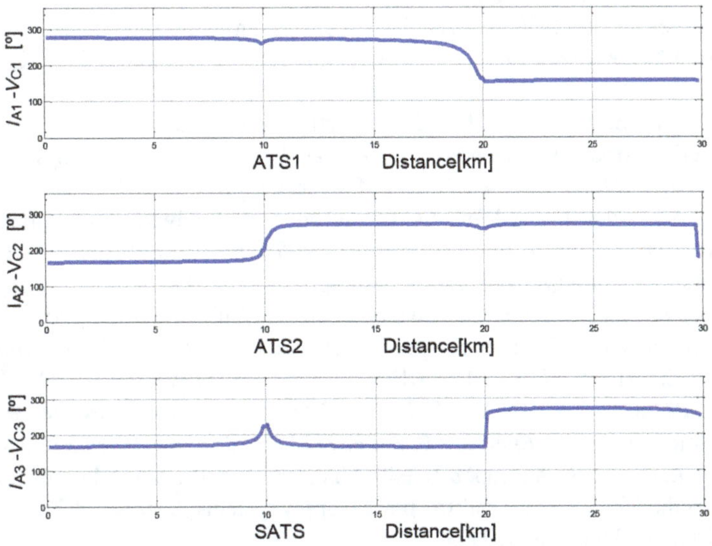

Figure 13: Phase angles between I_A and V_C at different ATS. Fault between feeder and rail. Two trains are placed at subsections A and C.

4 CONCLUSIONS

The influence of the high-speed trains consumption has been studied in the operation of a novel ground fault location method.

For this purpose a computer simulation model has been developed. Numerous simulations have been performed including one and two trains in the same section doing short circuits in the catenary and feeder along the 30 km of the section.

Although the power consumed by the high-speed trains slightly modify the theoretical 90° phase shift in case of ground fault, according to the results of all the simulations performed, the operation of the new ground fault identification method is correct, as long as the characteristic angle setting (previously set to ± 30°) of the directional overcurrent protection relay is properly modified [11]. The new value of the setting depends on the electrical section to be protected.

REFERENCES

[1] Pilo, E., Rouco, R. & Fernandez, A., A reduced representation of 2 × 25 kV electrical systems for high-speed railways. *Proceedings of the 2003 IEEE/ASME Joint Rail Conference.* Chicago, IL, pp. 199–205, 2003.

[2] Courtois, C., Why the 2 × 25 kV alternative? [autotransformer traction supply]. *50 kV Autotransformer Traction Supply Systems – The French Experience, IEE Colloquium*, pp. 1/1–1/4, 1993.

[3] Han, Z., Zhang, Y., Liu, S. & Gao, S., Modeling and simulation for traction power supply system of high-speed railway. *Power and Energy Engineering Conference (APPEEC)*,Wuhan, China, pp. 1–4, 2011.

[4] Richards, S.H., Application benefits of modern microprocessor distance protection for AC electrified railways. *Sixth International Conference on Developments in Power System Protection (Conf. Publ. No. 434)*: Nottingham, UK, pp. 338–341, 1997.

[5] Zhou, Y., Xu, G. & Chen, Y., Fault location in power electrical traction line system. *Energies*, **5**, pp. 5002–5018, 2012. DOI: 10.3390/en5125002

[6] AREVA T&D, *Network Protection & Automation Guide*, Alstom Grid: UK, pp. 354–363, 2002.

[7] Han, Z., Liu, S., Gao, S. & Bo, Z., Protection scheme for China high-speed railway. *10th IET International Conference on Developments in Power System Protection (DPSP 2010), Managing the Change*, Manchester, pp. 1–5, 2010.

[8] Millard, A., Taylor, I.A. & Weller, G.C., AC electrified railways-protection and distance to fault measurement. *Proceedings of the 1995 International Conference on Electric Railways in a United Europe*, Amsterdam, pp. 73–77, 1995.

[9] Wang, C. & Yin, X., Comprehensive revisions on fault-location algorithm suitable for dedicated passenger line of high-speed electrified railway. *IEEE Transactions on Power Delivery*, **27(4)**, pp. 2415–2417, 2012. DOI: 10.1109/TPWRD.2012.2207270

[10] Xu, G., Zhou, Y. & Chen, Y., Model-based fault location with frequency domain for power traction system. *Energies*, **6**, pp. 3097–3114, 2013. DOI: 10.3390/en6073097

[11] Serrano, J., Platero, C.A., López-Toledo, M. & Granizo, R., A novel ground fault identification method for 2 × 25 kV railway power supply systems. *Energies*, **8(7)**, pp. 7020–7039, 2015. DOI: 10.3390/en81010993

LIFECYCLE VS ELEMENT COSTS: A NEW APPROACH TO OPTIMIZE THE POWER SUPPLY SYSTEM DESIGN IN RAILWAYS INFRASTRUCTURES

M. SOLER, J. LÓPEZ, J.M. MERA & J. MAROTO
CITEF, Railway Technology Research Centre of the Universidad Politécnica de Madrid, Madrid, Spain.

ABSTRACT

Electrification systems applied to railway systems have high complexity to achieve the optimum in several terms. These terms could be separated in energy supply and budget costs, such as direct and indirect costs, all of them derived from the implantation of the electrification system along the railway line. Starting from the experience in several projects carried out in CITEF (Railway Research Center in Technical University of Madrid), a new research way has grown up. The main goal has been an expert system, which is able to improve substantially the design process of power supply systems in railways, such as in AC as DC. Bearing in mind the specific constraints in terms of budget and electric standards, the expert system is able to determine a set of possible design scenarios. Each of them has the number, type and location for the main elements (catenary, substation, autotransformer, etc.). In this article a step forward is presented. This new step would apply a better way to choose between elements, hence taking into account the relationship between lifecycle versus element cost or losses avoided in the infrastructure. This has been applied to the main elements like traction substations, autotransformers, catenary and impact on maintenance and environmental zones. This new concept for evaluation has been integrated in the objective functions. These functions are the managers for the guidance of the optimization method, in this case genetic algorithm AMGA-II, in order to achieve the final Pareto Front. This Pareto Front contains the several possibilities which the designers must evaluate to accomplish their goals to power supply system.

Keywords: AMGA-II, life-cycle costs (LCC), maintenance and environmental zones, optimization, power supply design (PSD).

1 INTRODUCTION

The electrical system that supplies energy to the railway traction is a very complex system, which must achieve several important keys. Any power supply design (PSD) must include concepts such as electrical validation according to the standards, power demanded by trains of the operation plan, avoiding zones with high impact on environmental and maintenance zones, considering areas to connect with the general electric grid and finally budget of the project. The choice of the components and position through the line is a task which has been planned as a research way in several studies and works.

In CITEF a methodology has been developed during recent years. This methodology helps to find optimal solutions for the complex system explained before. The work developed by Soler et al. [1] is applied to the DC power supply using an expert system that helps in the choice about the PSD related to the railway traction. This expert system works with Hamlet Simulator [2] developed in CITEF and integrating a multi-objective genetic algorithm NSGA-II [3]. The optimization algorithm from an initial configuration is able to converge to an optimum Pareto Front, which contains good proposals for the designer. The same research line

DOI: 10.2495/TDI-V1-N3-481-490

was applied to electrification systems in AC in [4]. The validation process has been improved with the integration of more electrical terms in the genetic algorithm. The integration of the methodology with other algorithms like AMGA-II [5] is another new step.

Definitely, this research and developing line has been an effort to accomplish a methodology which was able to help properly the designers in their task of defining a complex system, adequate to electrical standards, adhering to a limited budget and positioning critical elements that avoid critical zones. But obviously this methodology can grow up in different ways and areas of the PSDs. In this article, the introduction of elements selection for the supply system taking into account the life-cycle costs (LCCs) and the energy losses versus element costs, and concepts attending to the environmental and maintenance influence, is presented.

In engineering projects the studies to install critical elements is very important to analyse the LCC. Hence if this study is focused on power supply systems in railways, in the article presented in Rondón [6], the authors were working in a manner to evaluate the LCC from studying the overload of the autotransformer they were analysing. Chester and Horvarth [7] discuss the problem to a high-speed system in California, where they are comparing with other means of transport and adding the calculated demand of ridership and the type of energy consumed. Finally they have integrated their conclusions to the related environmental impact. Another interesting study is given by Abbes et al. [8], where the authors could combine a multi-objective optimization system, studies of LCC, embodied energy and calculus of probability of energy losses in order to supply correctly the energy at homes. The highlight is how the settings are interacting internally and what exactly they have taken into account to reach the calculus of the parameters. Another use of the multi-objective algorithms in LCC joined with environmental impact is addressed by Cerri et al. [9], who worked with a theoretical model using the NSGA-II. In the same line but focusing on railway systems, Salim and Xiaoqiang [10] presented their work. Reading this method is interesting to know how the environmental costs are minimized in direct connection using the planned optimization model. The stops in stations and times of circulations are very important to achieve this goal.

Studying the possibilities to improve the LCC connected with other parameters belonging to railway systems is needed to obtain the way to weight these costs. In the document extracted from Baumgartner [11], a measure to weight properly is obtained. Performing this initial evaluation some optimization works more linked with the power supply in railway systems are integrated, such as the study of comparing energy losses among several systems [12]. Next research way highlighted has become a different combination topic of interest to apply. Chuang et al. [13] model the performance of the traction substations in a MRT (massive rail transit), and they are using an optimization algorithm (dynamic programming) and are taking into account the ridership demand through the line, load flow, energy losses and computational time derived, finally finding a method to adapt the PSD to the needs and calculated costs. Furthermore, Hinow and Mevissen [14] have applied a genetic algorithm, trying to achieve an optimum schedule for the substation maintenance. This handles a new method to reduce significantly the LCC.

Throughout this article, the theoretical model of performance of the genetic algorithm integrated in our methodology has been modified. The goal is to include in the objective functions the capacity to get a PSD, taking into account the LCC, joined with environmental and maintenance impact and finally, energy losses versus traction elements costs. The optimization has been carried out with the multi-objective genetic algorithm AMGA-II, final results of which allow the designers to choose between a set of good possible designs in terms of electrical, budget, LCC, energy losses and low impact on environmental and maintenance.

2 METHODOLOGY – INCLUDING LCC IN ENVIRONMENTAL–MAINTENANCE AND ENERGY LOSSES VERSUS INSTALLING ELEMENTS

This topic is divided into two main parts. The first presents a short introduction about the later works related to the methodology, defining the most important features and skills to understand how it is working. Thereafter new introductions to include the LCC in some concepts that the methodology takes into account will be explained. It is important to mention that the new improvements have been carried out in DC systems.

2.1 Main concepts of the optimization of the PSD methodology

In this section the methodology carried out until now and published in papers that were mentioned previously is treated [1, 4].

2.1.1 Zonal discretization and weighting elements

The discretization of the layout of the railway line is a very important concept because it handles the distances like a bidimensional space and permits the critical zones stay along the line. This reason is one of the highlights of the methodology.

Initial step must be to associate the installation zones to the initial discretization. The critical point is related to the assurance to install a traction substation inside each installation zone.

The second step has a relationship with other important concept in this methodology. In railway electrical dimensioning design is mandatory taking into account several impact factors, which must be critical in terms of evaluating budget and time in construction and maintaining the project. In the recent years in CITEF, these critical factors have been linked with critical zones to study in this methodology:

- Environmentally protected zones
- Zones for maintenance of the railway infrastructure
- Zones related to the evaluation of difficulty of connection between general electric grid and traction substation

The ability of the system depends on the action of discretizing the map belonging to the initial route of the circulations of the trains. During this initial step, the length of the line is analysed. This length will be divided into zones in such a way that one traction substation will be able to be installed in each of this zone. In fact, this will be critical in the process of determining the positions of substations and if the participation of them is necessary in this zone or not.

Determining the cost or weight of each element and zone involved in each project is important to know if these costs are real or weighted.

2.1.2 Electrical validation of PSDs

The most important and critical thing to do would be finding a way to test a PSD but not the entire operation plan. The final Pareto Front filled by the genetic algorithm must be composed by individuals with good fitness, even global optimum. But obviously, these individuals must be validated in terms of electrical criteria. The process must be run into a finite and efficient computational time.

Hence while searching about possibilities, one can realize the first approximation to the problem was working with a pair of ways during the simulation. The first way was to attain the calculation of the voltage drops on each cycle. Second alternative was the calculation of the power supply from the traction substations, again on each cycle of the simulation. But both have a critical problem. The final results are highly connected with the calculation about the power supply system. This means that each scenario will have its own final results for both, and obviously that avoids a permanent and homogeneous concept to compare and validate any power supply scenario.

The solution embarked in this methodology is linked with the idea of maximum power demanded peak. This allows a steady-state parameter for any PSD, so this is hardly connected only with the mechanical part of the infrastructure. The initial simulation usually has the higher power demanded peaks at any snapshot belonging to this simulation. In cases of fail, simple or complete in any traction substation, is got the time when demanded power is higher, just affecting to traction substations, even to their rectifiers.

2.1.3 Genetic algorithm AMGA-II

The final consequences of analysing the problem were the need to integrate a powerful optimization algorithm, which was able to work with a multi-objective problem, non-analytical theory model and hardly non-linear model. The choice was AMGA-II. This genetic algorithm has a good computational time and is a raising evolution for a later algorithm, and of course a knowing process about its main configuration parameters has been necessary. These parameters must be met during each working process of the methodology, which creates a conjunction between the other parts of it and the heuristical work proposed for AMGA-II. Finally a set of Pareto optimal individuals is got and there will be the goal of AMGA-II.

2.1.4 Objective functions and restrictions

The objectives are divided into installation costs and operation costs. The first one combines the environmental and general electric grid connection areas with basic elements; the operational costs include the maintenance costs with losses in energy.

2.2 Integrating LCC versus direct and indirect costs and energy losses

The new approach to evaluate the PSD has been enhanced to adapt the LCC in maintenance, critical elements and interaction between energy losses. For this purpose, changes must be developed in the theoretical model represented by the objective functions in the evaluation of the genetic algorithm. In the following section innovations depending on both the objective functions are explained.

2.2.1 Including changes in installation objective function

To know the past objective function, the element costs and the impact on environmental and maintenance zones are important. New parameters and other calculation factors are also important in this task.

In an interesting paper referenced before (Chuang et al. [13]), the authors have included some equations in their model to add the influence of costs and inflation/interest rates [15] during life cycle. In this article, a first approximation taking this main idea has been developed.

To add the impact of the LCCs of the catenary and traction substations, determined by the designer:

– Overhead contact line direct costs

$$CostOhl = \sum_{o=1}^{nohl} \frac{budget_o}{LC_o} + \left(\frac{(1+\sigma)^{LC_o}}{(1+\tau)^{LC_o}} \right) * budget_o \tag{1}$$

– Traction substation direct costs

$$CostSs = \sum_{t=1}^{nss} \frac{budget_t}{LC_t} + \left(\frac{(1+\sigma)^{LC_t}}{(1+\tau)^{LC_t}} \right) * budget_t \quad , \tag{2}$$

where

nohl and *nss* are the total numbers for overhead contact lines and traction substations, respectively
budget is for the element, as catenary or traction substation *LC* is configured life cycle for the elements.
σ is expected inflation included in Fisher Equation.
τ is expected interests costs included in Fisher Equation.

Both equations are a powerful tool to adhere the life cycle of the elements in the calculation of the impact of these elements on the final budget of the PSD.

Once this first step was achieved, the application of the impact of the life cycle inside other important factors for the calculation of installation costs was the next step. In this work the focus on the impact on environmental costs during the years attending to the initial installation was established. After an initial knowledge of the system and a step forward including the planning to add a new measurement for the LCC in environmental zones, the final equation is presented:

$$CostCE_k = CostCE_k \times LC_t \quad , \tag{3}$$

where

LC_t is configured life cycle for traction substation.

Life cycle configured for each traction substations affects directly the environmental costs during the total life of this electrical element.

2.2.2 Including changes in operation objective function

Following the idea described in the previous point, the main differences between the later objective functions integrated in the methodology are pointed out.

The reader could examine the papers related with this newly presented work and then realize that in operation costs there are a clear distinction between maintenance impact costs and energy losses. The possibility to get more relevance to one of the calculations is integrated now by means of a probability value.

The equations to operate the calculus of maintenance are very similar to installation costs introduced in the previous section. The difference is the concept of maintenance costs of elements. This is the preconfigured value the designer must include to match the overhead contact line and traction substation with the expected cost in maintenance for a single year. Furthermore the equations are the same except changing $budget_0$ to $budget\_m_0$. Another relevant matter is the incorporation of the impact of the life cycle on the calculation of the maintenance zone, which finally is similar to the environmental zone changes explained earlier.

In energy losses calculation the impact of life cycle of the traction substation versus energy losses calculated in the simulation is included. That defines a new relationship between the losses calculated in one single cycle of time and an extrapolation of the life cycle of the traction substation:

$$Losses = \sum_{t}^{nsubst}\left(\sum_{r=1}^{nsubst} Ssubst_r - \sum_{u=1}^{trains} Pet_u\right) * \frac{LC_t}{porc\_LC}, \quad (4)$$

where

$porc\_LC$ is the total life cycle years calculated for the set of traction substations in the power supply system analysed.

So finally

$$f_2(x) = Factor_m * \sum_{v=1}^{nsubst}\left(CostCM_v\right) + Factor_l \alpha * Losses \quad , (5)$$

where

$Factor_m$ is the value in % assigned to maintenance in front of $Factor_l$ that represents the % assigned to losses calculated.

The rest of the variables are explained in the papers referenced later.

3 OUTCOMES

Applying a genetic algorithm to a complex, non-linear and non-analytic model is important in the set of tests. This set must validate the methodology in terms of searching an optimum of budget and is electrically convenient for the railway infrastructure. Here is shown a specific test using our methodology. This test is based on a real railway project, and its main characteristics are as follows:

- Geometry and elevation view: Double track along the whole route and 13 stations. Speed restrictions have also been taken into account (Fig. 1).
- Rolling stock: A train with 6 cars, maximum speed 80 km/h, maximum service acceleration 1 m/s^2 , maximum service deceleration 0.9 m/s^2.
- Signalling: The planned signalling system for this line was CBTC.
- Electrical scenario: The line is direct current at 1,500 V. Two types of overhead line systems have been chosen. Seven traction substations are considered to install on initial design.

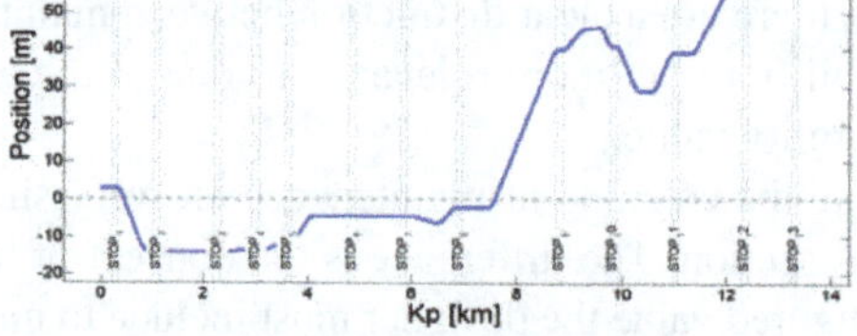

Figure 1: Elevation of the line.

– Trains with a 90s of headway. Stopping times in stations are 15–20 seconds, depending on the station stop. In case of any rectifier fail, the operation will be degraded as well.

Finally, the main characteristics of the machine used for this optimization process are i5-2400 3,1 GHz 4Gb RAM.

Once the project has been configured in the methodology, and the main parameters belonging to the genetic algorithm, the next is to test the system. Several tests have been launched with differences in some parameters of configuration (Table 1). Next figures are the Pareto Front belonging to the three main tests.

Figure 2 shows the three more representative tests. The results brought by the tests are very interesting in the way to get the key to understand the performance of new skills. Obviously the tests have similar specific data. These similarities have given the opportunity to get clear

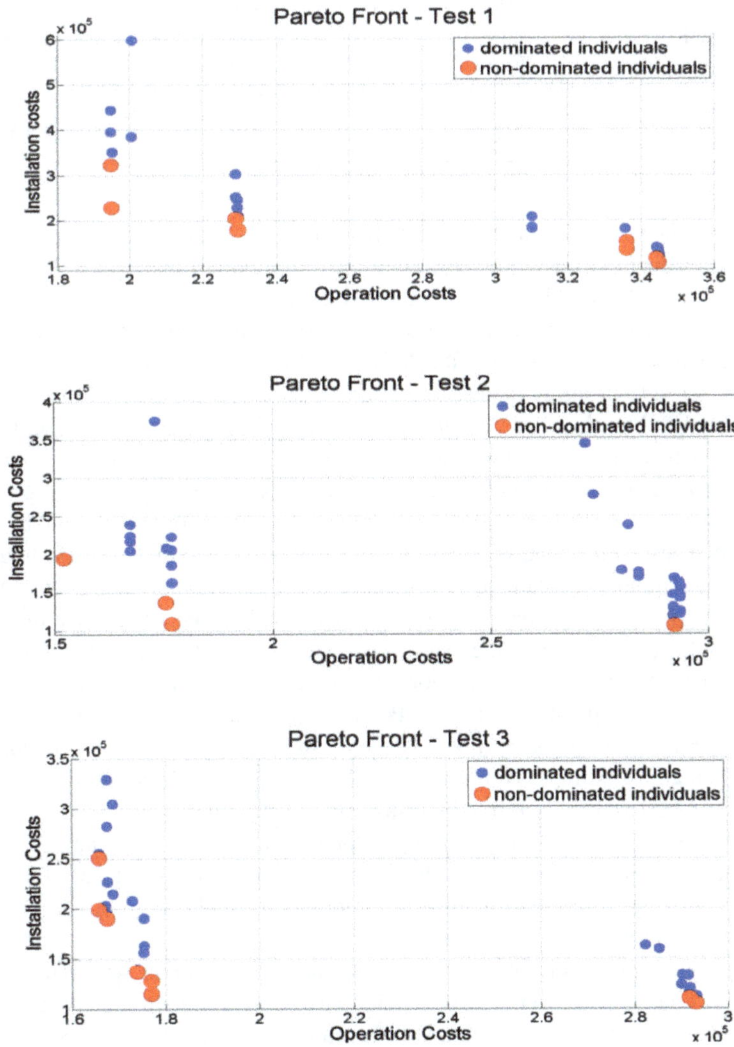

Figure 2: Pareto Front in three different cases.

Table 1: Configuration of main parameters related to calculation of LCC.

	Test 1 (cost/LC/CM)	Test 2 (cost/LC/CM)	Test 3 (cost/LC/CM)
Substation 1	1E5/5/40	1E5/45/4	1E5/45/4
Substation 2	4E5/15/6	4E5/45/6	4E5/45/6
Substation 3	1E5/10/5	1E5/40/5	1E5/40/5
Substation 4	4E5/20/30	4E5/50/3	4E5/50/3
Substation 5	7E5/80/3	2E4/50/5	2E6/20/25
Substation 6	4E5/20/20	2.4E4/50/25	2.4E6/20/2
Substation 7	4E5/10/5	4E5/40/5	4E5/40/5
Maintenance 1	2000	2000	2000
Maintenance 2	500	500	500
Maintenance 3	20	3000	3000
Maintenance 4	200	3200	3200
Environmental 1	2000	2000	2000
Environmental 2	200	200	200
Environmental 3	20	320	320
Environmental 4	200	3200	3200

the differences when the new parameters are changed. The AMGA-II has been parameterized with 120 final evaluations and 25 desired solutions. Mutation probability is 0.2 and crossing probability 0.7. The threshold for removing traction substation is 0.28.

Focusing on the three final Pareto Fronts, two very clear behaviours were found. On the one hand, independently the configuration AMGA-II tries to locate two different final designs. There is a better side in terms of installation and another part is better in operation terms. The reason is related to the relative location of traction substation 5. In any case, the tests are useful to remove this traction substation and that allows considerable saving of money in global budget. On the other hand, obviously, the algorithm takes into account the energy losses and maintenance costs and awards the possibility to keep the substation 5. Due to that the energy losses are less, even with a relative low life cycle. This part belongs to the best part of individuals bearing in mind the installation costs. Moreover, the factors of maintenance and energy losses integrated in operation objective function (explained in Section 2) have their influence in the final results. For these tests, the values have been 0.7 for $Factor_l$, and naturally, 0.3 for $Factor_m$. Hence the energy losses gain more relevance inside the operation costs.

One of the choices represented in final Pareto Fronts has been collected to analyse the electric restrictions, such as correct power supply to the demand of trains and avoiding critical voltages. The final example is on the Pareto Front of case 2. It is examining a proposed scenario for the best fitness in terms of installation costs, that is, in net costs of elements involved in this PSD and analysing the impact on environmental zones and LCC for these elements. The main characteristic is the choice of the second type of catenary for the railway line and the removing of the sixth traction substation. The reallocation of the rest of substation is another highlight.

Observing Figs. 3 and 4, it is interesting to conclude that even in the worst situations, when substation 7 or 5 falls, the power system is able to avoid the limits of the normative in minimum voltages in catenary and maximum voltages in return earth rail. Despite the voltage rise till 152 Vcc in situations like permanent regime, 360 s and 3600 s, the voltage is under limits.

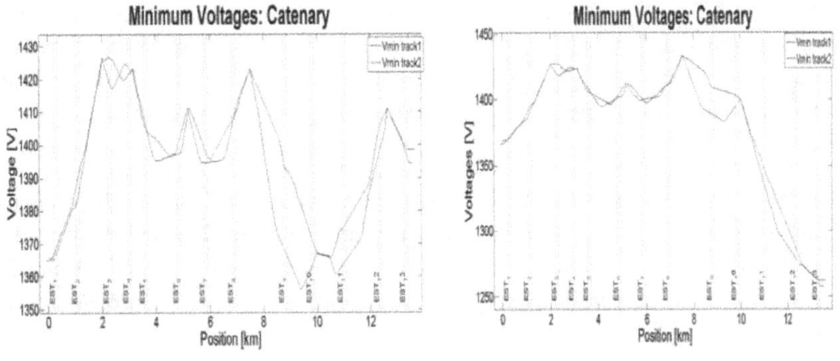

(a) (b)

Figure 3: Minimum voltages in catenary in cases (a) degraded in ss 5 and (b) degraded in ss 7.

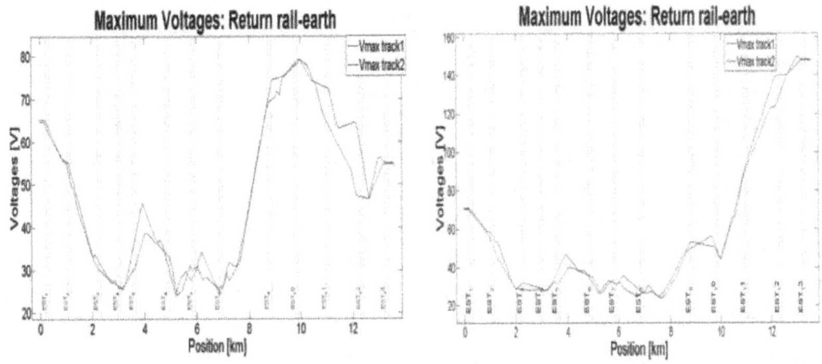

(a) (b)

Figure 4: Maximum return voltages between rail and earth in cases (a) degraded in ss 5 and (b) degraded in ss 7.

Focusing on the allocation and therefore the subrogation to LCC and environmental costs is important to say that the methodology helps the optimization to remove the substation in worst zone, in terms of environmental costs. On the other hand if it sticks out the difference between LCC in substation 4 till 7, it is important to note that the methodology has tried to remove substation 5 or 6, discarding from the beginning the others, due to the life cycle, 40 versus 20.

4 CONCLUSIONS

The satisfactory results obtained during the optimization process in the past using our methodology have not been a barrier to try to go a step forward. In this article an improvement that introduces the concepts of LCC joined with energy losses, main elements (traction substation and overhead contact line) and maintenance and environmental zones has been presented. Once the modifications have affected directly the objective functions, on the core of the genetic algorithm, then the need is to test deeply the new skills. In this article one of the tests has been detailed and the effectiveness of the optimization methodology in terms of proposing a set of final best choices has been described. This set has considered budget conditions and electrical restrictions, and it is proved that now we are able to work towards PSD in terms of LCC linked with direct and indirect costs, taking into account the energy losses during the life cycle of elements, and depending on the relevance in this matter, we are able to define a better solution as well. Next ways of research must be directed to enhance this technique and to include some studies of energy regeneration during optimization process.

REFERENCES

[1] Soler, M., López, J., Mera, J. M. & Tapia, S., Expert system using multi-objective optimization of the direct current railway power supply system. *Transport,* pp. 1–12, 2015.

[2] Soler, M., López , J. & Mera, J. M., Simulation system for the optimization of a block distribution under the ERTMS-1 signalling system, de *Computer in Railways XII*, WIT Press, pp. 61–73, 2012.

[3] Deb, K., Amnit , P. & Sameer, A., A fast and elitist multiobjective genetic algortihm NSGA-II. *IEEE Transactions on Evolutionary Computation,* **6**(**2**), pp. 182–197, 2002. DOI: 10.1109/4235.996017.

[4] Soler, M., López, J., Mera , J.M. & Maroto, J., Methrodology for Multiobjective Optimization of the AC Railway Power Supply System. *IEEE Transactions on Intelligent Transportation Systems,* **15**(**5**), pp. 2531–2542, 2015. DOI: 10.1109/TITS.2015.2412460.

[5] Tiwari, S., Fadel, G. & Deb, K., AMGA2: improving the performance of the archive-based micro-genetic algorithm for multi-objective optimization. *Engineering Optimization,* **43**(**4**), pp. 377–401, 2011. DOI: 10.1080/0305215X.2010.491549.

[6] Rondón, I., Calache ,V. & Rojas, F., Metodología para estimar la vida útil del autotransformador AT-4, 700 MVA de la subestación Guayana 'B' debido a sobrecargas. *Universidad Ciencia y Tecnología,* **17**(**64**), 2014.

[7] Chester, M. & Horvarth, A., Life-cycle assessment of high speed rail: the case of California. *Environmental Research Letters,* **5**(**1**), 2010. DOI: 10.1088/1748-9326/5/1/014003.

[8] Abbes, D., Martinez ,A. & Champenois, G., Life cycle cost, embodied energy and loss of power supply probability for the optimal design of hybrid systems. *Mathematics and Computers in Simulation,* **98**, pp. 46–62, 2014. DOI: 10.1016/j.matcom.2013.05.004.

[9] Cerri, D., Taisch, M. & Terzi, S., Multi-objective Optimization of Product Life-Cycle Costs and Environmental Impacts. *Advances in Production Management Systems. Competitive Manufacturing for Innovative Products and Services,* pp. 391–396, 2012.

[10] Salim, V. & Xiaoqiang, C., A genetic algorithm for railway scheduling with environmental considerations. *Environmental Modelling and Software,* **12**(**4**), pp. 301–309, 1997. DOI: 10.1016/S1364-8152(97)00026-1.

[11] Baumgartner, J.P., *Prices and Costs in the Railway Sector*, Zurich: Institut des Transport et de planification, 2001.

[12] Açıkbaş ,S., & Turan Söylemez, M., Energy loss comparison between 750 VDC and 1500 VDC power supply systems using rail power simulation, de *Computers in Railways IX*, WIT Press, pp. 951–960, 2004.

[13] Chuang, H., Chen, C., Lin, C., Chen, Y. & Ho, C., Optimal Expansion Planning of Traction Substations for an Electrified Mass Rapid Transit System. de *International Conference on Power System Technology*, pp. 1-7, 2006.

[14] Hinow ,M. & Mevissen, M., Substation maintenance strategy adaptation for life-cycle cost reduction using genetic algorithm. *IEEE Transactions on Power Delivery* , **26**(**1**), pp. 197–204, 2011. DOI: 10.1109/TPWRD.2010.2065247.

[15] Fisher, I., *The Theory of Interest*, New York: The Macmillan Company, 1930.

INCREASING RAILWAY ENERGY EFFICIENCY: A THREE-LEVEL METHOD

R.M. PAULUSSEN[1], G.F. TEN HARVE[2], T. PLOEG[3] & A. ZOETEMAN[2]
[1]ARCADIS, Amsterdam, The Netherlands.
[2]ProRail, TU Delft, The Netherlands.
[3]DNV GL, Arnhem, The Netherlands.

ABSTRACT

Energy efficiency in railway operation has been a topic of high interest for years. Lowering energy costs can make railway operation not only more sustainable, but also more competitive towards other modalities. Energy efficiency can be reached in two ways: system efficiency and operational efficiency. Innovation can be found in combining the right system measures with the right operational measures. Introducing the right system measures can even lead to energy efficiency independent of operational strategies, such as coasting and regenerative braking.

Rail infra providers struggle with the question which innovations – system, operational or both – will bring them closer to their targets. To determine which measures will contribute best, the authors developed a new universal method based on energy efficiency at three levels: individual rolling stock performance, rolling stock performance in networks and operational strategies.

The authors used this three-level method to study the energy efficiency effect of a system change from the current 1,500 V_{DC} to a 3 kV_{DC} traction power supply system, including operational strategies. The results of the study revealed that in the Netherlands this system change will result in a return on investment (ROI) within 15 years and a reduction of energy consumption ranging up to 20%.

Keywords: energy efficiency, return on investment, simulation study, social cost–benefit analysis, system changeover.

1 INTRODUCTION

Energy efficiency in railways has been a major topic in the past years. Due to the high investment costs in infrastructure and rolling stock, operational strategies are in favour for application. However, will this kind of strategies still work when system changeover is applied?

In the Netherlands, the current traction voltage is 1,500 V_{DC}. In the past a changeover towards a 25 kV_{AC} system was considered in order to be able to meet the future increasing traction power demand. However, the idea of a changeover from 1,500 V_{DC} towards a 25 kV_{AC} system was put aside because of high costs and difficulty in changing a DC system into an AC system.

The introduction of increased train frequencies leads to high investments in the existing 1,500 V_{DC} network. With that, a question arises if a changeover to a 3 kV_{DC} network could be an alternative: are the benefits – lower energy losses and higher performance – feasible and will it deliver enough return on investment (ROI)?

To increase energy efficiency in train operation Dutch Railways has introduced operational strategies like coasting to make train drivers aware of energy efficient ways of driving. The efficiency of such operational strategies depends strongly on individual train driver's

DOI: 10.2495/TDI-V1-N3-491-500

behaviour and the fit into the timetable. Will these strategies still work after a changeover from 1,500 V_{DC} towards 3 kV_{DC}?

This article deals with the energy efficiency part of the changeover and the relation between system voltage and operational strategies. To determine which measures – which voltage in combination with which operational strategies – will contribute best to energy efficiency, the authors developed a new universal method based on energy efficiency at three levels:

1. Individual rolling stock performance. The performance of rolling stock on energy consumption and energy recuperation is defined, thereby determining its maximum performance.
2. Rolling stock performance in networks. Insight is given in the performance of rolling stock in networks by studying its interactions with those networks. The difference between the competence of individual rolling stock and the performance of rolling stock in a network defines the energy opportunity gap.
3. Operational strategies. Different operational models in combination with different system topologies are studied to determine which minimizes the energy opportunity gap the most.

In the next sections these three levels will be discussed in relation to the possible changeover from 1,500 V_{DC} towards 3 kV_{DC}.

2 INDIVIDUAL ROLLING STOCK PERFORMANCE

This section deals with the first step of increasing energy efficiency for individual rolling stock: the possible contribution of rolling stock to regenerative braking energy, and the relationship between available power and system voltage. As individual rolling stock is part of a network and operational strategies, the type of rolling stock and service performed determine the influence on the energy consumption and regeneration: freight trains and intercity services operate in more or less continuous power consumption mode. Stop services, on the contrary, are accelerating or decelerating most of the time and therefore are of great interest in increasing energy efficiency by reusing the regenerative braking energy.

2.1 Regenerative braking

In order to determine the contribution of rolling stock to its own regenerative braking possibilities, a few boundaries of these possibilities must be clarified. Individual rolling stock represents a total mechanical braking power P_m which is equal to

$$P_m = F \cdot v = m \cdot a \cdot v = \frac{1}{\eta} U \cdot I_{\max} = \frac{P_{ED,\max}}{\eta} \tag{1}$$

The electrodynamic braking power (P_m) is limited at a certain velocity by a certain electrical power $P_{ED,\max}$. The relation between the mechanical and electrical braking power is determined by the efficiency η.

So, to optimize electrodynamic regenerative braking, deceleration should be adapted to the ability of the train system to transform mechanical braking force into electrical power. As the mass of the train is to be considered constant during braking, only the deceleration (a) and velocity (v) are left to maximize the electric regenerative braking energy.

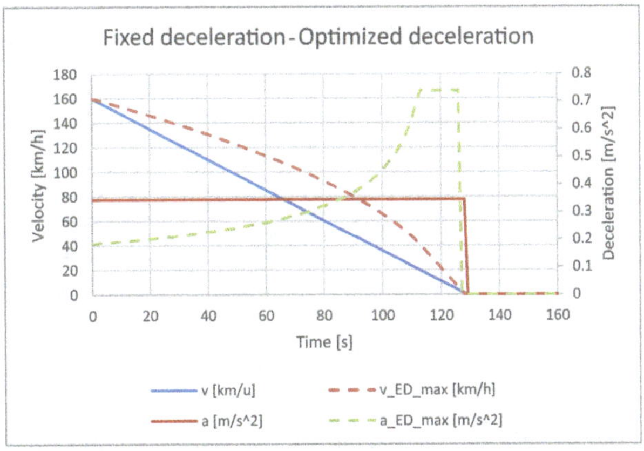

Figure 1: Individual rolling stock, constant deceleration (*a*) versus optimized deceleration (a_ED_max, dashed lines) for electrodynamic braking.

This may lead to the first conclusion that braking with a constant deceleration leads to adding up mechanical braking to electrodynamical braking at high speed and to underachievement of electrodynamical braking at lower speed. As an example of this mechanism, a combination of Sprinter Light Train (SLT) of Dutch Railways is presented: 2,805 kW of maximum mechanical braking power at 40 km/h. To utilize the entire electrodynamic braking force during braking, the deceleration must be speed dependent (optimized). Speed-dependent and fixed decelerations, having the same braking time, are shown in Fig. 1.

The difference in braking energy – from 160 to 0 km/h – between fixed (0.345 m/s^2) and optimized deceleration for this train set is in favour of optimized deceleration: 65 kWh (100%) versus 83 kWh (128%). Based on the measurements by Dutch Railways the SLT stopping services are modelled with a constant braking deceleration of 0.5 m/s^2 [1], leading to 50 kWh (100%) versus 83 kWh (166%).

Introducing a constant deceleration of 0.5 m/s^2 for this rolling stock will lead to utilization of 60% of the available electrodynamic braking energy. For intercity services – based on Dutch Railway measurements of VIRM rolling stock [2] – the deceleration is modelled with 0.35 m/s^2, which leads to an utilization of 87% of the available electrodynamic braking energy.

Based on this first step in the first level of energy efficiency it can be concluded that the characteristic of the individual train, together with the way a parameter of deceleration is used, in the real world as well as in simulation tools, determines the regenerative braking energy available to the network.

2.2 System voltage

In the Netherlands, the system voltage of the conventional rail system is 1,500 V_{DC}. All trains in the network have to comply with the current limitation of 4,000 A according to EN50388. This means that train performance is decreased when a train initially wants to draw more current than 4,000 A. In fact, the maximum power of the VIRM and SLT rolling stock is only available above 1,900 V_{DC}. In Fig. 2 this voltage dependency of the tractive effort is presented for a combination of 10 coaches SLT.

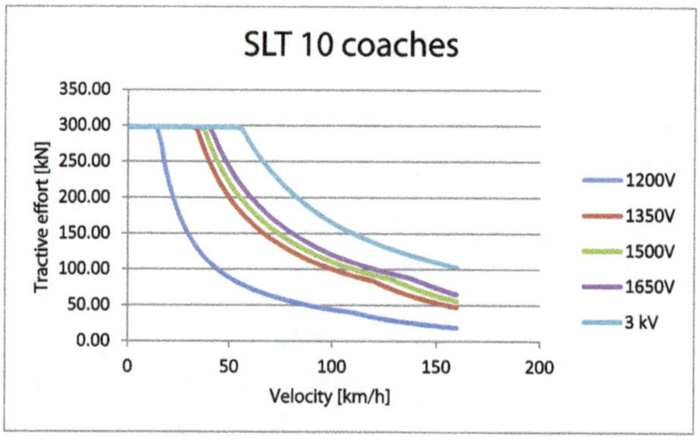

Figure 2: Tractive effort – speed characteristics depending on train voltage.

This formation of SLT rolling stock runs without power limitation in a 3 kV_{DC} system. When operated under 1,500 V_{DC}, however, its performance is limited by the train voltage in the network. This performance limit leads to an extended power draw from the network when accelerating. The difference in mechanical power available between 1,200 V and 3 kV is 1,157 kW versus 4,462 kW. At velocities above 110 km/h this difference will increase further.

Based on this second step in the first level of energy efficiency it can be concluded that the tractive effort characteristic of the individual train is voltage dependent and must be considered when comparing different power supply systems.

3 ROLLING STOCK PERFORMANCE IN NETWORKS

This section deals with the performance of rolling stock in a network by studying its interactions with that same network. The difference between the performance of individual rolling stock and rolling stock in a network defines the energy opportunity gap, which is network dependable. The main question in this section is, what is responsible for the energy opportunity gap?

Although the capability of the network depends on many parameters, only three of them are regarded to be responsible for the efficiency of the network. First, the operational voltage of the network and the voltage-current limitation of the regenerative braking power. Both accommodate the reach of the braking energy in the network. Second, the network configuration consisting of paralleling points, single or more tracks in parallel and the overall cross section of the catenary and return system. These characteristics of the network determine the efficiency of the reach of the braking energy in the network. Finally, the operation of the network, which can be clarified by an example. If two trains arrive at the same time at the same station, the regenerative power of these trains must be transported elsewhere in the network. If the first train leaves and is able to consume the regenerative power from the second train the regenerative power is used within a small circuit, avoiding high losses. So, operation of the network in combination with the network topology can influence the energy consumption of the network. This third parameter introduces the question whether it is possible to uncouple the operation of the network and the ability to increase energy efficiency. The benefit of this uncoupling would be the 'freedom' for train drivers to implement the driver's strategy best fit for the occasion.

To illustrate the train–network interaction described above, several case studies (based on simulations) are described. In order to validate the simulation model, the results of these case studies are compared with measurement data. The mean deviation turned out to be 4%, which is very small. The maximum deviation of 14% was found on parts of the network that include freight services. For freight services there is a lack of information about the use of diesel or electrical traction. The model uses only electrical freight trains.

3.1 BASIC ASSUMPTIONS

To obtain a fair comparison between two systems – 1,500 V_{DC} and 3 kV_{DC} – other parameters than the system voltage ought to be equal or known unequal. For both systems regenerative braking will be applied.

Section 2 already made clear that train performance depends on the power supply system applied. The train characteristics are not adapted to be equal for both power supply systems but their qualities are used in the simulation: trains are operated with a margin of 5% to comply with the timetable. This 5% extra time is uniformly applied to all modes – acceleration, driving, deceleration – of the train between two stations. When a train is delayed, the margin is dropped up to the station where the delay is absent. The recuperation voltage at 1,500 V_{DC} is limited for the current rolling stock: BR189, 1,900 V_{DC}, VIRM, 1,925 V_{DC} and SLT, 1,950 V_{DC}. According to EN50163 the maximum recuperation voltage for a 1,500 V_{DC} system is 1,950 V_{DC}. The recuperation voltage for a 3 kV_{DC} system is 3,900 V_{DC}. This voltage is applied to the 3 kV train characteristics. The substation impedance at 3 kV_{DC} is four times higher than that at 1,500 V_{DC} for a comparable substation. All substations in the existing 1,500 V_{DC} network are used as substations in the 3 kV_{DC} network.

3.2 CASE STUDIES

The simulation model used in the case studies is based on part of the network in the Netherlands: the connection between Utrecht and Den Bosch including all lines connecting to Den Bosch. This model represents the complete network [3].

The energy delivered by the substations is related to the 1,500 V_{DC} system without recuperation: 1,500 V_{DC} recuperation 85%, 3 kV_{DC} recuperation 95% and 3 kV_{DC} recuperation 75% (see Fig. 3). The difference in recuperation energy between 1,500 V_{DC} and 3 kV_{DC} can be explained by two factors: a reduced recuperation voltage for VIRM and BR189, and a reduced susceptibility for recuperative energy by the 1,500 V_{DC} system. The first factor plays a minimal role, because the VIRM and BR189 represent intercity and freight services and therefore represent a minority in recuperation energy in the network. The stop service (SLT) on the contrary has no diverting limitation. The susceptibility for recuperation energy is demonstrated in Fig. 4 for both power supply systems.

Figure 4 demonstrates the limited ability of the 1,500 V_{DC} system to deliver the demanded power and to receive the regenerative power. It also illustrates the ability of the 3 kV_{DC} system not only to deliver a higher power demand, but also to receive the complete recuperation power. Measurements of Dutch Railways confirm that up to 50% of the regenerative power of intercity services between Utrecht and Den Bosch could not be fed back into the network.

A further increase of the regenerative power is thus only feasible in the 3 kV_{DC} network. To investigate if an increase of density of rolling stock will increase the susceptibility of the 1,500 V_{DC} for regenerative energy a high frequency timetable between Utrecht and Den Bosch was introduced (PHS 8/8), using the same infrastructure as the 2013 timetable [4]. In Fig. 5 the result of the simulation study is presented in comparison with the 2013 timetable.

The efficiency of the 1,500 V_{DC} system is between 88% (PHS 8/8) and 89% (timetable 2013) compared with 94% (PHS 8/8) and 95% (timetable 2013) for the 3 kV_{DC} system. The increased number of trains has not led to an increase in recuperation energy in the 1,500 V_{DC} system (relative).

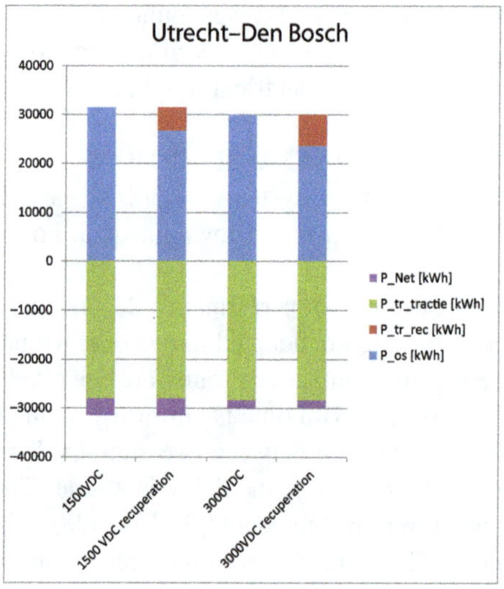

Figure 3: Energy consumption of 1 h of the complete network. Timetable 2013.

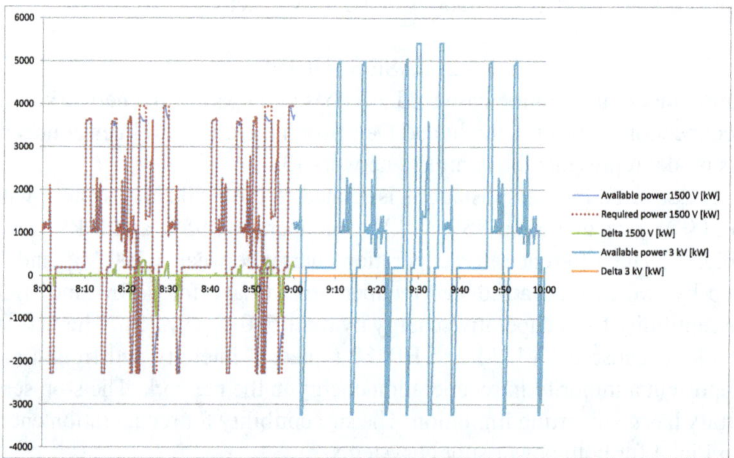

Figure 4: Train results 1,500 V_{DC} (8:00–9:00) 3 kV_{DC} (9:00–10:00). Timetable 2013.

Besides the Utrecht–Den Bosch network, three regional lines (Heuvellandlijn, Betuwelijn and Valleilijn) were studied [4]. Is the advantage of the 3 kV_{DC} over the 1,500 V_{DC} network on the conventional lines also valid on regional lines? One of this regional lines, the Betuwelijn – mostly a single-track layout – is crossing the conventional line Utrecht–Den Bosch at Geldermalsen. Figure 6 demonstrates the result of individual regenerative power of rolling stock, which leads again to the conclusion that a 1,500 V_{DC} network is not very susceptible to regenerative energy.

Although the timetable on a single-track layout is blocking the possibility to regenerate energy close by, the 3 kV_{DC} system is able to provide the regenerative energy towards the line Utrecht–Den Bosch: 17% less energy supplied by the substations on the Betuwelijn compared with the 1,500 V_{DC} variant.

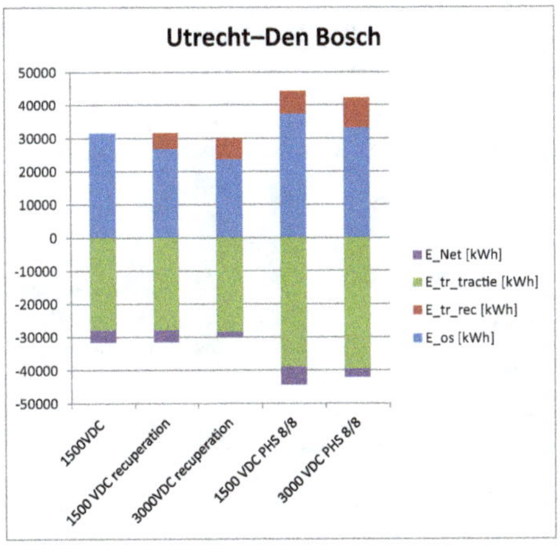

Figure 5: Simulation results of the energy consumption of 1 h of the complete network. Timetable 2013 and increased frequency (PHS 8/8).

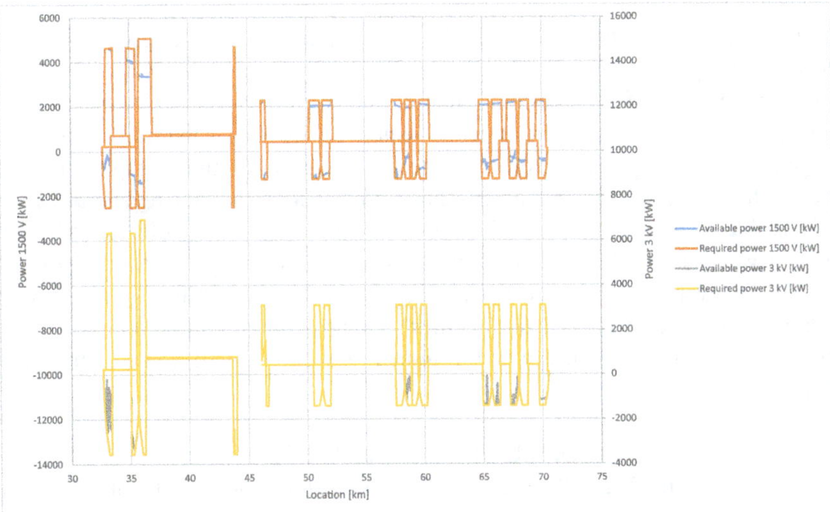

Figure 6: Train performance, above 1,500 V_{DC}, beneath 3 kV_{DC}.

The Valleilijn is a regional line of 18 km single-track line length without own substations, connecting two conventional lines. The absence of substations enables regenerative braking for both power supply systems, although the difference between the power supply systems on regenerative braking is still present: the regenerative power under 1,500 V_{DC} is limited. The tractive effort performance under 1,500 V_{DC} is also limited on major parts of the Valleilijn. For that reason both power supply systems perform with the same energy efficiency. The result of the regional line in the complete network is demonstrated in Fig. 7. Because of the small amount of energy on the regional line compared to the complete network, the overall result is equal to conventional lines.

The Heuvellandlijn is a double-track regional line of 22 km with substations, connecting two conventional lines. The power supply systems do not limit the performance – tractive and

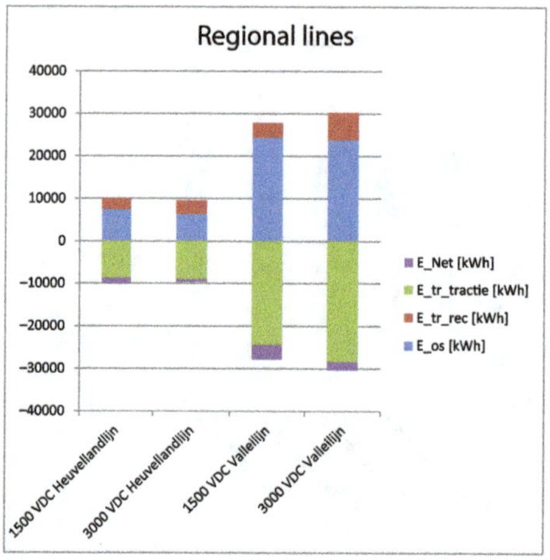

Figure 7: Simulation results of the energy consumption of 1 h of the regional networks. Timetable 2013.

regenerative – on this regional line. Figure 7 shows that both power supply systems perform with the same energy efficiency.

The main conclusion of this section is that a 1,500 V_{DC} power supply system negatively influences the performance of rolling stock. The network decreases the tractive effort as well as the regenerative energy and therefore creates an energy opportunity gap. Increase of the frequency in the timetable only strengthens these effects. Increasing the regenerative energy – which is possible according to Section 2 – does not increase the energy efficiency because the 1,500 V_{DC} network is already overloaded by the regenerative energy supplied by the trains.

The 3 kV_{DC} network, on the other hand, offers complete susceptibility for not only tractive effort but also regenerative energy and therefore avoids an energy opportunity gap. Furthermore, increasing regenerative energy as mentioned in Section 2 will lead to a higher energy efficiency.

4 OPERATIONAL STRATEGIES

The results of Sections 2 and 3 give rise to the question whether operational strategies should differ for the two power supply systems. In a 1,500 V_{DC} system operational strategies aimed at increasing the regenerative energy will not lead to a higher energy efficiency of the system. In a 3 kV_{DC} system, however, there is still room for an increase of regenerative energy. Operational strategies to increase this energy will lead to a higher energy efficiency of the system.

The previous two sections also revealed that the energy opportunity gap is not only network dependable but also dependent on the use of the individual rolling stock. In order to get a clear view of their effect on the energy efficiency of the two power supply systems, several parameters (coasting, a wide variety of decelerations, speed restriction and catenary coupling) were studied. In Fig. 8 comparisons are made between coasting, recuperation and the combination of coasting and recuperation.

In the current situation in the Netherlands, coasting is an attractive strategy to increase energy efficiency. (In 2013 about 50% of the rolling stock was capable to regenerate power back to the network.) The result of coasting is equal to having all trains equipped with regenerative braking facilities. The combination of coasting and regenerative braking increases

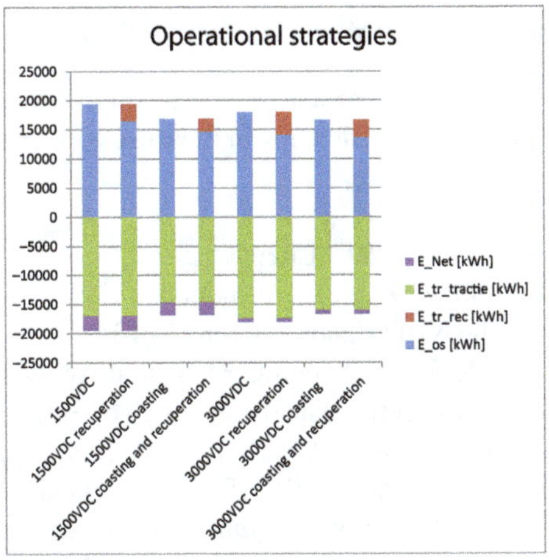

Figure 8: Simulation results of the energy consumption of 1 h of operational strategies.

the efficiency furthermore. The coasting strategy of Dutch Railways is – based on the status of the rolling stock – a conscious choice in reducing energy consumption in railways. The outcome of Sections 2 and 3 demonstrates that obtaining higher levels of regenerative energy in a 1,500 V_{DC} system will not lead to an increase in energy efficiency of the system. The combination of coasting above 100 km/h and regenerative braking below 100 km/h is a well-working strategy for a 1,500 V_{DC} system.

Despite the success of this strategy in a 1,500 V_{DC} system, a 3 kV_{DC} system needs another strategy to maximize the increase of energy efficiency. The 3 kV_{DC} system is able to couple high performance – substantial higher power per train compared with 1,500 V_{DC}, and reduced travelling time – with high efficiency (due to its high susceptibility for regenerative energy).

Introducing the 3 kV_{DC} system in the Netherlands will disconnect energy efficiency from the operational strategies (regenerative braking or coasting combined with regenerative braking). This conclusion is based on the 60% regenerative energy of Section 2. If this regenerative energy can be increased, regenerative braking will be preferable to coasting.

5 SOCIAL COST–BENEFIT ANALYSIS

This method has been used for an indication of the social cost–benefit analysis of the introduction of a 3 kV_{DC} traction power supply system in the Netherlands [5]. Fifty percent of the benefit consists of energy efficiency. The cost of the system changeover is calculated at 400 million euros for infrastructure and a couple of 100 million euros for rolling stock [6]. Based on the energy efficiency calculations, the ROI period is 15 years [2].

6 CONCLUSION

This article described a three-level method to obtain energy efficiency in railways. The application of this method on the possible system changeover in the Netherlands shows that a system changeover should also lead to a 'changeover' in operational strategy: energy efficiency based on coasting in a 1,500 V_{DC} system has to be replaced by regenerative braking in a 3 kV_{DC} system. To improve the efficiency furthermore the way of braking has to be improved by introducing optimized deceleration for regenerative braking. For stop services in the Netherlands an improvement of 40% of the total regenerative energy is feasible. So, operational strategy has to be tuned to the traction power supply system. A changeover from

1,500 V_{DC} to 3 kV_{DC} – based on the outcome of conservative energy efficiency calculations – will give ROI within 15 years.

Using the method discussed in this article, railway companies can easily determine the most efficient way to increase energy efficiency in their network and perform a well-founded social cost–benefit analysis.

REFERENCES

[1] Steenis, N.H. van, *Monitoring Train Performance in Case of Low Adhesion,* Enschede: University of Twente, The Netherlands, 11 June 2010.

[2] Dutch Government, Guideline for Social Cost Benefit Analysis. Ministry of Economic Affairs & Ministry of Infrastructure and Transport, The Hague, 2000.

[3] Paulussen, R.M., *Electrical Energy Simulations on Behalf of Introduction of 3 kV_{DC} Traction Power Supply,* Amersfoort: ARCADIS Netherlands BV, 26 September 2014.

[4] Paulussen, R.M., *Electrical Energy Simulations on Behalf of Introduction of 3 kV_{DC} Traction Power Supply: Regional Lines,* Amersfoort: ARCADIS Netherlands BV, 27 October 2014.

[5] Zoeteman, A., Harve, F. ten, & Ploeg, T., *Societal Cost benefit Analysis 3 kV and 1500V Traction Power Supply*, Utrecht: ProRail, 2014. .

[6] Zoeteman, A., Harve, F. ten, & Ploeg, T., *Migrating to DC 3 kV traction power supply for Dutch Railways – A feasibility analysis.* Elektrische Bahnen 114 (2016) Heft 1–2.

[7] Harve, G.F. ten., *Sustainable Traction Supply,* Utrecht: ProRail, 2012.

[8] Paulussen, R.M., *Electrical Energy Simulations on Behalf of Introduction of 3 kVDC Traction Power Supply: PHS-study,* Amersfoort: ARCADIS Netherlands BV, 26 September 2014.

[9] Welvaarts, J., *3000V Traction Energy Supply: Travel Time and Recuperation Effects,* Utrecht: Lloyd's Register Rail, 2014.

FREIGHT TRAIN SCHEDULING WITH MINIMUM ENERGY CONSUMPTION

T.H. LE & B. JAUMARD
Computer Science and Software Engineering Department, Concordia University, 1455 de Maisonneuve Blvd. West, Montreal, QC, Canada H3G1MB

ABSTRACT
We consider energy-efficient scheduling of freight trains on a line with single-track sections, as is the case on many freight railways routes in North America. We determine the train speed that minimizes the energy consumption, taking into account the departure and arrival times and exploiting the waiting times. The proposed energy consumption model takes into account the ground topography, the speed, the number of axles the axle load and the type of locomotives. Some experiments are conducted in order to estimate the energy gain when adapting the train speed, using Canada Pacific Railway (CPR) data.
Keywords: freight trains, optimization modeling

1 INTRODUCTION

Freight train, for its high energy efficiency and safety advantage compared to other land and air transportation options, plays a very important role in transportation in North America. In America, the annual cost for train industry is up to 9 billion USD [1]. This article studies the fuel consumption within the context of freight train schedule.

In order to deal with the fuel consumption, there are few main approaches, among them simulation is the more popular. Bai et al. [2] take into account the tractive power required and also the energy loss due to braking. Then by using simulation, they investigate several driving profiles, i.e. the way that trains should be driven (accelerating, coating, braking, etc.) so that the energy consumption is reduced. However, the simulation is done only in a limited environment of about 30 km long from Hefei and Chang'anji, China, and it is not clear how the train profiles behave in larger instances. Similarly in Chang and Morlok [3], the authors, using simulation, evaluate energy consumption with respect to different train driving profiles. The factors being considered are the tractive power, the tangents and the curves. In another vein, Howlett et al. [4] use control theory to derive the optimal switching points and consequently optimal driving strategies for freight trains. However, the article, as well as others taking the simulation approach, do not take into account the scheduling of the trains. This can be addressed with the optimization approach. Ghoseiri et al. [5] apply a simple formula to calculate the friction and deduce from it a formula for the fuel rate consumed by a train. The authors utilize the Pareto frontier method and formulate a non-linear programming model to deal with basic constraints such as headway (i.e. safety) constraints and deadlock constraints. However, these constraints are modelled using many binary variables which render the solution of the problem difficult and consequently, the size of data instances that can be solved is small, only a few trains and stations.

Our article addresses the problem of building train schedule in single-track railway while minimizing the fuel consumption. The article is organized as follows. First, we discuss factors that contribute to the train fuel consumption estimation and then a fuel consumption

DOI: 10.2495/TDI-V1-N3-501-510

formula; taking into account the train weight, train speed is derived. Next, a mathematical model is built, which aims to minimize the fuel consumption using the formula derived from the first step and satisfies the operational constraint. We then describe our two-phase method to solve the problem together with some numerical results.

2 ESTIMATING FUEL CONSUMPTION

2.1 Assumptions

In this project, we assume that the amount of fuel required to produce a unit of force is the same, regardless of the speeds of trains [5]. Therefore, we study the fuel assumption by studying the resistance of the trains.

The resistance of the train is composed of three components: train rolling resistance, grade resistance and curve resistance. Out of the three, train rolling resistance is the most important. The grade and curve resistances are both dependent only on the weight of the trains. For simplification, in this project we ignore the grade and the curve resistance as they do not depend on speed and we cannot control these two resistances by varying the speed. In the remaining part of this article, we discuss only train rolling resistance and subsequently the fuel consumption resulting from the rolling resistance.

2.2 Train rolling resistance

Train rolling resistance was studied by Davis in the 1920s and is still used today, e.g. in Refs [3, 5–7]. It takes the following form:

$$\text{Train rolling resistance} \quad R = A + BV + CV^2, \tag{1}$$

where R is train rolling resistance, A component independent of train speed, B component dependent on speed, C drag coefficient based on the shape of the front of the train and other features affecting air turbulence and V the train speed.

The fuel consumption estimation based on the rolling resistance can be derived as follows. Let P denote the horsepower produced by the locomotives of a given train. At constant speed, the fuel consumption per hour depends linearly on the horse power, according to [6]

$$F = a + bP \quad gal\,/\,hr \tag{2}$$

Let TE denote the tractive effort. We also have

$$P = TE.V \quad \text{and} \quad F = a + bTE.V. \tag{3}$$

Assuming the horsepower TE is approximately equal to the train resistance $TE = R = (A + BV + CV^2)$. After some algebraic transformations and scaling, the fuel consumption takes the following form:

$$E = \left(\frac{a'}{V} + (A + BV + CV^2)\right)d \quad gal \tag{4}$$

3 TRAIN SCHEDULING AND FUEL CONSUMPTION OPTIMIZATION MODEL

3.1 Generalities

Although the energy consumption is very important in train operations, other aspect should be taken into account in order to maintain the quality of service. One of the other requirements is that train schedule should be reasonable with respect to the travelling time; i.e. the trains should not take too much time to travel from their origin to their final destination. The trains must, on the other hand, satisfy the operational constraints, e.g. safety constraints, no deadlock constraints [8]. These concerns are to some extent independent of the fuel consumption. Therefore, we propose a heuristic in which the optimization model is performed in two steps as follows.

The first step is to build a train schedule using the algorithm described in [8] which minimizes the average travelling time. The input of the optimization model includes the train departure times (which are planned by the train company), the speed limits (which implicitly take into account the network features such as grade, curvature, etc.), the number of siding tracks, etc. The schedule produced by the model will specify the arrival and departure of each train at each station while satisfying the operational constraints as specified in [8]. In other words, this first step builds an initial feasible schedule that minimizes the average travelling time.

In the second step, a more detailed schedule is built based on the schedule obtained from the first step. In the detailed schedule, the arrival and the departure times of the trains at each station are kept the same as the ones from the first step but the speed of each train in each part of its initiatory is determined by the mathematical model so that the fuel consumption of trains is minimized. The fuel consumption to be minimized is calculated using eqns (1) and (4).

The advantage of this approach is that we can reuse the mathematical model in [8] which leads to a scalable train scheduling algorithm. This model is also quite flexible and we can easily incorporate many types of operational constraints. Another important advantage is that in the first step, we work with all the trains, ensuring they have to satisfy all the operational constraints whereas in the second step, we work with each train individually and also each segment, finding a detailed schedule that minimizes its fuel consumption. This helps increasing significantly the scalability when solving the model.

As an output of the optimization model, we get the average speed in the segments, connecting two successive sidings/stations p and p'. In order to compute the energy consumption in that segment $[p, p']$, we may need to divide $[p, p']$ into sub-segments, on which the speed is constant. We next discuss how to do it on a base case that is depicted in Fig. 1. Generalization to more sub-segments is straightforward.

3.2 Outline of energy consumption optimization in a segment

Let I be the set of indices indicating the sub-segments. Let v_i be the speed of train t on ℓ_i, and v_{opt} the optimized average speed of the train between p and p', as output by the optimization model.

$$\min \sum_{i \in I} \left(\frac{a'}{v_i} + A + Bv_i + Cv_i^2 \right) \ell_i \tag{5}$$

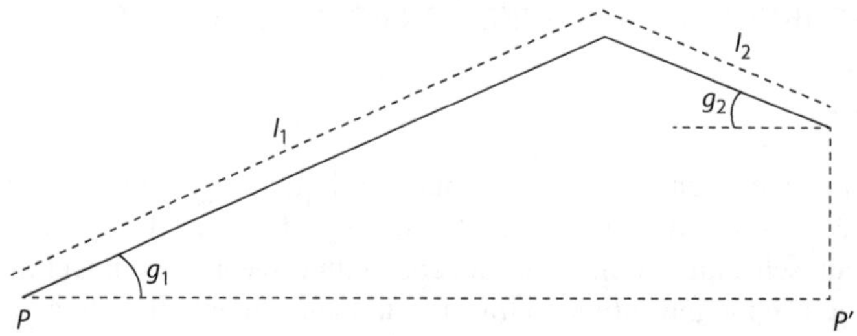

Figure 1: The case of two grade distributions in one segment.

$$\text{subject to: } \sum_{i \in I} \frac{\ell_i}{v_i} = a_t^{p'} - d_t^p \quad i \in I, \tag{6}$$

$$v_i \in \mathbb{R} \quad i \in I\ . \tag{7}$$

In terms of solving techniques, the model above is in a separable form with respect to variables v_i, i.e. the objective and the left-hand side of the constraints are in the form of $\sum_{i \in I} f_i(v_i)$. Therefore, we can linearize the terms in the objective and the constraints, namely $E_i = \left(\frac{a'}{v_i} + A + Bv_i + Cv_i^2\right)\ell_i$ and $\frac{\ell_i}{v_i}$. Moreover, they are all convex and we are minimizing the objective; therefore when we linearize using λ-method, the resulting linearization leads to a linear programming problem.

3.3 Time domain formulation

The formulation above with respect to speed has one disadvantage, that is the constraint (6) is non-linear and we need to linearize it. However, while linearizing a constraint, we could face infeasibility unless we linearize v_i with a sufficiently small granularity. Then, however, the number of variables in the resulted linearization problem might become too large. One way to avoid this issue is, instead of modelling the problem using velocity variables v_i, we can use time index variables to reformulate the problem.

Let t_i be the time it takes to travel l_i then $v_i = \frac{\ell_i}{t_i}$ and the model can be written as follows:

$$\min \sum_{i \in I} \left(\frac{a'}{\ell_i} t_i + A + B\frac{\ell_i}{t_i} + C\frac{\ell_i^2}{t_i^2}\right)\ell_i \tag{8}$$

$$\text{subject to: } \sum_{i \in I} t_i = a_t^{p'} - d_t^p \tag{9}$$

$$t_i \in \mathbb{R} \qquad i \in I. \tag{10}$$

It is easy to verify that the function $E_i = \left(\frac{a'}{l_i} t_i + A + B\frac{\ell_i}{t_i} + C\frac{\ell_i^2}{t_i^2} \right)$ is still convex and separable in t_i. Therefore, if we are to apply linearization using λ-method, the resulting formulation is still a linear programming problem.

Let t_i be linearized by the set of k_i points $\tau_i^1, \tau_i^2, \ldots, \tau_i^{k_i}$ with associated λ variables $\lambda_i^1, \lambda_i^2, \ldots, \lambda_i^{k_i}$. Besides, let τ_i^1 and $\tau_i^{k_i}$ be equal to lower bound and upper bound of t_i, respectively. We then have the following linear programming problem:

$$\min \sum_i \sum_{j=1}^{k_i} E_i\left(\tau_i^j\right)\lambda_i^j \ell_i \tag{11}$$

$$\text{subject to: } \sum_i t_i = a_t^{p'} - d_t^p \tag{12}$$

$$t_i = \sum_{j=1}^{k_i} \tau_i^j \lambda_i^j \quad i \in I, \tag{13}$$

$$\sum_{j=1}^{k_i} \lambda_i^j = 1 \quad i \in I, \tag{14}$$

$$\lambda_i^j \geq 0 \quad i \in I, j \in J. \tag{15}$$

3.4 Safety constraint

Assuming that, from the schedule produced from the first phase, in segment $[p, p']$, train m' depart after m but the time m' spend $[p, p']$ overlaps with that of m. In other words, there is a time that both m and m' are on segment $[p, p']$. We need to make sure that m and m' maintain the safety distance in all point i.

In general, in any given segment $[p, p']$ there might be k trains $m_1, m_2,, , m_k$ in which m' overlaps with m_2 and m_2 overlaps with m_3 and so on. If this is the case, we need to maintain the safety distance between m_i and m_{i+1} in segment $[p, p']$. Our experiments with the data show that k usually takes the value of 2.

Let the set of sub-segments ends of $[p, p']$ be indexed by point i. Since train m cannot stop in the middle of a segment, departure and arrival times at each point i inside the segment are equal; i.e. we only need to be concerned with the departure time at each point i. Denote by d_i^m the departure time at point i of train m. Let M be the set of trains that overlaps $m_1, m_2, \ldots m_k$ at segment $[p, p']$; i.e. train m_j overlaps train m_{j+1}. We then have the following model:

$$\min \sum_{m \in M} \sum_{i \in I} \left(\frac{a'}{\ell_i} t_{mi} + A + B\frac{\ell_i}{t_{mi}} + C\frac{\ell_i^2}{t_{mi}^2} \right) \ell_i \tag{16}$$

$$\text{subject to: } t_{mi} = d_{i+1}^m - d_i^m \qquad m \in M, i \in I, \tag{17}$$

$$\sum_i t_{mi} = a_m^{p'} - d_m^p \qquad m \in M, \tag{18}$$

$$d_i^{m_j} \geq d_i^{m_{j+1}} + \text{safety\_distance} \qquad i \in I, \tag{19}$$

$$d_i^m, t_{mi} \in \mathbb{R} \qquad m \in M, i \in I. \tag{20}$$

Observe that the safety distance is then ensured by constraint (19).

3.5 Taking advantage of the waiting time in the second phase

In the area where there are only single tracks, when two trains meet, one of them has to slow down and wait on the siding for the other train to pass. The train schedule built from the first phase specifies where trains meet and how much time trains have to wait. Figure 2 shows an example of a train meeting. Train 1 goes from station C to D to E and Train 2 in the opposite direction from station E to D to C. They meet at D at 08:12 and Train 1 has to take the siding and wait from 08:10 to 08:15.

As described above, the second phase will find a refined schedule for the trains so that the overall fuel consumption is minimized. For a train that waits at a siding such as Train 1 in the example, the wait time can be exploited to reduce speed and therefore fuel consumption as follows. In the original schedule obtained from phase 1, Train 1 passes the station C at 08:05 , arrives at D at 08:10 and waits there until 08:15 then departs from D and arrives at E at 08:22. Train 2 passes E at 08:07, D at 08:12 and C at 08:17. We adjust the timeline of Train 1 such that the Train 1 passes C at 08:05 (same as the original schedule) and arrives at D at 08:12 (the time Train 2 passes D). Train 1 then departs from D at 08:12 and arrives at E at 08:22. The adjusted timeline of Train 1 is depicted in dash line in Fig. 2. The timeline of Train 2 is kept the same.

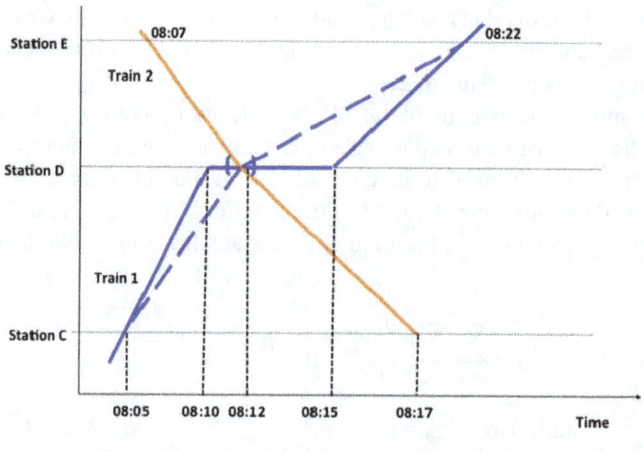

Figure 2: Adjust timeline of a waiting train.

As can be seen, the adjustment above does not affect the feasibility of the schedule; therefore, in the second phase, the scheduling constraints will still be satisfied. Consequently, no binary variable is needed and the model in the second phase is a continuous linear problem and is highly scalable.

3.6 Combining successive segments

In addition to the improvement discussed in Section 3.5, it turns out that the schedule of the train can be adjusted so that the resistance and therefore fuel consumption can be further reduced. The schedule output in the first phase is optimized with respect to the travel time. Travelling times, i.e. the speed, may vary quite significantly from one segment to the next. So, if the optimization process is performed over several consecutive segments, then speed differences can be smoothed out and the overall saving could be larger than the combined savings on each individual segment. In that case, since the travelling time of each segment may be adjusted, the minimum travel time constraint in each segment should be added to the model so that the schedule is still feasible.

In principle, the more segments are combined, the more smoothness of the speeds between segments and therefore the more fuel consumption saving can be obtained. However, in order to maintain feasibility of the schedule, successive segments are combined, for any given train, until only it meets another train on opposite direction. For example, consider Train 2 in Fig. 3, travelling from station F going through E then meeting Train 1 at D and continuing its trip through C then B and A (not shown in Fig. 3). The successive segments that will be included into the set are F-E and E-D but not D-C. If segment D-C was included then the arrival and departure times of Train 2 at D would be modified and consequently Train 2 and Train 1 might no longer meet at D but inside either segment E-D or D-C and the schedule's feasibility no longer holds. The departure time of Train 2 at the first station of the segment set (F in this case) and the arrival time at the last one (D) will be fixed to those obtained from phase 1 (07:55 and 08:12 in Fig. 3). The departure and arrival at other stations are flexible, assuming the minimum travelling time constraint is satisfied.

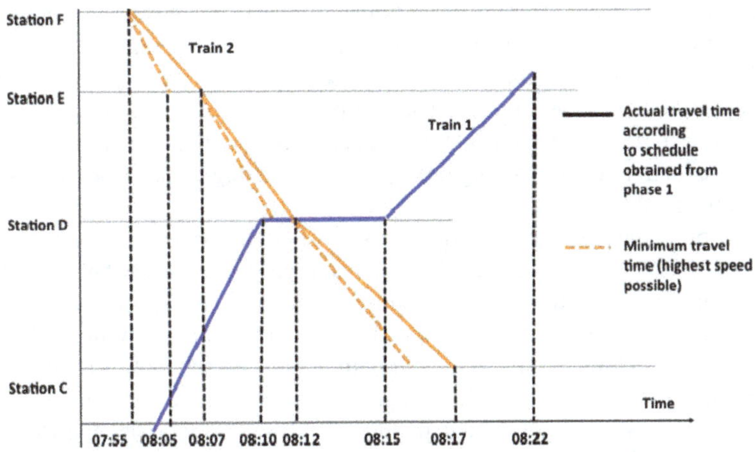

Figure 3: Combining consecutive segments of one train.

Let S be the set of segments, i_s be the sub-segment index of segment s, t_{smi_s} be the time train m spends at sub-segment i_s of segment s then the objective is modified as follows:

$$\min \sum_{s \in S} \sum_{m \in M} \sum_{i \in I_s} \left(A + B \frac{\ell_{i_s}}{t_{smi_s}} + C \frac{\ell_i^2}{t_{smi_s}^2} \right) \ell_{i_s}. \tag{21}$$

The following constraints must be satisfied in addition to the constraints (17)–(19):

$$t_{smi_s} = d_m^s - d_m^s \geq min\_travel_s \qquad s \in S, m \in M, \tag{22}$$

$$\sum_{s \in S} t_{smi_s} = a_m^{s^{last}} - d_m^{s^{first}} \qquad m \in M, \tag{23}$$

$$d_m^s, t_{smi_s} \in \mathbb{R} \qquad s \in S, i \in I. \tag{24}$$

4 NUMERICAL EXPERIMENTS

As an initial step, we focus on the division Calgary to Field of the railway network of Canadian Pacific Railway. This part of the network consists essentially of single tracks with about 42 stations. We evaluate the energy consumption saving in 5 instances with a number of trains varying from 10 to 18.

Table 1 shows the result of our experiments when we fixed the departure times as planned. The first and second columns correspond to the number of trains and their average travelling time obtained from phase 1 of the model. The third column is the fuel consumption obtained from the first phase without any optimization. This quantity is the sum of the consumption in each segment. To estimate each segment's consumption we assume the speed is constant and equal to the average speed in that segment, which can be deduced from the arrival and departure times obtained from the first phase. The fourth and fifth columns describe the fuel consumption and the percentage of the saving when we optimize the fuel consumption with the inclusion of waiting times. Similarly, the last two columns show the fuel consumption and the percentage of the saving in case we combine consecutive segments and include the waiting times. As can be seen from the last two columns of Table 1, the proposed approach shows significant saving of 14% to 16%. Including waiting time in the optimization contributes the most to the reduction of energy consumption or about two-thirds of this reduction; the remaining one-third of the saving comes from the combination of several segments.

Table 2 compares the two scenarios. In the first one, we fix the departure times of the trains as planned and in the second one, we allow the departure times to vary between 30 minutes before and 30 minutes after the planned departure time. Since the second scenario has more flexibility, it can reduce the average travelling time between 8 and 43 minutes depending on the number of trains. However, as the travelling time reduces, the fuel consumption tends to increase, as the last two columns of Table 2 show. Also, as can be seen in columns 6 and 9 of Table 2, the savings in case of fixed departure time are usually higher and vary much less than those in case of flexible departures.

Table 1: Fuel consumption – fixed departure times.

Train	Average travelling time	Without optimization	Fuel consumption (in gal) With optimization			
			Include wait times	Saving	Combine segments	Saving
10	10:18	25587.9	22982.5	10.18%	21909.1	14.38%
12	10:19	30055.7	26983.3	10.22%	25483.2	15.21%
14	10:28	34188.8	30523.8	10.72%	29140.5	14.77%
16	10:28	38572.3	34485.4	10.6%	32363.7	16.1%
18	11:02	41977.7	37501.7	10.66%	34960.8	16.72%

Table 2: Fuel consumption – fixed vs flexible in departure time.

Train	Average trav. time		Fuel consumption (in gal) Fixed			Flexible			Diff. Fle & Fix	
	Fix	Fle	W/o opt.	With opt.	Saving	W/o opt.	With opt.	Saving	(in gal)	%
10	10:18	9:32	25587.9	21909.1	14.4%	26522.4	23792.2	10.29%	1883.1	8.6
12	10:19	10:11	30055.7	25483.2	15.2%	30408.8	26739.2	12.07%	1256	4.9
14	10:28	10:08	34188.8	29140.5	14.8%	34656.8	29067.8	16.13%	-72.7	-0.2
16	10:28	10:13	38572.3	32363.7	16.1%	38743.9	33877.8	12.56%	1514.1	4.7
18	11:02	10:19	41977.7	34960.8	16.7%	42324.8	36296.8	14.24%	1336	3.8

5 CONCLUSION

In this article, we develop a two-phase approach to address the problem of scheduling trains, satisfying different operational constraints while minimizing their fuel consumption. In the first phase, the train schedule is built with the objective of minimizing the travel time. In the second phase, the obtained schedule is optimized with respect to the fuel consumption. Some improvements are proposed so that the consumption is further minimized. Experiments are performed on reasonable size of real-life dataset of 42 stations and up to 18 trains with the savings between 10% and 16% depending on the instances. The results show that our method is scalable, and hence, could be used as a tool to evaluate different train scheduling scenarios.

REFERENCES

[1] Vantuono, W.C., A closer look at LNG. *Railway Age,* October 2013, p. 14.
[2] Bai, Y., Mao, Y., Zhou, F., Ding, Y. & Dong, C., Energy-efficient driving strategy for freight trains based on power consumption analysis. *Journal of Transportation, Systems Engineering and Information Technology,* **9**, pp. 43–50, 2009.
[3] Chang, D. & Morlok, E., Vehicle speed profiles to minimize work and fuel consumption. *Journal of Transportation Engineering,* **131**, pp. 173–182, 2005.

[4] Howlett, P.G., Vu, X., Pudney, P.J., Local energy minimization in optimal train control. *Automatica,* **45**, pp. 2692–2698, 2009.

[5] Ghoseiri, K., Szidarovszky, F. & Asgharpour, M.J., A multi-objective train scheduling model and solution 2004. *Transportation Research Part B: Methodological,* **38(10)**, pp. 927–952, 2004.

[6] Radford, R.W., Fuel consumption of freight trains hauled by diesel electric locomotive. *Journal of Engineering for Industry,* **105**, pp. 75–87, 1983.

[7] AREMA, *American Railway Engineering and Maintenance-of-Way Association Manual,* Maryland, USA, 2003.

[8] Jaumard, B., Le, T., Tian, H., Akgunduz, A. & Finnie, P., A dynamic row/column management algorithm for freight train scheduling. *12th Workshop on Algorithmic Approaches for Transportation Modelling, Optimization, and Systems (ATMOS),* Ljubljana, Slovenia, pp. 1–12, 2012.

STRAY CURRENT CORROSION MITIGATION, TESTING AND MAINTENANCE IN DC TRANSIT SYSTEM

S.A. MEMON[1] & P. FROMME[2]
[1]Arup North America, Houston, Texas.
[2]University College London, London, UK.

ABSTRACT

Stray current corrosion in direct current (dc) transit systems occurs because of the mechanism of current transfer between metals and a conductive electrolyte such as concrete, soil and water. Stray current reactions can be considered as a special case in that the anode (point of current discharge) may be at a considerable distance from the cathode (point of current pickup). The risk of stray current corrosion arising from the operation of dc-powered transit system is difficult to eliminate completely. However, suitable design of dc traction power systems and structures carrying the railways can significantly reduce the risk of corrosion both to the transit system structures and third-party structures.

Stray currents can cause safety risks, thus making the design of stray current mitigation, testing and maintenance an important element of the holistic design for a dc transit system. Based on the results of the literature research, interviews with over two dozen dc rail transit systems, and testing of dc rail tracks, this paper presents and analyses various mitigation methods currently in use in the industry to control stray current corrosion. Recommendations for the testing (monitoring) and maintenance procedures to keep the stray current leakage and the related corrosion in control are then presented.
Keywords: corrosion, direct current transit system, leakage current, rail potential, rail-to-earth potential, stray current, stray current corrosion, stray current testing, substation spacing, testing and maintenance of dc operating transit system.

1 INTRODUCTION

Stray current corrosion has been a cause of concern for dc rail transit agencies, electrolysis committees, utility owners and providers all across the globe. Over the years various stray current control and collection methods have been adopted by different transit agencies with varying results. The mitigation methods usually depend on various factors including but not limited to geographical location of the transit system, physical environment (weather), soil investigations, economic considerations and other elements. These methods are seldom defined in the transit agency design criteria especially when it is an older agency. In general most of the transit agencies in Europe and often outside of Europe follow the BS EN 50162:2004 [1] & BS EN 50122-2:2010 [2] – BSI (British Standards Institution) standards.

In the US an ASTM designation G 165 –99 (ASTM [3]), which was issued in 1999 as a standard practice for determining rail-to-earth resistance, is also used by the transit agencies to carry out the rail-to-earth testing. Besides the above-mentioned standards there are no other national standards or guidelines for stray current leakage and corrosion control that can be used by the transit agencies in the US (National Research Council [4]). In the absence of any defined standards or guidelines the transit agencies occasionally end up choosing from a vast array of mitigation methods. These mitigation methods are adopted based on the knowledge of previously completed projects instead of the agency choosing them based

DOI: 10.2495/TDI-V1-N3-511-519

on applicability and understanding of project-specific factors, surveys, studies and systems design carried out for that specific transit system.

Thus to understand the stray current design process, used by various transit agencies, an initial desk study of numerous dc-powered transit systems was performed. This was used to collect pertinent information on design criteria, performance specifications, constructability issues and physical environment as it relates to stray current mitigation decisions. Based on the findings of the literature review a mix of 30 transit agencies (including US and international agencies) were contacted. A questionnaire was emailed to these transit agencies to gather information on their existing stray current mitigation and collection procedures, methods of testing and measurement, criteria for acceptable levels including rail-to-earth potential, and agency-specific existing and/or previous issues.

This paper highlights the most common mitigation, testing and maintenance methods narrowed down during the study that have been successfully adopted by these agencies in effectively controlling the stray current corrosion.

2 STRAY CURRENT CORROSION

The operating current for the electric traction power supply flows through the overhead catenary system or the third rail to the vehicle and returns to the substation through the return circuit. The return circuit includes numerous conductors that help complete the path of the return current to the substation. Running rails are typically the most widely used conductors for the return of electric current. Since perfect insulation does not exist and that rail has a finite resistance, the return current leaks into the earth and finds its way to the substation via the path of least resistance. This current that leaks into the earth by taking the path of least resistance is called stray current and the resulting corrosion is called stray current corrosion.

In dc-powered rail transit systems the stray current will follow any path of least resistance (rather than the rail), if not designed properly, to return to the substation. This can cause significant corrosion to the metallic structures along the rail transit system where the current leaves the conductor and potentially to third-party buried infrastructure. Hence measures need to be taken to contain stray current at the source by providing suitable insulation and/or isolation of the rail to prevent the current flow into the earth.

Some transit systems use the fourth rail system for the return of current which is typically an insulated conductor fourth rail, electrically isolated from the running rails and the surrounding soil. This fourth rail collects the current and returns it to the substation. However, fourth rails are not often used due to limited transit system construction budgets.

3 MITIGATION AND CONTROL METHODS

Early project surveys (including the stray current specific line items), commonly referred to as baseline surveys, are an integral part of the initial design of stray current corrosion mitigation. Baseline surveys consist of the following important elements (Moody [5]):

- Utility location information and coordination – this includes collecting/measuring voltage potential on existing utility structures.
- Soil corrosion characteristics – this includes existing ground testing.
- Atmospheric corrosion characteristic – this includes weather variations.

A baseline survey plays a vital role in the design and mitigation of stray current in dc-powered transit system which helps develop a planned approach with foreseeable milestones from inception to the operation stage of the transit system (Memon [6]).

Depending on the findings of the baseline survey and traction power design during the design of the dc-powered rail transit system, steps must be taken to incorporate stray current controls. These steps should start with controlling the stray current at the source which is commonly referred to as 'control at source'. This should then subsequently be followed by mitigation of the stray current, the collection of the stray current leakage and then finally the ongoing planned maintenance and testing of the tracks.

Contingent on the type of track this 'control at source' is accomplished by frequently spaced substations, cross-bonding, rail resistivity, use of insulated track fasteners, rail boots, and improved coatings, maintaining a continuous electrical path, and isolating the rail in the yards (Fromme & Memon [7]).

3.1 Track type

Embedded track design requires a more complex level of electrical isolation as compared to ballasted track and thus demands a more aggressive approach to control the stray current leakage. The following are some of the key elements specific to embedded track design that must be evaluated at an early design level to avoid potential stray current leakage issues:

- Drainage at and around the tracks
- Rail boot construction (joints in particular)
- Grade crossings
- Insulating fasteners
- Special construction like bathtub isolation
- Collection mats

3.2 Substation spacing

This technique reduces the feeding distances and the amount of current to be returned to any one point, resulting in the reduction of track voltage drop, thereby reducing the amount of current which will stray away from the rails. Technology papers by the Bureau of Standards on leakage of currents from electric railways, issued in 1916, by Stratton, M'Collum and Logan (Stratton et al. [8]), explain the importance of reducing the feeding distance to minimize the stray current leakage for both grounded and ungrounded systems with the use of mathematical expressions.

Considering the marked effect on the reduction of the stray currents and overall potentials due to the reduction in the feeding distance, average feeding distances between the substations have been reduced to two to three miles on a typical dc-operated light rail system (Fig. 1).

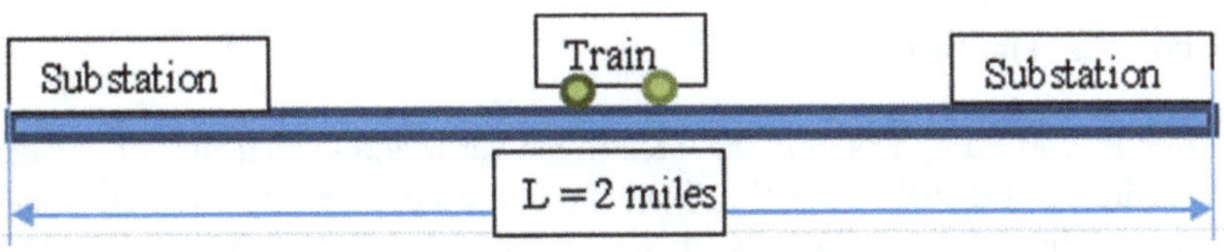

Figure 1: Substation spacing.

3.3 Rail resistivity and cross-bonding

Rail resistivity is mostly based on the type of rail used and its cross section and weight. Different size and type of rail are provided by different manufacturers. Based on the design loads, design life, cost and other track requirements the transit agency decides which vendor and rail section to select.

The use of heavy rail sections and suitably bonded rail joints were one of the earliest implemented mitigation methods for the control of stray current. The evolution of rail sections and steel along with other metal compositions has continued throughout the years and across the globe. With time rail sections have been improved in cross section, length and the method of joining the two sections of rail. Increasing the cross-sectional area or size of the rail is achieved by using standard size rails ranging from 90 to 120 pounds (115 RE tee is the commonly used rail with a longitudinal resistance of around 40–80 mΩ/km) (Cotton & Charalambous [9]).

Welded joints are used to provide conductivity equal to or greater than the continuous rail. These joints are less subject to failure compared to other forms of rail joints. Welding of rail lengths is thus used as the standard form of construction, especially in the embedded rails. This has not only proven instrumental in the reduction of stray current but also improved the performance of rail.

Cross-bonding between single-track and parallel-track rails is installed to ensure rail connectivity and to equalize the current flow between the rails, thus reducing voltage drop (rail potential). A frequent cross-bonding of tracks at an average of 500 to 1,000 feet is recommended.

3.4 Resistance to earth and rail isolation

Resistance of the ground immediately in contact with the rail depends primarily on the type of ground material that is in contact with the rail. Measures are taken to insulate the track from the earth to reduce the stray current process, thus reducing the corrosion of the base of the rails and other grounded steel structures for the sections of the rail that were embedded in the ground in urban areas.

As the current return path to the adjacent substation increases, the rail potential increases and similarly does the stray current. The track type along the track and in the immediate vicinity of the substation plays an important role in keeping the stray currents under control. It is a general understanding in the industry that the ballast section will have a high rail-to-earth resistance and thereby a reduced stray current as compared to embedded track.

Tracks are regularly maintained to keep the vegetation out of the tracks, retaining the track clean, dry, and dirt- and salt-free to help preserve the resistivity of the rail-to-earth high by keeping them insulated from earth. Well-drained broken stone ballast or gravel ballast is used in non-embedded sections for its much higher resistance to stray current as compared to concrete, whereas for embedded tracks, the rail is typically isolated from the ground by a sealing compound or rail boot (Fig. 2).

3.5 Isolation of rail in yards

Stray current is reduced by isolating the track in the yards and storage areas by detaching it from the sections of the main track (Szeliga [11]). To protect the staff from electrical shock in the yards and shops the running rails are directly grounded and earthed back to the nearest substation via insulated cables. This results in excess leakage of stray current since the only

Figure 2: Rail boot (Memon & Fromme [10]) and sealing compound replacing rail boot.

way back for the current to the substation is through the ground which contributes to the current leakage. Therefore, yard tracks are isolated from the main track and are provided with a dedicated substation.

3.6 Ungrounded or floating system

A floating system has no deliberate connection to earth and the stray currents are restricted by high rail-to-earth resistance using rail boot, rail coating and rail fasteners. However, this could potentially result in increased running rail voltage, as compared to the grounded system, causing safety concerns for the general public and transit agency staff. Though these safety concerns present a downside to the system, this system is preferred over the other systems because surveys conducted have shown that the floating system with ungrounded substations provides the best earthing system and keeps the potential within limits. Safety concerns are addressed with the use of overvoltage protection equipment and platform insulation procedures.

3.7 Drainage and insulating fasteners

Maintaining proper drainage around the rail boot and the tracks is vital. Proper drainage measurements must be taken during construction with careful design of water flow at low elevations and other critical locations. In newer constructions it is recommended to provide a rubber boot flangeway to isolate the rail contact. Similarly, the use of insulating rail fasteners is also important. This is especially vital where the use of insulated clip-type fasteners for rail to tie connection and continuous welded sections are common.

3.8 Collection mats

For embedded tracks operating in city streets, stray current corrosion is a major concern to the track owners and utility and other infrastructure owners in the vicinity of the transit system. Stray current collection systems are sometimes necessary to control the current leakage on these systems. The goal of insulating the rail is to control the current at the source and minimize the stray current leakage. Thus the need for a collection mat can be eradicated if the level of stray current being produced by the transit system is controlled by rail insulation. Research shows that there will be a certain amount of stray current leakage in spite of these control measures. This leakage usually happens after a few years of track life or even earlier if the tracks are not maintained and tested routinely (based on the survey results).

Collection mats are used to intercept and retain stray current on embedded track sections which are laid on concrete slabs with steel reinforcement such as those in tunnels and viaduct. A stray current collection mat in these instances provides a low-resistance path to intercept and retain the stray current leaving the rails. These collection mats are continuously bonded together along their length to provide the stray current with a low-resistance path (Fig. 3) (Tzeng & Lee [12]).

Figure 3: Reinforcement steel and collection mat for embedded track.

4 STRAY CURRENT TESTING AND MAINTENANCE

Mitigation methods installed to control stray current corrosion require periodic testing and maintenance to ascertain their safe performance and effectiveness. Therefore, it is important to conduct regular inspection and testing of the tracks including the inspection and testing of any mitigation techniques installed by the transit system to ascertain that the stray current leakage is within limits and the mitigation measures are operational as designed (Memon [6]).

The transit agencies interviewed, in general, agreed on the need and importance of a well-thought-out testing and maintenance plan tailor-fitted to their geographical and system necessities.

4.1 Coordination and communication

The development of a coordinated effort to sustain effective stray current control requires education, communication and cooperation of all stakeholders and concerned parties. Communication is the foundation of the effort to achieve and maintain effective stray current control. Thus it is essential that regular exchange of information be maintained between the interested parties to develop an overall sustainability of effective and efficient stray current control.

Based on the results of the transit agency survey it was observed that the stray current issues were significantly less onerous in those transit systems which had corrosion committees located in reasonable geographic proximity. This shows that coordination and communication with the corrosion engineers and utility owners are the key elements in keeping the corrosion-related problems under control.

4.2 Stray current testing

It is good transit industry practice to perform track-to-earth and utility-to-earth resistance testing prior to any revenue service operation to document the reference point characteristics and stray current activity of the system. These track-to-earth resistance values are used by the transit agencies and corrosion design consultants to achieve the control at source and thereby design and maintain the stray current mitigation and collection techniques. Based on the findings of the survey, following are some of the most commonly performed tests by the transit agencies (Memon & Fromme [10]):

- Track-to-earth resistance survey – measures the resistance and helps in locating and isolating the track work discontinuities.
- Track slab current measurement – provides an insight into the magnitude and direction of possible current leakage from the rails.
- Structure (utility pipe)-to-soil potential measurement – helps identify whether the structure is influenced by stray current and the direction of the current flow.
- Audio frequency testing – helps in pinpointing the local low-resistance areas and is used in conjunction.

5 STRAY CURRENT MAINTENANCE

The key fundamental maintenance essentials to be carried out on the system must include the following:

- Visual inspection – this is the first level of defence in identifying stray current leakage areas and providing mitigation for them.

- Rail isolation from conductive elements – this can be spotted during visual inspections and/or the testing of the system.
- Maintaining clean and dry tracks – it is imperative to control vegetation and design the tracks such that they drain properly and the water does not stand close to the rail boot and the tracks.
- Testing regime – working out a regular testing schedule to perform testing of the tracks in coordination with the corrosion consultant.

6 CONCLUSION

Of the many transit agencies interviewed (with testing being done at some of these) most of the ballasted section results showed that stray current leakage is under control and track-to-earth resistance values are within the limiting values based on the agency's criteria. However, unlike ballasted sections, most of the embedded sections showed higher track-to-earth resistance values along at least one segment of the track. Some of these track sections were using a concrete flangeway which holds earth and debris that in return provides a path for the stray current. This along with the fact that the concrete repairs with time had spilled over the rail boot at many locations, which resulted in the creation of a contact with the rail thereby annulling the purpose of the rail boot isolation.

Drainage of storm water seems to be another common issue with several transit agencies. These drainage correction measurements must be taken into account during initial construction with careful design of water flow at low elevations and other critical locations. In newer constructions it is recommended to provide a rubber boot flangeway to isolate the rail contact with the earth.

Another issue observed was the inconsistency of the testing methods and frequency of testing. Testing methods should be adapted based on the age of the transit system, location of the tracks, type of the track bed, the type of structure under investigation and the source of leakage.

Lastly, it would be easier to implement most of the above-mentioned isolation, mitigation and collection options on a newer transit system with proper foresight and planning. However, not all the options and recommendations discussed here will apply to older systems or systems that are currently going through extension of their existing systems. In such instances it will be the responsibility of the design engineer/consultant in consensus with the transit agency/owner to design the system that will keep the stray current corrosion to a minimum. The key to achieve a leakage-free transit system is to follow the logical sequence of the design process and then maintain a stringent maintenance and testing regime.

REFERENCES

[1] Protection against corrosion by stray current from direct current systems. BS EN 50162:2004 – BSI (British Standards Institution), 2004

[2] Railway applications – Fixed installations – Electrical safety, earthing and the return circuit. Part 2: Provisions against the effects of stray currents caused by DC traction systems. BS EN 50122-2:2010 – BSI, 2010.

[3] Standard Practice for Determining Rail-to-Earth Resistance. American Society for Testing and Material (ASTM) Designation: G 165–99, 1999

[4] Electrolytic Corrosion in DC Powered Transit Systems, IIT Research Institute, Chicago, IL, 8 Reports Prepared for the National Cooperative Transit Research and Development Program (NCTRP) Project 48–1, TRB, National Research Council, Washington DC, 1985–1987.

[5] Moody, K.J., A cookbook for transit system stray current control. Presented at NACE Corrosion, Paper No. 14, New Orleans, LA. 1993.
[6] Memon, S. A., Understanding stray current mitigation, testing and maintenance on DC powered rail transit systems. *Proceedings of the 2013 Joint Rail Conference*, Knoxville, TN, USA, April 15–18, 2013.
[7] Fromme, P. & Memon, S.A., Use of rail boot and collection mat to control the electrolysis of rail and utilities in DC powered transit agencies. *Proceedings of the 2014 Joint Rail Conference*, Colorado Springs, CO, USA, April 2–4, 2014.
[8] Stratton, S.W., M'Collum, B. & Logan, K.H., Technology papers of the bureau of standards – leakage of currents from electric railways. Washington Government Printing Office. 1916.
[9] Cotton, I. & Charalambous, C., Influence of soil structures on corrosion performance of floating-DC transit systems. *IET Electric Power Applications*, **1(1)**, pp. 9–16, 2007. DOI: 10.1049/iet-epa:20050449.
[10] Memon, S.A. & Fromme, P., Stray current corrosion and mitigation in DC transit systems. *IEEE Electrification Magazine,* **2**, pp. 22–31, 2014. DOI: 10.1109/MELE.2014.2332366.
[11] Szeliga, M. J., *Stray Current Corrosion: The Past, Present and Future of Rail Transit Systems*, NACE International: Houston, TX. ISBN 1-877914-57-6, 1994.
[12] Tzeng, Y.S. & Lee, C.H., Analysis of rail potential and stray currents in a direct-current transit system. *IEEE Transactions on Power Delivery*, **25(3)**, pp. 1516–1525, 2010. DOI: 10.1109/TPWRD.2010.2040631.

CHARACTERISTICS OF CORRUGATION IN SHANGHAI METRO AND SUGGESTED MAINTENANCE THRESHOLD FOR AMPLITUDE

WANYI LIU[1], HAIFENG LI[1,2], YUDE XU[1], WANQING ZHANG[3], SIHAN YAN[1], FENGYU LI[1] & JUNXING QIU[1]

[1]Key Laboratory of Road and Traffic Engineering of the Ministry of Education, Tongji University, China.

[2]Department of Transportation Engineering, Tongji Zhejiang College, China.

[3]Vibration and Noise Research Department, Shanghai Marine Diesel Engine Research Institute, China.

ABSTRACT

To find out the relationship between the track structure parameters and rail corrugation type and put forward a reasonable maintenance threshold for the amplitude of rail corrugation, surveys on rail corrugation by rail static and dynamic tests were carried out on Shanghai Metro recently. Based on the in situ data, the wavelength and amplitude of rail corrugation were calculated and analysed by different categories according to their service conditions. Investigation data and its analysis show that different track structure parameters, such as ballast, fastener, curve radius, etc., can affect the amplitude and wavelength in pretty different ways. In order to know the development trend in rail corrugation, statistical significance on the distribution of amplitude and dynamic data is calculated, the distribution of amplitude at certain wavelengths is figured out, and accordingly a reasonable threshold of the amplitude maintenance is put forward.

Keywords: in situ tests, maintenance threshold, rail corrugation, subway, track structure.

1 INTRODUCTION

Rail corrugation is a normal phenomenon in the world but appears now to be substantially understood, and lots of researches have been done on its cause and treatments. Rail corrugation is quasi-sinusoidal irregularity where the wavelength is less than 1 m [1]. Grinding corrugated rail reduced the power spectrum of railhead roughness by about 2 orders of magnitude of wave numbers up to about 25 cycles m^{-1} and by about 1 order of magnitude of higher wave numbers [2]. According to Sato et al. [3], for five out of corrugation's six categories, not only characteristics but also causes and even its treatments are identified. According to their opinion for the sixth category (roaring rail-type corrugation), a wavelength-fixing mechanism yet had to be pursued. Torstensson [4] monitored rail roughness in a curve with radius 120 m in the Stockholm Metro and pointed out that roughness growth rate increased with the number of days after grinding until the measurement that was taken 300 days after grinding. Besides, corrugation does not consistently show uniformly increased magnitudes in the same sections of the curve as before grinding. Wang X. [5] did continue tests on corrugation's development of two different base plates, figuring out that the wavelengths of stiffness 90 MN/m and 60 MN/m are 80–100 mm and 160–185 mm, respectively. Cao [6] analysed the corrugation's characteristics of Shanghai Metro 1 and found out there existed huge differences between the two transition curves.

DOI: 10.2495/TDI-V1-N3-520-528

Rail corrugations can increase vibrations and noises and multiple the costs for maintenance. In order to find out the characteristics of corrugation with different track structures and put forward a reasonable criterion for amplitude controlling, some surveys are carried out. Based on experimental data, characteristics of corrugation with various structures are analysed and amplitude thresholds which can be applied on maintenance are offered.

2 SOURCE AND PRE-PROCESSING

In this paper, data of sections with severe corrugation and their track structures and line conditions are recorded. Rail surface irregularities are measured with RMF 2.3E, which can identify wavelength of 10 to 300 mm. This research covered 6 lines and 8 sections and some of them are measured continuously. Line conditions are shown in Table 1.

In dynamic in situ tests, stress and acceleration are measured for a train with different speeds. Measuring spots are shown in Fig. 1.

The measured rail corrugation data may contain some interference and some pre-processed works need to be done before performing a detailed analysis, which includes the following:

(a) Categorizing data based on working conditions: one working condition means these sections share the same track structure and similar curve radius.

(b) Eliminating trend term: due to error in the measure instrument and the influence of environment, rail irregularity shows trends of upward or downward movement. Raw data need to be fitted with least square method.

Table 1: Line conditions.

Line	1, 2, 3, 8, 9, 11
Ballast	Monolithic track bed, normal bearing block, ballast, long sleeper buried, floating slab
Fastener	Vanguard, DTIII-2, WJ-2A, elastic rod-II, absorber, LORD
Curve radius (m)	350, 400–450, 500, 700, 800, 1200

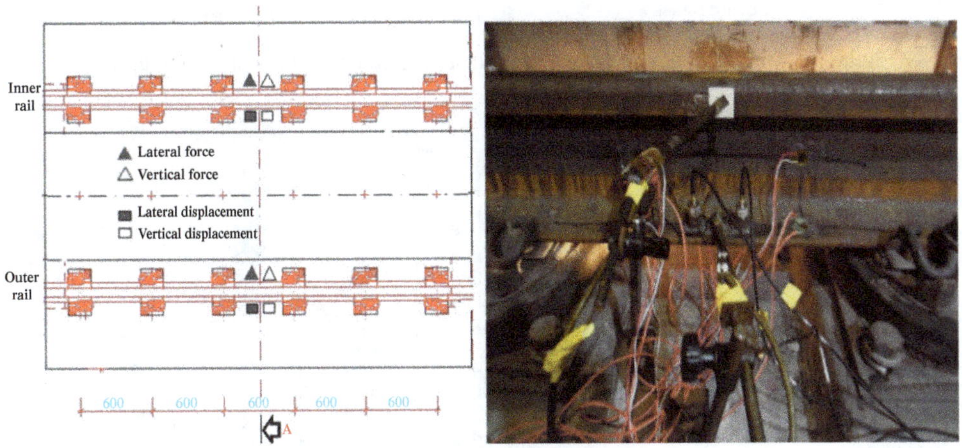

Figure 1: In situ tests.

(c) Deleting effects of rail joints: identifying where these rail joints are and then deleting data of a certain range.
(d) Deleting abnormal data.

Data after pre-processing can be used to analyse corrugation's distribution and relationship with different working conditions.

3 RELATIONSHIP BETWEEN CORRUGATION AND TRACK STRUCTURES

3.1 Relationship between corrugation and ballasts

As shown in Fig. 2(a), amplitude of corrugation where monolithic track bed is utilized is smaller than that where floating slab is used as ballast. And in Fig. 2(b) when elastic-II and curve with radius 400–450 m is applied, amplitude of long sleeper buried is smaller than normal bearing block and ballast. In other conditions amplitude of monolithic track bed is smaller than long sleeper buried. In short, when we consider corrugation mainly, no matter which kind of fasteners and radius is used, monolithic track bed's amplitude is relatively short while ballast's is long

Figure 3(a) and (b) shows influences of ballast on wavelength when using DTIII-2 or elastic-II as fastener. When fastener is DTIII-2, wavelength of floating is shorter than that of monolithic track bed. When fastener is elastic-II, wavelength of long sleeper buried is the shortest and the wavelength of ballast is pretty long. In brief, the wavelength of floating slab is relatively short and the wavelength of ballast or normal bearing block is pretty long.

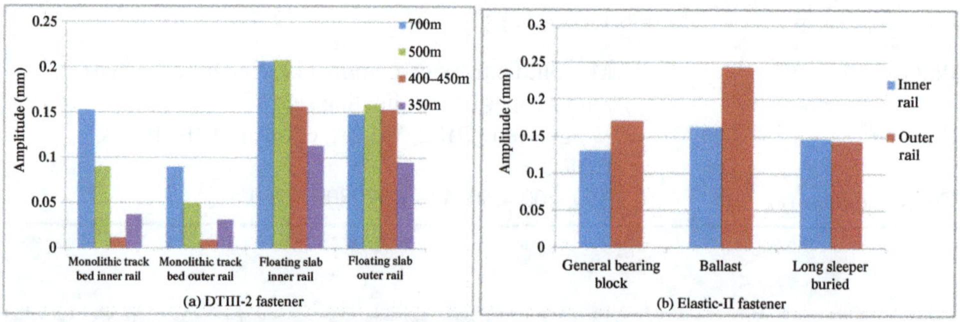

Figure 2: Amplitudes with different ballasts where certain fasteners are applied.

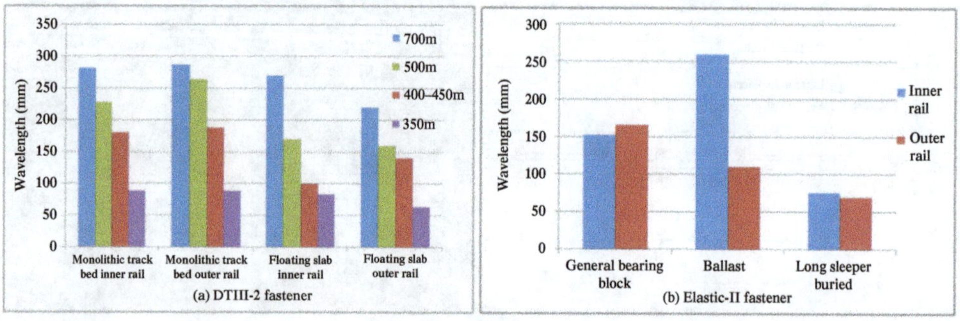

Figure 3: Wavelengths with different ballasts when certain fasteners are applied.

3.2 Relationship between corrugation and fasteners

Figure 4 shows what will happen to different fasteners' amplitude when we use (a) floating slab and (b) monolithic track bed as ballast. When ballast and curve radius are the same, the amplitude of vanguard or absorber is smaller than the amplitude of DTIII-2.

What Fig. 5 shows is the wavelength of different fasteners when we use (a) floating slab and (b) monolithic track bed. We can easily find that wavelength of vanguard is the longest and that of absorber is the shortest.

3.3 Relationship between corrugation and curve radius

Actually, curve radius influences amplitude in different ways under different working conditions. Figure 6 shows that amplitude reaches its minimum where the curve radius is about 400–450 m when (a) monolithic track bed and DTIII-2 or (b) absorber is applied.

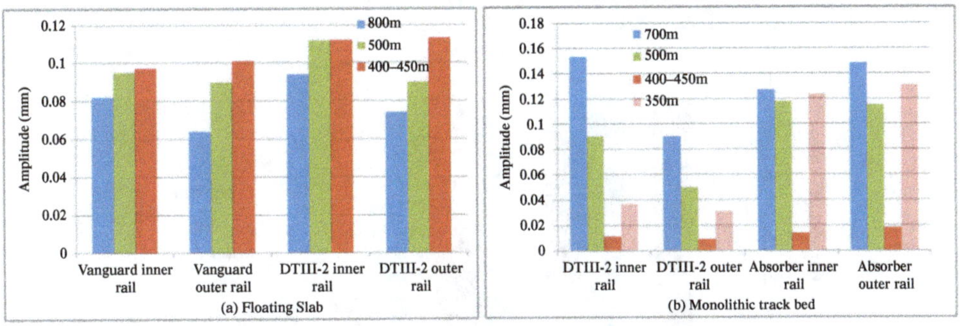

Figure 4: Amplitudes with different fasteners where certain ballasts are applied.

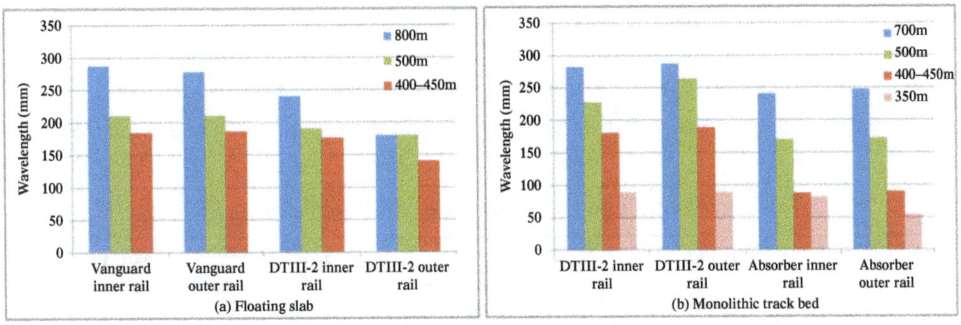

Figure 5: Wavelengths with different fasteners where certain ballasts are applied.

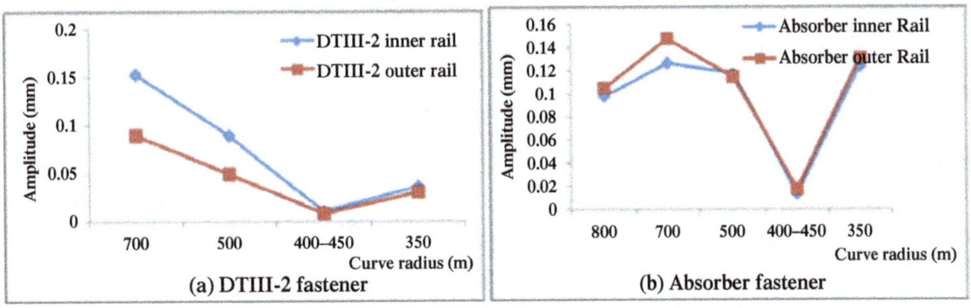

Figure 6: Amplitudes with different radius where monolithic track bed is applied.

Figure 7 points out that wavelength decreases with decreased curve radius. Actually, wavelength shares the same rule when other kinds of ballast are applied.

4 CORRUGATION CONTROLLING AND MAINTENANCE

4.1 Distribution of amplitude

Statistical analysis was conducted on the corrugation of Shanghai Metro 8 and 11, contributions of which are shown in Figs. 8 and 9 and Table 2.

According to Figs. 8 and 9 and Table 2, contribution of amplitude of different wavelengths is standard normal distribution. If we set triple standards as the threshold, most of the portion is below 3%, which is pretty reasonable for the operation.

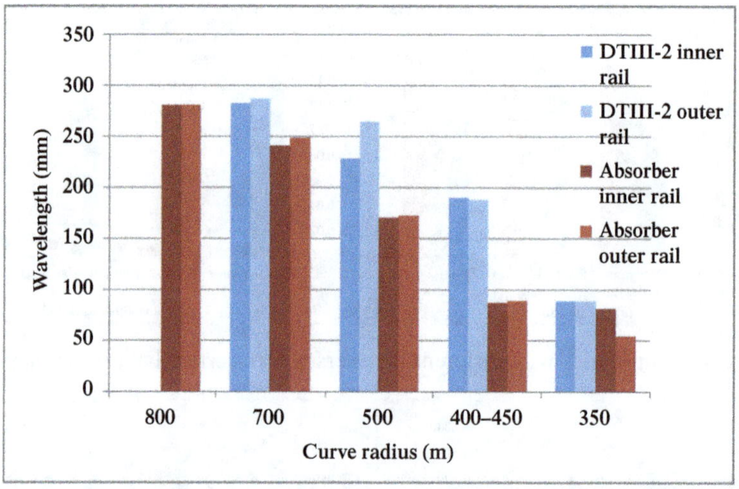

Figure 7: Wavelengths with different radius where monolithic track bed is applied.

Table 2: Contributions of amplitude of different wavelengths.

Main wavelength (mm)	0–100		100–200		200–250		250–300	
	Inner rail	Outer rail	Inner rail	Outer rail	Inner rail	Outer rail	Inner rail	Outer rail
Average (mm)	0.01	0.01	0.01	0.01	0.00	0.00	0.01	0.01
Triple standard (mm)	0.42	0.43	0.44	0.50	0.31	0.34	0.36	0.37
Range of amplitude (mm)	Portion of amplitude (%)							
<–0.4	0.31	0.44	0.5	1	0.04	0.09	0.34	0.15
(–0.4, –0.3)	1.19	1.05	1.8	3.05	0.67	1.23	0.99	1.21
(–0.3, 0.3)	96.08	96.44	94.86	92.81	98.92	97.57	96.58	96.88
(0.3, 0.4)	1.66	1.65	2.28	2.56	0.3	0.63	1.38	1.41
>0.4	0.76	0.42	0.57	0.58	0.08	0.48	0.72	0.36

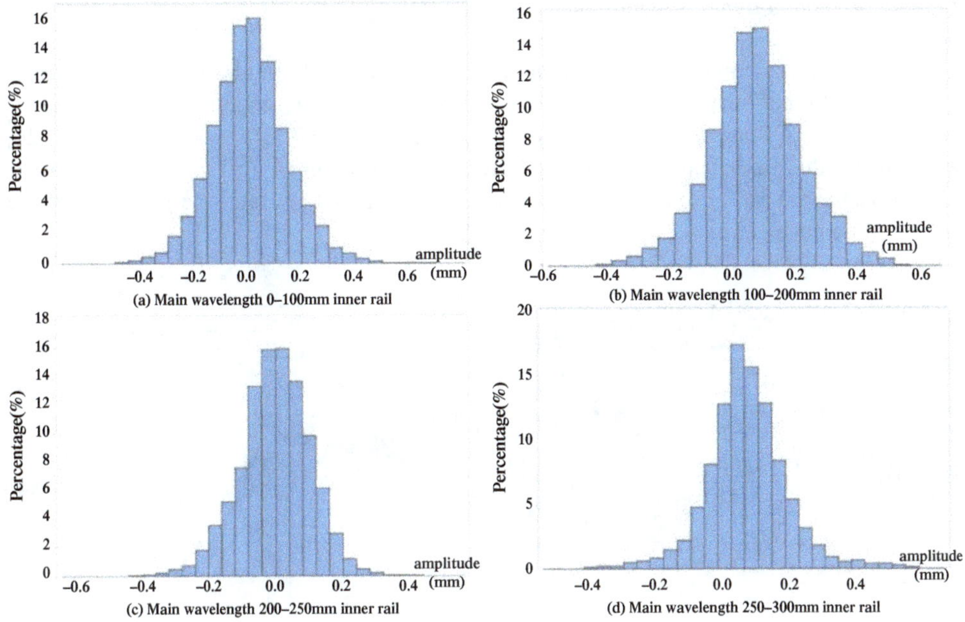

Figure 8: Distribution of amplitude of different wavelengths on inner rail.

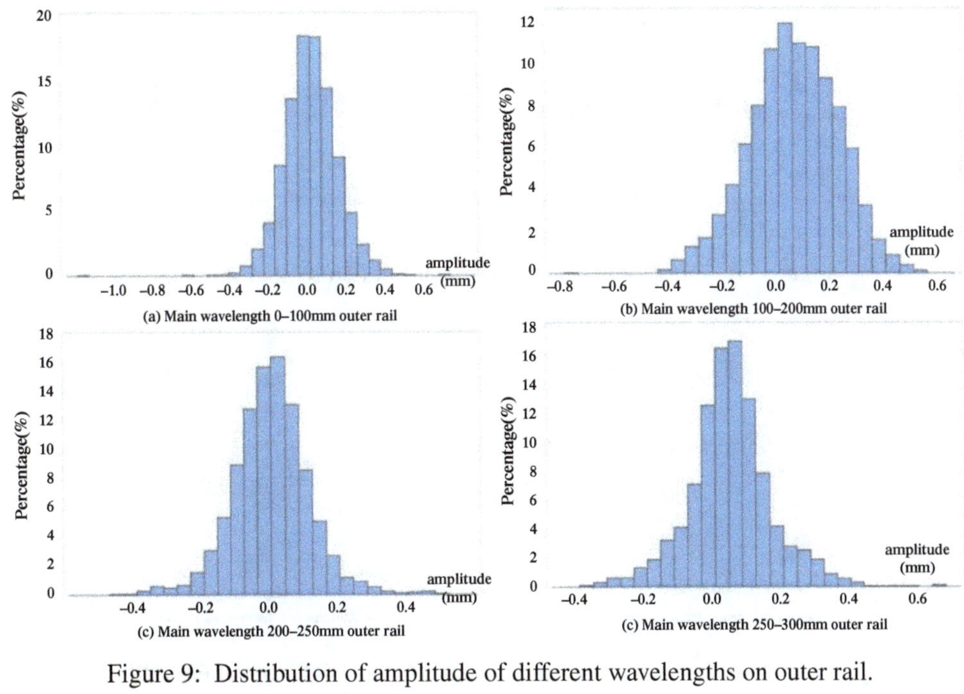

Figure 9: Distribution of amplitude of different wavelengths on outer rail.

4.2 Distribution of dynamic data

The effect of rail grinding is considered in this research (Fig. 10). Figure 11 shows the waveform graph of inner/outer rails before/after grinding on a curve at section Xujiahui to Shanghai Natatorium of Shanghai Metro 11. This curve lies at about XK35 + 494 – XK35 + 620, when monolithic track bed and absorber fasteners, with radius of 350 m, are used. Figure 11 demonstrates vertical forces before/after grinding in about 7 s after the train passed the measured point (Fig. 12).

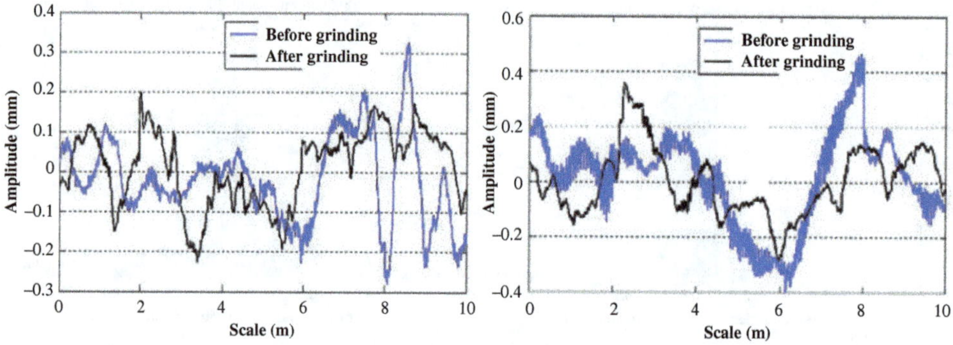

Figure 10: Waveform graph of inner/outer rail before/after grinding.

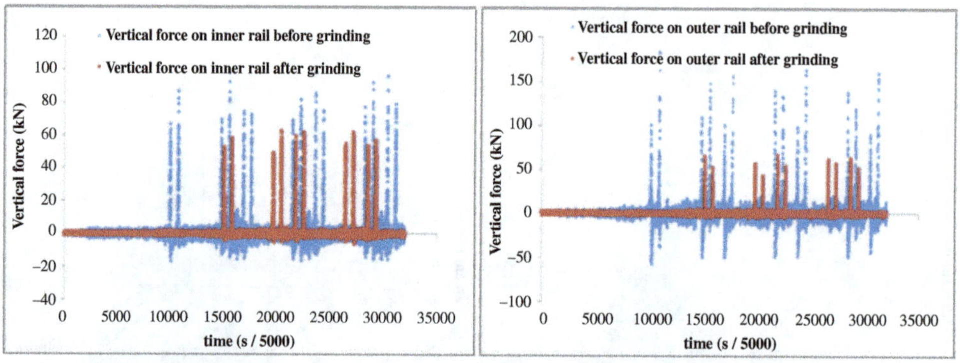

Figure 11: Vertical forces of inner/outer rail before/after grinding.

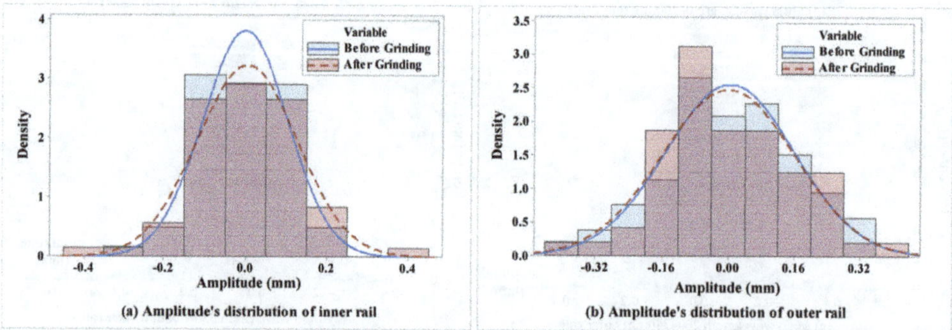

Figure 12: Distribution of amplitude of inner/outer rail before/after grinding.

As shown in Fig. 11, vertical forces decreased after grinding. In Table 3, the numbers of points amplitude of which is bigger than 0.3 mm or smaller than –0.3 mm decreased, while the number of points amplitude of which is in the range of –0.3 to 0.3 mm increased. Although there is no proof to show that that decreasing of amplitude is the only cause of decreased vertical forces, we can draw the conclusion that decreased amplitude is helpful for decreasing vertical force.

4.3 New threshold for corrugation amplitude

According to in situ data, after grinding, amplitudes that are smaller than –0.3 mm or bigger than 0.3 mm disappeared a lot, which is pretty similar to the triple standard in the contributions of amplitude of different wavelengths. Existed threshold is pretty severe and may increase maintenance work. Here we suggest using the conservative values of triple standard as shown in Table 4, which haven't been verified by applying them on maintenance now.

Table 3: Amplitude contribution before/after grinding.

	Before grinding		After grinding	
	Inner rail	Outer rail	Inner rail	Outer rail
Average (mm)	–0.001	0.006	0.005	0.002
Triple standard (mm)	0.314	0.468	0.371	0.482
	Number of points amplitude			
<–0.4 mm	0	0	1	0
(–0.4, –0.3) mm	0	3	0	2
(–0.3, –0.2) mm	3	4	2	2
(–0.2, –0.1) mm	7	9	5	15
(–0.1, 0) mm	22	16	28	12
(0, 0.1) mm	17	10	19	13
(0.1, 0.2) mm	12	16	15	8
(0.2, 0.3) mm	1	6	1	6
(0.3, 0.4) mm	0	2	0	1
>0.4 mm	0	0	1	1

Table 4: Suggested values for amplitude threshold of different wavelengths.

Main wavelength (mm)	0–100	100–200	200–250	250–300
Triple standard (mm)	0.43	0.46	0.32	0.36
Suggested value (mm)	±0.4	±0.4	±0.3	±0.3
Overrun portion (%)	1.07	1.07	1.09	3.43

5 CONCLUSIONS

By analysing rail corrugation data and dynamic in situ test data and plotting the contribution of them, we arrived at the following conclusions:

1. Ballast/fastener/curve radius can affect the amplitude and wavelength of corrugation. When preventing corrugation is the primary purpose, our suggestion is using monolithic track bed and floating slab as the ballast and vanguard as the fastener, and the curve radius can be about 400–450 m.
2. The distribution of amplitude at certain wavelengths is approximately normal standard distribution. Using triple standard as the threshold for corrugation amplitude is reasonable.

ACKNOWLEDGEMENTS

This paper has received the financial aid from International Exchange Program for Graduate Students, Tongji University.

REFERENCES

[1] Grassie, S.L. & Kalousek J., Rail corrugation: characteristics, causes, and treatments. *Proceedings of the Institution of Mechanical Engineers Part F: Journal of Rail and Rapid Transit*, **207(16)**, pp. 57–68, 1993.

[2] Grassie, S.L., Short wavelength rail corrugation: field trials and measuring technology. *Wear*, **191**, pp. 149–160, 1996.

[3] Sato, Y., Matsumoto, A., & Knothe, K., Review on rail corrugation studies. *Wear*, **253**, pp. 130–139, 2002.

[4] Torstensson, P.T. & Nielsen, J.C.O., Monitoring of rail corrugation growth due to irregular wear on a railway metro curve. *Wear*, **267(1–4)**, pp. 556–561, 2010.

[5] Wang, G., Chen, G., & Wu, P., Influence of sleeper support stiffness and damping on wear-type rail corrugation on a tight curve (in Chinese). *Journal of Vibration and Shock*, **30(2)**, pp. 99–103, 2011.

[6] Cao, L., Xu, Y., Zhou, Y. & Zhan, G., Characteristics of passenger evacuation in urban mass transit (in Chinese). *Urban Mass Transit*, **13(2)**, pp. 46–46, 2010.

PRELIMINARY RESULTS AND CONCLUSIONS FROM THE EXPERIMENTAL MONITORING OF THERMAL REGIME OF RAILWAY TRACK STRUCTURE

L. IŽVOLT, P. DOBEŠ & M. PITOŇÁK
Department of Railway Engineering and Track Management,
Faculty of Civil Engineering, University of Žilina, Slovakia.

ABSTRACT
The article describes the preliminary results and conclusions of experimental monitoring of the thermal regime of railway track structure performed on the experimental stands (railway track models in 1:1). The purpose of this study was to optimize the structural thickness of the protective layer in the subgrade structure. The first part of the study describes the experimental stand design and methodology for monitoring the heat and humidity variations in the railway track structure. The following parts describe the results of verification of the thermal characteristics of the material incorporated into the model of the railway track and of its subgrade.
Keywords: climatic factors, experimental monitoring of thermal regime, railway track, subgrade dimensioning.

1 INTRODUCTION

The quality and safety of railway tracks are clearly dependent not only on the technical level and state of train sets but also on the state and quality of the railway superstructure and substructure. They compose one technical–physical unit, affecting the required operational capability and railway track safety. Due to this, the Department of Railway Engineering and Track Management (DRETM) of the University of Žilina has been paying long-term attention to the problem of railway substructure reliability under the influence of various factors, with specific focus on climatic factors [1, 2]. The course of winter periods, particularly over the past 20–30 years, shows different trends than in the past. As a result, there is a necessity to update the input design parameters of subgrade structure dimensioning, or possibly to amend the dimensioning methodology stated in [3] as it reflects the research results from the 1960s.

In the Slovak Republic, as well as in neighbouring countries, the weather characteristics of the winter period considerably influence the dimensions of the subgrade structure. Because of this, it is necessary to identify and include the influence of changing climatic conditions on the railway substructure into the design methodology of subgrade structure dimensioning for non-traffic loads (influence of climatic factors). Concerning foreign research, in the past, this problem was analysed in Germany, Norway, Japan and Canada. At present, the most commonly used research output is from UIC ORE research, which is carried out under the auspices of Trondheim University. This research took into account the climatic conditions of the 1960s and the 1970s. The importance of studying the problem of the thermal regime of railway track structure (subgrade) can be underlined by the fact that new construction

DOI: 10.2495/TDI-V1-N3-529-539

materials and structural elements are applied within the rail track modernisation, and as it was stressed before, the climatic conditions for present and future railway track operation have been changed.

2 EXPERIMENTAL MONITORING OF THERMAL REGIME AND CHANGES IN HUMIDITY OF THE RAILWAY TRACK STRUCTURE

The problem of monitoring the thermal regime of railway track structure has been solved at the DRETM, Faculty of Civil Engineering of the University of Žilina (FCE, ŽU - UNIZA) for at least two decades. At present, the research also focuses on verification of thermal conductivity coefficients, λ, and variations in humidity of various materials applied in the subgrade structure and on monitoring the influence of winter period courses (the number of frost periods and size of the frost index) and the average annual temperature on the depth of subgrade freezing. This study is carried out using experimental stands of railway track models and software support for mathematical modelling of the thermal regime and variations in humidity of railway track structure. The following section of the article will briefly describe the experimental basis that DRETM uses for monitoring the thermal regime and variations in humidity at the experimental stands with railway track models.

2.1 Experimental stand DRETM I

The experimental measurements at the railway track model in the scale 1:1 were started as early as in 2003. The structure of the railway track model, the so-called experimental stand DRETM I (see Fig. 1), was built as a track section run on terrain level. The railway superstructure consists of rail skeleton and ballast bed (BB) fraction 31.5/63 mm of total thickness 500 mm. The railway substructure composes the subgrade with in-built MACRIT GTV/50 geocomposite, where the protective layer of subgrade surface is formed by 450 mm thick layer of crushed aggregate fraction 0/31.5 mm and its soil material is sandy clay. The thickness of the protective layer was dimensioned in compliance with the then in force regulation [4], where the design value of air frost index for Žilina was I_F = 550°C.day. To monitor the temperatures in the individual structural layers of this experimental stand, we installed 11 resistance thermometers, Pt100, whose arrangement can be observed in Fig. 1. An important piece of input data in the process of monitoring the effects of non-traffic loads is the air temperature that is being measured 2.00 m above the level of the surrounding terrain and serves as a determination of the frost index value, I_F.

We are aware of the fact that snow decreases the effects of frost on a subgrade and acts as an insulation layer. To monitor its effects, we removed the snow cover completely during winter or, conversely, the snow cover of maximum height approx. 0.12 m, which takes into account the real operational conditions, is kept.

At this experimental stand during each winter period, the following characteristics were monitored and determined:

- mean daily temperature, θ_s,
- minimal temperature in the individual structural layers of the stand, θ_{min},
- depth of structural freezing in the given period, D_F,
- air frost index, I_F,
- mean daily surface temperature, θ_{ss},
- frost index at the track bed surface, I_{FS}.

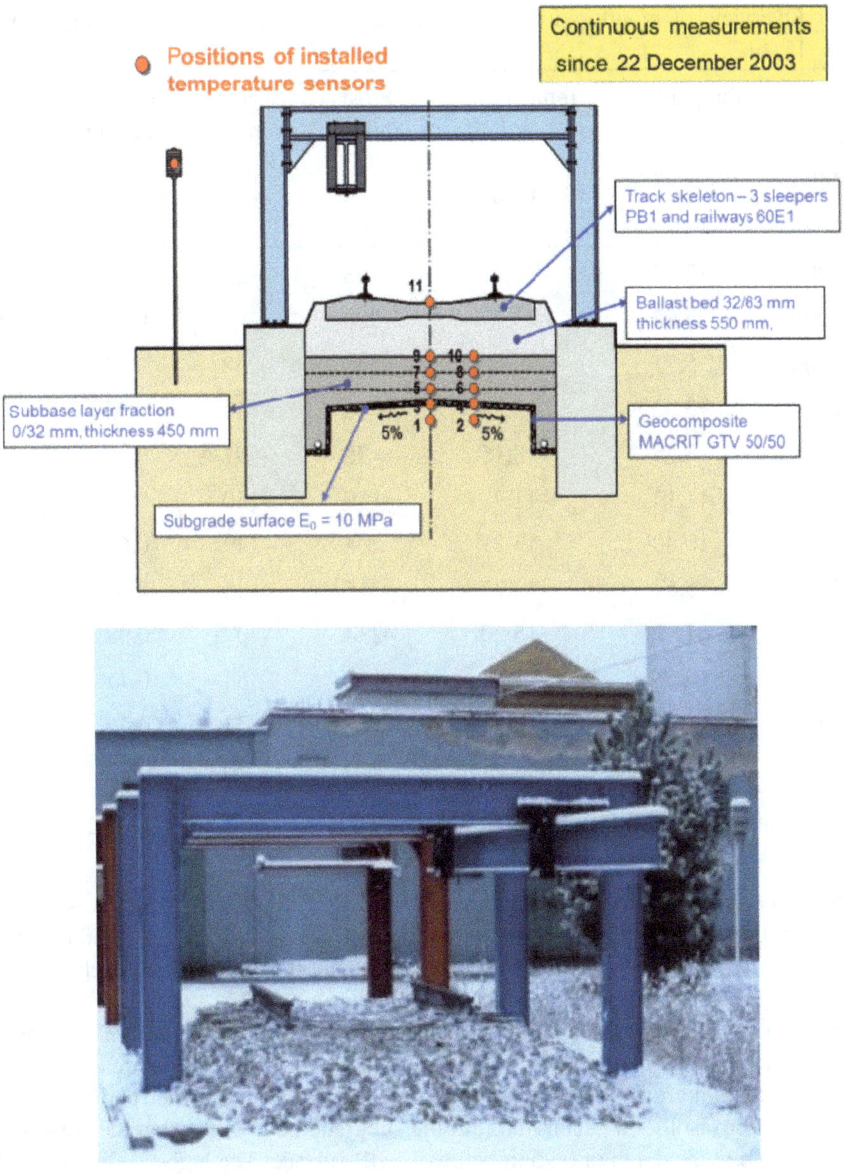

Figure 1: Experimental stand DRETM I.

This experimental stand, besides being used for monitoring the development and course of the subgrade thermal regime during individual winter periods, also served for monitoring the variations in deformation resistance. This was facilitated by a steel frame with a mobile beam and set for measuring the static and dynamic modulus of deformation.

The courses of 12 winter periods were recorded from 2003 to 2015. Their monitored characteristics are stated in Table 1. The winter period, the above-mentioned parameters were stated for, is limited by the first and last frost period ($\theta_s \leq -0.1$°C), while the frost period must last for at least 4 days.

Table 1: Overview of monitored climatic quantities at the experimental stand DRETM I [5, 6].

Winter period	Temperature, $\theta_{s,max}$ (°C)	Temperature, $\theta_{s,min}$ (°C)	Frost index, I_F (°C.day)	Surface frost index, I_{FS} (°C.day)	Depth of freezing, D_F (m)	I_F for $D_{F,max}$ (°C.day)
2003/2004[a]	13.19	−12.31	−168	− 58	0.46	− 29
2004/2005	7.97	−13.37	−228	−143	0.70	−165
2005/2006[a]	5.71	−16.68	−388	−248	0.82	−334
2006/2007	13.02	− 5.26	− 16	− 14	0.33	− 9
2007/2008	8.55	− 8.15	− 96	− 91	0.48	− 20
2008/2009	11.13	−11.78	−150	−130	0.58	−145
2009/2010	10.40	−13.27	−204	−149	0.60	−145
2010/2011	5.67	−12.27	−215	−145	0.61	−215
2011/2012	4.93	−15.20	−238	−207	0.98	−184
2012/2013[a]	10.68	−11.45	−162	− 54	0.44	−107
2013/2014	11.21	−10.50	− 27	− 33	0.46	− 23
2014/2015	9.14	− 8.95	− 40	− 20	0.34	− 15

[a]Parameter values acquired for snow cover of maximum thickness approx. 0.12 m.

Table 1 clearly shows that the coldest winter period in the last 12 years was the winter period of 2005/2006, when the air frost index reached the value I_F = −388°C.day, although the subgrade freezing depth in this winter period was not the lowest, as it reached the value D_F = 0.82 m (freezing of subgrade surface did not occur).

From the point of view of subgrade freezing, the winter of 2011/2012 can be considered the most unfavourable winter period, when the depth of structural freezing reached the value 0.98 m for a maximum frost index of, I_F, only −238°C.day. This winter period is the only one when the zero isotherm penetrated into the subgrade below the subgrade surface level, which is located at the depth of 0.95 m below the BB surface. The higher value of the depth of subgrade freezing at the lower frost index compared to the winter period 2005/2006 is a consequence of snow removal from the BB surface during the entire winter period.

The most favourable winter period, not only from the point of view of subgrade freezing but also due to the reached air frost index I_F, was the winter period of 2006/2007, when the subgrade freezing depth reached the value of only 0.33 m and the maximum air frost index −16°C.day, which means that there was no freezing of the entire BB structural thickness.

For individual dimensioning of subgrade structure (the thickness of protective layer of subgrade surface), the repeatability of winter periods is important. If the winter period of 2003/2004 is compared to the winter period of 2012/2013, or the winter period of 2009/2010 to the winter period of 2010/2011, it can be ascertained that the differences in frost indices (on the BB surface, I_{FS}) and depths of subgrade freezing, D_F, are only minimal. Table 1 also clearly shows that the air frost index, I_F, is not the only parameter affecting the depth of subgrade freezing, D_F. The depth of subgrade freezing, D_F, is, besides the occurrence of snow cover on the rail track surface, also influenced by the average annual air temperature, θ_m (the amount of heat accumulated in the subgrade), and – certainly – by the humidity of subgrade materials before the first frost period.

2.2 Experimental stand DRETM II

In order to consider the results from the Experimental stand DRETM I and the efforts to load the subgrade structure by the maximum effects of frost, an experimental stand DRETM II was designed and constructed in 2012. The concept and design of the experimental stand was based on the long-term experience gained from the experimental measurements of the subgrade thermal regime at the experimental stand DRETM I. As at the time of construction of the experimental stand DRETM II the software for mathematical modelling of the subgrade thermal regime was available, besides higher number of thermal sensors, humidity sensors were also embedded to acquire one of the input parameters for mathematical modelling of the thermal regime using *SoilVision* software (see [7]).

The experimental stand DRETM II represents the railway track model run on the embankment in the scale 1:1 – Figs. 2 and 3. The superstructure of this model consists of the track skeleton and BB fraction 31.5/63 mm of overall thickness 500 mm. From the structural point of view, this experimental stand is divided into two structural parts. The first part (half) of the experimental stand (measuring profile 1) is the entire earthwork at the height of approximately 1.00 m built of crushed aggregate fraction 0/63 mm (reclaimed rail ballast) and its subgrade is formed by clay with an addition of river gravel (identified by the geotechnical investigation). The second part (half) of the experimental stand (measuring profile 2) consists of a protective (subbase) layer of crushed aggregate fraction 0/31.5 mm at the thickness of 450 mm, placed on the subgrade surface. The earthwork, 550 mm high, is built of crushed aggregate, a fraction 0/63 mm and foundation soil also of clay with an addition of river gravel. The thickness of the protective layer was equal to the experimental stand DRETM I dimensioned in accordance with the regulation [4], to enable the comparison of structures

Figure 2: In the foreground: snow-covered experimental stand DRETM II on the Žilina university campus.

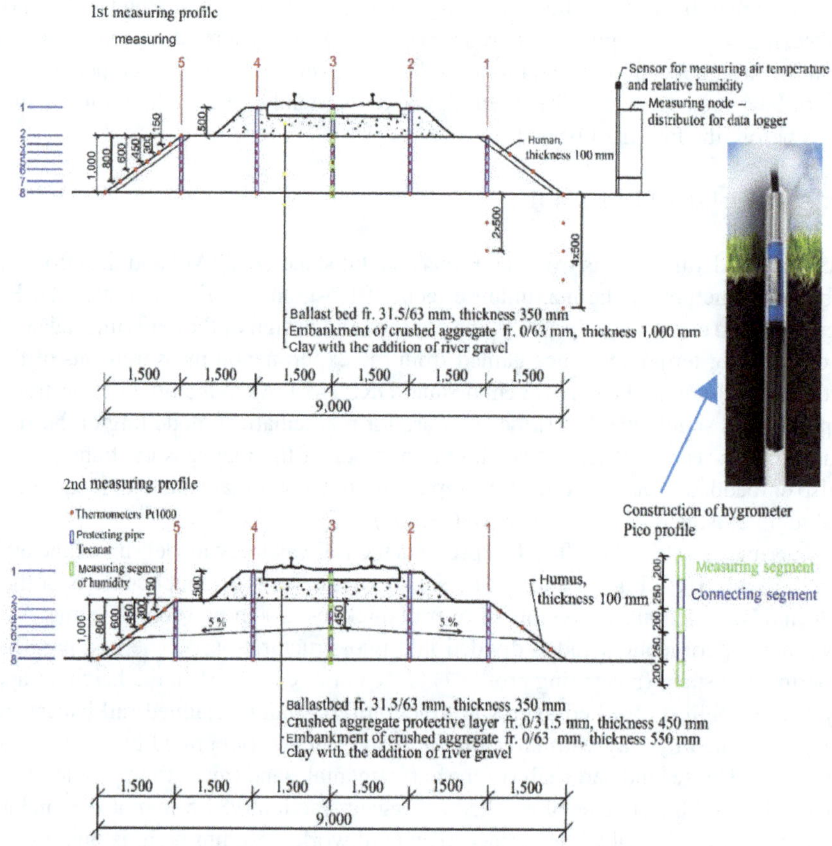

Figure 3: Cross sections of the experimental stand DRETM II.

from the point of view of thermal regime and the position of zero isotherm. The slope treatment was finished with topsoil application at the thickness of 0.10 m and the slopes were grassed over.

To monitor the thermal regime in the experimental stand DRETM II, 100 resistive temperature sensors, type Pt1000 (measuring range from –50°C to +400°C), were embedded. The temperature sensors were placed in each measuring profile on eight horizontal levels, in a mutual distance of 150 mm (the distance of the last two levels is 200 mm) and in five rows, while some thermal sensors were also placed on the surface of both embankment slopes, or below the level of the original terrain.

Besides the thermal sensors, the experimental stand also featured protective insulated pipes for humidity sensors of the type *PICO-T3P* (Fig. 3), enabling non-destructive humidity measurement in the range from –15°C to +50°C with a precision of ±2% [8]. To prevent damage to the probe tube in contact with sharp material, we placed the probe tube into a special protective pipe *TECANAT* at a diameter of 40 mm. The placement of the *TECANAT* pipes of humidity sensors is similar to the thermal sensors, thus in five rows, while the thermometers were placed outside the pipes. The structure of the experimental stand included 11 protective pipes (5 in each measuring profile and 1 pipe placed in the foot of the embankment to measure the humidity of the embankment subbase). The arrangement and position of individual thermal and humidity sensors in the experimental stand DRETM II is shown in Fig. 3.

In the vicinity of the experimental stand the sensor *Comet T3419* was installed, which served for acquiring the data about the air temperature and humidity in the height of 2.00 m above the terrain. The air temperatures can be recorded in the range of –30°C to +105°C with a precision of ±0.4°C and the air humidity in the range of 0–100%, with a precision of ±2.5%.

As experimental stand DRETM II serves the same purpose as the previous structure (experimental stand DRETM I), identical parameters, stated in paper, are monitored. In the case of this experimental stand, there was also added the parameter of inbuilt material humidity. This parameter, which substantially influences the thermal regime of the rail track structure and also the depth of subgrade freezing, is necessary for mathematical modelling of the subgrade thermal regime using the *SoilVision* software (closely characterized in [7]).

The experimental measurements of the thermal regime and variations in humidity in the rail track structure model and its subgrade at the experimental stand DRETM II have been carried out since 2013. That is why only two winter periods have been evaluated, winter of 2013/2014 and winter of 2014/2015. The required thermal characteristics of these two winter periods are determined in the same way, as with the experimental stand DRETM I. Tables 2 and 3 show the measurement results that were acquired in the case of no snow cover on the surface of the experimental stand during all the winter periods.

Equally as with the experimental stand DRETM I, the winter period, the above-mentioned parameters were stated for, is limited by the first and last frost period ($T_s \leq -0.1$°C), while the frost period must last for at least four days and the winter period can consist of several frost periods. Their monitored characteristics are stated in Table 1.

Table 2: Parameters acquired by evaluating of the monitored quantities at the experimental stand DRETM II – 1st measuring profile without the protective layer.

Winter period	Temperature, $\theta_{s,max}$ (°C)	Temperature, $\theta_{s,min}$ (°C)	Frost index, I_F (°C.day)	Surface frost, index I_{FS} (°C.day)	Depth of freezing, D_F (m)	I_F for $D_{F,max}$ (°C.day)
2013/2014	10.45	–11.70	–38	–22	0.43	–32
2014/2015	8.50	–10.80	–77	–32	0.42	–75

Table 3: Parameters acquired by evaluating of the monitored quantities at the experimental stand DRETM II – 2nd measuring profile with a protective layer.

Winter period	Temperature, $\theta_{s,max}$ (°C)	Temperature, $\theta_{s,min}$ (°C)	Frost index, I_F (°C.day)	Surface frost index, I_{FS} (°C.day)	Depth of freezing, D_F (m)	I_F *for* $D_{F,max}$ (°C.day)
2013/2014	10.45	–11.70	–38	–22	0.42	–32
2014/2015	8.50	–10.80	–77	–32	0.41	–75

As manifested in Tables 2 and 3, the so far acquired measurement results at the experimental stand DRETM II from two latest winter periods confirm that not only the air frost index, I_F, influences the depth of subgrade freezing, D_F, but it is also influenced by the course and intensity of the frost period, the air frost index was based on reaching approximately the same depth of subgrade freezing, D_F, for double value of air frost index I_F. The comparison of individual set parameters for both monitored and evaluated winter periods is given in Figs. 4 and 5 show the course of daily mean air temperatures (at the ballast bed surface) and subsequently the frost indexes, I_F, (of the ballast bed surface I_{FS}) for winter periods of 2013/2014 and 2014/2015 are determined. Figures 7 and 8 show the course and the position of the zero isotherm in the subgrade for both winter periods.

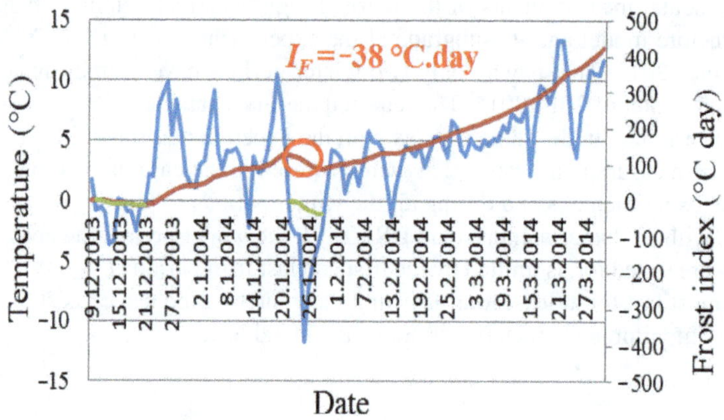

Figure 4: The stated value of the air frost index I_F – winter 2013/2014.

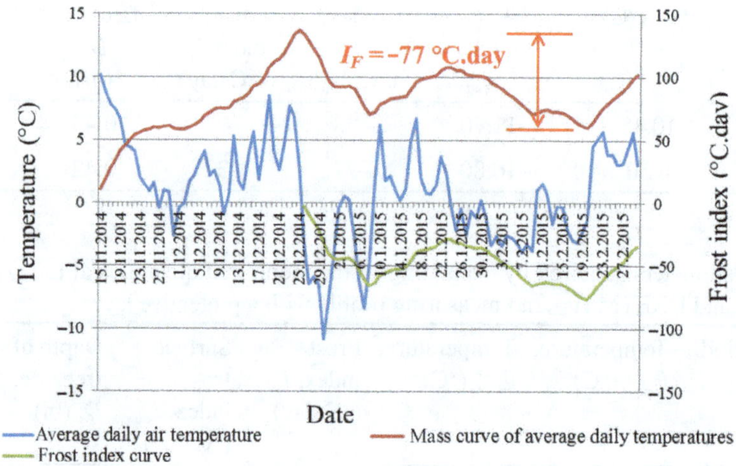

Figure 5: The stated value of the air frost index, I_F – winter 2014/2015.

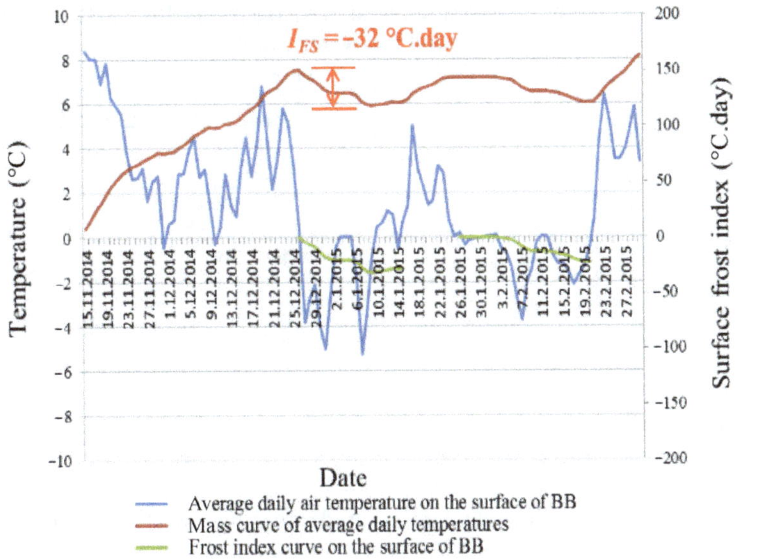

Figure 6: The stated value of the surface frost index, I_{FS} – winter 2014/2015.

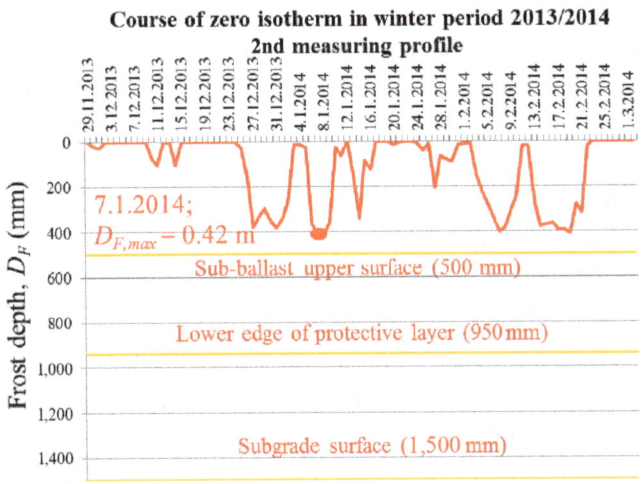

Figure 7: The stated value of the depth of subgrade freezing – winter period of 2013/2014.

Also the results of experimental measurements of two evaluated winter periods at the experimental stand DRETM II provide the same conclusions, as mentioned in relation to the DRETM I, that the depth of subgrade freezing is not only influenced by the frost index, I_F, but also by other factors, stated in the final section of the article.

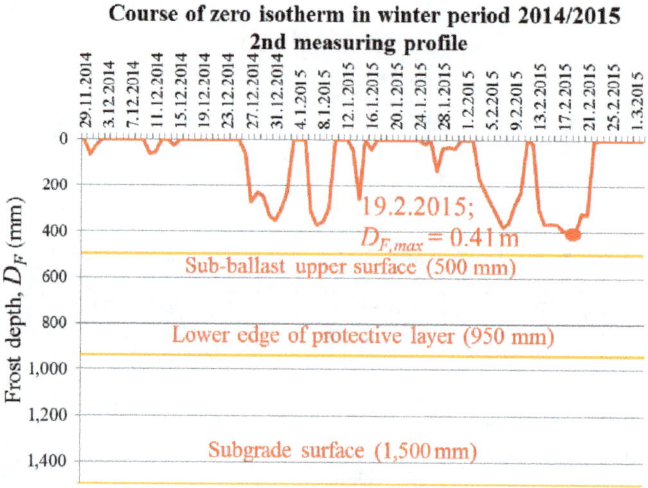

Figure 8: The stated value of the depth of subgrade freezing – winter period of 2014/2015.

3 PRELIMINARY RESULTS FROM THE EXPERIMENTAL MONITORING OF THERMAL REGIME OF RAILWAY TRACK STRUCTURE

The experimental measurements as well as the mathematical modelling of the thermal regime of the rail track (subgrade) confirmed that the air frost index is not the only parameter that influences the depth of rail track freezing. In the design methodology of the determination of structural thickness of the protective layer, it is also necessary to consider other characteristics, namely:

- length and number of frost periods,
- course of a dominant frost period during the winter,
- occurrence and thickness of snow cover,
- temperature of individual structural layers of railway track and its subbase in the beginning of the winter period, i.e. before the first frost period,
- humidity of materials of individual structural layers of the railway track, or the subbase in the case of a railway track run on an embankment, in the beginning of the winter period,
- average annual air temperature.

The measurement of the humidity of materials in the individual structural layers of rail track (subgrade) was justified and it is necessary to consider the determined range of actual measured humidity in the dimensioning methodology of subgrade structure for non-traffic load.

ACKNOWLEDGEMENT

The presented results and conclusions are the results of solving the VEGA 1/0275/16 Design optimization of sleeper subgrade due to non-traffic load aspect [9].

REFERENCES

[1] VEGA 1/0756/12, Experimental monitoring and mathematical modelling of thermal regime of railway subgrade structure. Scientific Research of Department of Railway Engineering, Faculty of Civil Engineering, University of Žilina, Žilina, available at http://svf.uniza.sk/kzsth. VEGA – Ministry of Education, Science & Research of the Slovak Republic, available at http://www.minedu.sk, Bratislava, SK, 2012–2015. Accessed 9 November 2015. Accessed: November 9, 2015.

[2] Ižvolt, L., Dobeš, P. & Pitoňák, M., Some experience and preliminary conclusions from the experimental monitoring of the temperature regime of subgrade structure. Proceedings of the Computer in Railways XIV – Railway Engineering Design and Optimization, WIT Press, available at http://www.witpress.com, *WIT Transactions on the Built Environment*, **135,** pp. 267–278, 2014. DOI: 10.2495/CR140211.

[3] TNŽ 73 6312, The design of structural layers of subgrade structures, GR ŽSR, Slovakia, 2005.

[4] S4 Railway substructure, Regulations of Slovak Railways ŽSR, 1999.

[5] Kupčuliak, P., Design and evaluation of railway subgrade structure of aspect of climatic load factors, Doctoral dissertation, Department of Railway Engineering and Track Management, Faculty of Civil Engineering, University of Žilina, 142 p., 2011.

[6] Dobeš, P., Optimization of the subgrade design for non-traffic load, Doctoral dissertation, Department of Railway Engineering and Track Management, Faculty of Civil Engineering, University of Žilina, available at http://svf.uniza.sk/kzsth, 136 p., 2015. Accessed: October 12, 2015.

[7] Hodas, S., Ižvol, L. & Dobeš, P., Preliminary results and conclusions from mathematical modelling of thermal regime of railway track structure. Proceedings of the COMPRAIL 2016 – 15th International Conference on Railway Engineering Design & Operation, Madrid, Spain, WESSEX Institute of Technology, available at http://www.wessex.ac.uk, WIT Press, available at http://www.witpress.com, Southampton, UK, 12p., 2016. Accessed: April 18, 2016.

[8] Trime Pico-profile Manual, available at http://www.imko.de/en/products/soilmoisture/soil-moisture-sensors/trimepicoipht3. Accessed: December 1, 2015.

[9] VEGA 1/0275/16, Design optimization of sleeper subgrade due to non-traffic load aspect. Scientific Research of Department of Railway Engineering, Faculty of Civil Engineering, University of Žilina, Žilina, available at http://svf.uniza.sk/kzsth. VEGA – Ministry of Education, Science & Research of the Slovak Republic, available at http://www.minedu.sk, Bratislava, SK, 2016–2019. Accessed: February 23, 2016.

AN EFFICIENT CONTACT MODEL FOR DYNAMIC INTERACTION ANALYSIS OF HIGH-SPEED TRAIN AND RAILWAY STRUCTURE INCLUDING DERAILMENT DURING AN EARTHQUAKE

M. TANABE[1], K. GOTO[2], T. WATANABE[2], M. SOGABE[2], H. WAKUI[2] & Y. TANABE[3]
[1]Kanagawa Institute of Technology, Japan.
[2]Railway Technical Research Institute, Japan.
[3]Laboratory for Computational Mechanics Inc., Japan.

ABSTRACT

A computational model to solve for the dynamic interaction of a high-speed train and railway structure including derailment during an earthquake is given. An efficient mechanical model to express contact–impact behaviours between wheel and rail before derailment, between wheel and the track structure after derailment, and between the car body and railway structure during and after derailment is presented. The motion of the train with nonlinear springs and dampers is modelled in multibody dynamics. The railway structure is modelled with various finite elements. The combined dynamic response of the train and railway structure during an earthquake is obtained by solving the nonlinear equations of motions of the train and railway structure in the modal coordinates subjected to the interaction between the train and railway structure. Numerical examples of a high-speed train running on the railway structure after derailment during an earthquake are demonstrated.

Keywords: contact, derailment, dynamic interaction, FEM, multibody dynamics, post derailment, railway structure, train.

1 INTRODUCTION

When a high-speed train runs on the railway structure during an earthquake, there is a radical dynamic interaction between the train and railway structure. The huge impact force between the wheel and rail may cause lifting of the wheel, derailment and post-derailment behaviours such as the contact of wheel on the track structure, and also the contact of the car body of the train on the railway structure. It is very important to solve the dynamic interaction between the train and the railway structure including derailment and post-derailment behaviours to build an earthquake-safe railway system.

Regarding the wheel–rail contact, Pombo *et al.* [1] presented a generic contact detection formulation that allows the lead and lag flange contact in the rail direction. Ju [2] studied the method to express the vertical stiffness between the wheel and rail as a power function of changing contact force and developed a nonlinear moving wheel element to solve the contact problem of a moving train. Montenegro *et al.* [3] presented an algorithm to express the nonlinear vehicle–structure interaction by constraint equations and to solve them by Lagrange multipliers based on the Newton method due to the nonlinearity.

Regarding the derailment behaviour in the dynamic interaction between the train and the railway structure during an earthquake, Xiao *et al.* [4] studied the derailment criteria based on

DOI: 10.2495/TDI-V1-N3-540-551

the derailment coefficient that is defined as the ratio of the lateral force to the vertical force of wheel to predict the derailment and evaluate the vehicle running safety for the various operation speeds of the train and earthquake motions. Sumami *et al.* [5] studied the way to solve the impact behaviour between the derailed wheel and sleeper based on a wheel–sleeper impact model and investigated the derailed vehicle motion.

However, very little work related to the computational method to solve the dynaimic interaction including derailment and post-derailment behaviours during an earthquake has been reported so far.

In this article, a computational model to solve for the dynamic interaction of a high-speed train and railway structure including derailment and post–derailment during an earthquake is given. An efficient mechanical model to express contact–impact behaviours between wheel and rail before derailment, between wheel and railway structure after derailment and between the car body of the train and railway structure during and after derailment is presented.

The motion of the train with nonlinear springs and dampers is modelled in multibody dynamics. The railway structure is modelled with various finite elements. The modal reduction is applied to equations of motion of the train and railway structure. The combined dynamic response of the train and railway structure during an earthquake is obtained by solving nonlinear equations of motions of the train and railway structure in the modal coordinates subjected to the interaction between wheel and rail before derailment and between the train and railway structure during and after derailment.

Numerical examples of a high-speed train running at high speed on the railway structure after derailment during an earthquake are demonstrated.

2 MECHANICAL MODEL OF A HIGH-SPEED TRAIN AND RAILWAY STRUCTURE

2.1 Mechanical model of a Shinkansen train

A Shinkansen train (high-speed train in Japan) is modelled in multibody dynamics with rigid components of car body, truck and wheel-set connected by nonlinear springs and dampers as shown in Fig. 1. Assuming that the train runs at a constant speed, the equation of 3D motion of the train with n cars connected is derived with $31n$ degrees of freedom and written in a familiar matrix form as [6]

$$M^V \ddot{X}^V + D^V(\dot{X}^V) + K^V(X^V) = F^V \tag{1}$$

where X and F^V are displacement and load vectors of the train, M^V is the mass matrix, and $D^V(\dot{X}^V)$ and $K^V(X^V)$ are the damping and internal force vectors that are nonlinear depending on the velocity and displacement vectors of the train and railway structure.

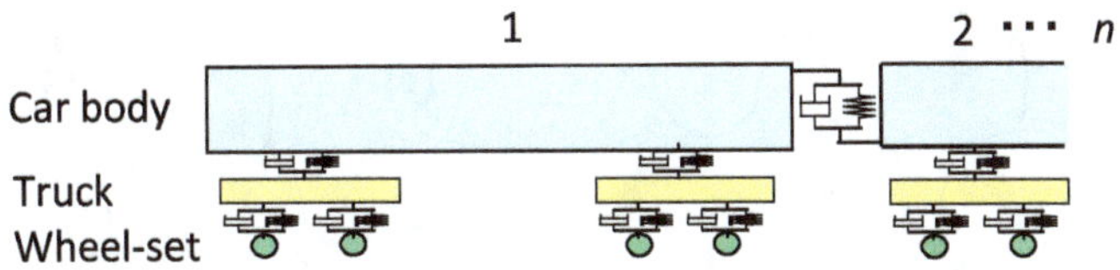

Figure 1: Mechanical model of a Shinkansen train.

2.2 Mechanical model of a railway structure

A railway structure is modelled with various finite elements such as truss, beam, shell, solid, mass and damper elements, and also nonlinear spring element that is to model simply a concrete structure with an elastic–plastic behaviour appeared under cyclic loads during an earthquake [6]. The whole equation of motion of a railway structure is written in the same matrix form as the train as

$$M^b \ddot{X}^b + D^b(\dot{X}^b) + K^b(X^b) = F^b \tag{2}$$

where X^b and F^b are displacement and load vectors of the railway structure, M^b is the mass matrix, and $D^b(\dot{X}^b)$ and $K^b(X^b)$ are the damping and internal force vectors that are nonlinear depending on the velocity and displacement vectors of the train and railway structure, respectively.

3 INTERACTION BETWEEN WHEEL AND TRACK STRUCTURE

3.1 Contact between wheel and rail before derailment

Assuming that the yawing and rolling of the wheel-set are relatively small for the contact behaviour between wheel and rail before derailment, two-dimensional geometries of the cross-sections of wheel and rail are considered, and the contact–impact behaviour in the normal direction on the contact surface between wheel and rail is modelled simply in two modes of the contact in the vertical and transverse directions and also in the combination as shown in Fig. 2 [6].

Of the vertical mode of the contact, the contact displacement between wheel and rail in the vertical direction, δ_Z, is expressed here as a function of displacements of rail, z_R, and wheel, z_W, in the vertical direction, the relative displacement on the contact surface between wheel and rail in the transverse direction, d_y, and also rail irregularity in the vertical direction, ε_z, as follows [6,7]:

$$\delta_Z = \delta_Z(z_R, z_W, d_y, \varepsilon_z) \tag{3}$$

The contact displacement, δ, in the normal direction on the contact surface between wheel and rail is obtained from the contact angle at the contact position. The contact force on the contact surface between wheel and rail in the normal direction, H, is expressed as a function of δ and d_y as follows:

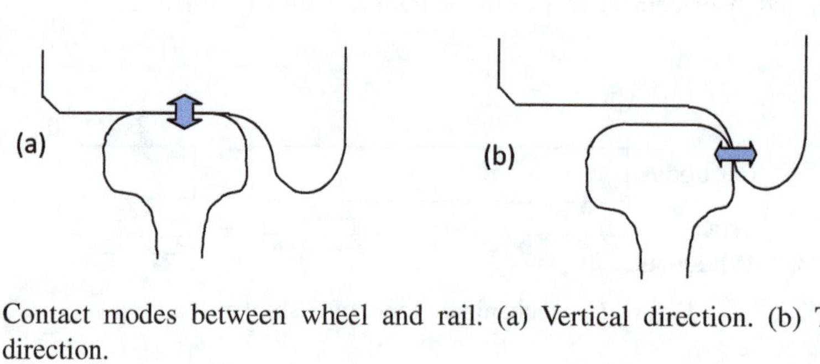

Figure 2: Contact modes between wheel and rail. (a) Vertical direction. (b) Transverse direction.

$$H = H(\delta, d_y) \tag{4}$$

Of the transverse mode of the contact between wheel and rail, the contact displacement in the transverse direction, δ_y, is also expressed as a function of d_y, δ_Z, and the rail irregularity in the transverse direction, ε_y, as

$$\delta_y = \delta_y(d_y, \delta_Z, \varepsilon_y) \tag{5}$$

The contact force is obtained in the same manner as the contact mode in the vertical direction described above.

Regarding the tangential and longitudinal directions on the contact surface between wheel and rail, constitutive equations to describe the relationship between creep forces and slipping rates of wheel are given [8]. The creep force in the tangential direction, Q_c, and yaw moment, T_c, due to the creep force in the longitudinal direction on the contact surface between wheel and rail are described mathematically here as functions of slipping rates of wheel in the longitudinal and tangential directions, S_x and S_t, and also of the spin rate around the normal vector on the contact surface, S_n, as [9]

$$Q_c = Q_c(S_x, S_t, S_n) \tag{6}$$

$$T_c = T_c(S_x) \tag{7}$$

3.2 Derailment criterion

When the relative displacement between wheel and rail in the transverse direction, d_y, exceeds derailment criteria, u_F, in the field side and, u_G, in the gauge side, as shown in Fig. 3, derailments in the field and gauge sides are initiated, respectively. After the derailment of the wheel from rail during an earthquake, the wheel touches down on the track structure and the post-derailment behaviour of the wheel on the track structure starts.

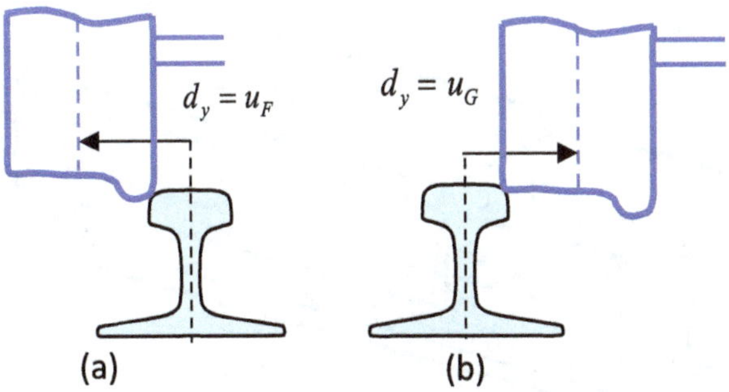

Figure 3: Derailment criterion of left wheel. (a) Field-side derailment. (b) Gauge-side derailment.

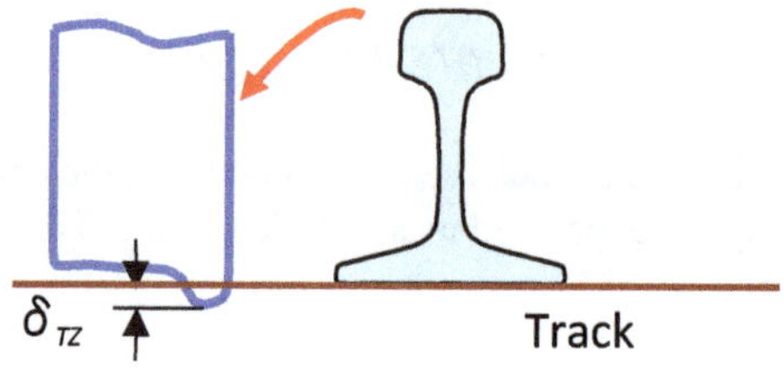

Figure 4: Vertical relative displacement between wheel and track structure, δ_{TZ}.

3.3 Contact-impact between wheel and track surface after derailment in the vertical direction

When a wheel touches down on the track surface after derailment (Fig. 4), the impact force of the wheel on the track structure on the contact surface, Q_{TZ}, is given here as a function of the relative displacement between wheel and the track structure in the vertical direction, δ_{TZ}, depending on material properties of the contact between wheel and the track structure as

$$Q_{TZ} = Q_{TZ}(\delta_{TZ}) \tag{8}$$

4 INTERACTION BETWEEN THE CAR BODY AND RAILWAY STRUCTURE

4.1 Contact censor point C between the car body and railway structure

In the pre- and post-derailments during a strong earthquake, the car body of the train may contact and impact on the railway structure. An efficient contact model to express the contact-impact behaviour between the car body and railway structure is given where a number, n, of contact censor points C_i ($i = 1, n$) are given on the surface of the car body to detect the contact between the car body and railway structure effectively as shown in Fig. 5. When a censor point C contacts on the railway structure, the contact force is created between the car body and railway structure. Censor points C_i ($i = 1, n$) needed are given depending on contact areas on the car body to express contact forces descretized (Fig. 5).

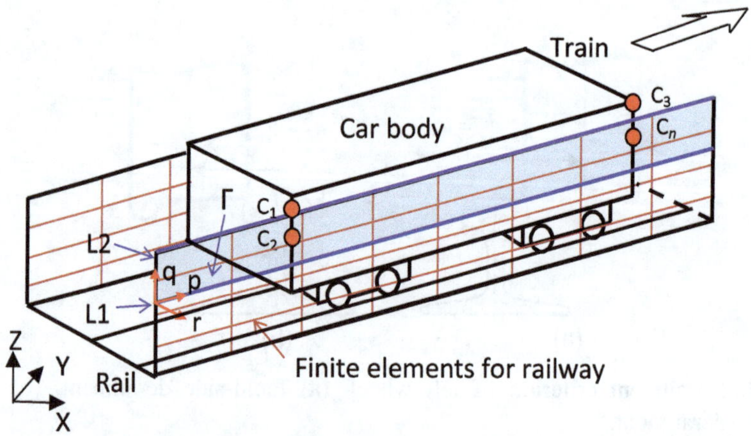

Figure 5: Contact between censor points C_i ($i = 1, n$) on the car body and contact surface Γ on the railway structure.

4.2 Definition of a contact surface Γ on the railway structure

A railway structure is modelled with various finite elements such as shell and solid elements. To make the contact calculation between the car body and railway structure effective, a contact surface Γ, that is a possible surface for censor points C_i (i = 1, n) to contact on the railway structure, is defined on the finite element mesh of railway structure by using two lines of L1 and L2 in the rail direction as shown in a dark colour in Fig. 5. Lines L1 and L2 are the lower and upper bounds of the contact surface for the censor points to contact on the railway structure. The contact point C on the car body can contact on the railway structure only if the point is within the surface Γ. The displacement of the surface, Γ, is approximated from nodal displacements of finite elements in the surface, Γ.

4.3 Contact displacement, δ_C, and force, F_C, between the car body and railway structure at censor point C

Local coordinates p, q, r on the contact surface Γ on the railway structure are defined as shown in Fig. 5, where, p is the rail direction, q is the vertical direction and r is the normal direction on the surface Γ. The contact displacement between the car body and railway structure, δ_C, at the point C in the normal direction is expressed as a function of displacements of the car body, u_C, and the contact surface, u_Γ, in the normal direction at the point C as

$$\delta_C = \delta_C(u_c, u_\Gamma, \varepsilon_\Gamma) \tag{9}$$

where, ε_Γ is the irregularity in the normal direction on the contact surface Γ of railway structure. The contact force, F_C, is obtained here as a nonlinear function of contact displacement, δ_C, between the car body and railway structure at C as

$$F_C = F_C(\delta_C) \tag{10}$$

The relationship of the contact displacement and contact force between the car body and railway structure at point C is obtained from experiments or the finite element static contact analysis between the car body and railway structure. The contact force, F_C, is applied to the car body of the train at the point C and to the railway structure at the contact point on the contact surface Γ.

5 NUMERICAL EXAMPLES

The combined dynamic response of the train and railway structure during an earthquake is obtained by solving nonlinear equations of motions of the train and railway structure in eqns. (1) and (2) subjected to the interaction between wheel and rail before derailment, between wheel and the track structure after derailment and between the car body and railway structure during and after derailment as discussed in Sections 3 and 4. However, a modal reduction to displacement vectors of the train and railway structure is applied to solve the combined response of a large-scale railway system during an earthquake effectively [6]. Equations of motions of the train and railway structure are solved in modal coordinates for each time increment. However, since the equations are strongly nonlinear, iterative calculations are made during each time increment until they converge within a small tolerance specified.

The contact model between the train and railway structure presented here has been implemented in the simulation program DIASTARS [6] for dynamic interaction analysis of a Shinkansen train and railway structure including derailment and post-derailment during an earthquake.

5.1 Simulation of a Shinkansen car running on the railway structure with side walls attached during an earthquake

The simulation of a Shinkansen car running at a speed of 300 km/h on the railway structure with side walls attached during a sinusoidal seismic wave with the frequency of 1 Hz, the maximum amplitude of 3.8 m/s^2 and wave number of 5 given from the base in the transverse direction has been conducted (Fig. 6). Figure 7 shows the distance between the car body and side walls of the railway structure. The railway structure is assumed to be stiff and the side walls are modelled with stiff shell elements. The seismic wave of 1 Hz is a severe frequency for the car as it is close to the resonance frequency of the rolling motion of the car body, truck and wheel-set which leads to derailment easily. Four censor points C_i (i = 1,4) at four corners of the car body are given at the height of the centre of gravity of the car body. The

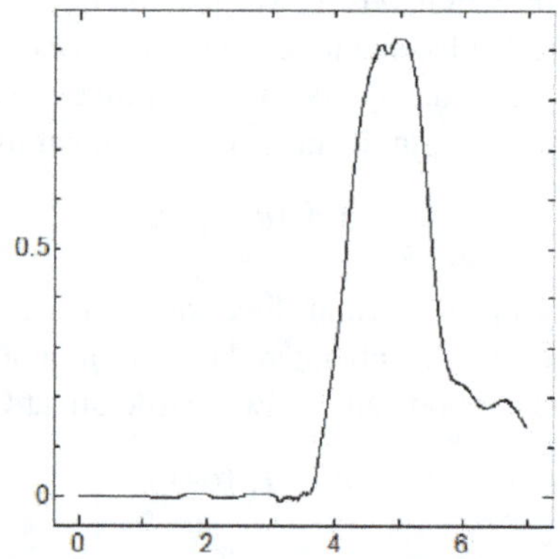

Figure 6: Contact displacement at C_3.

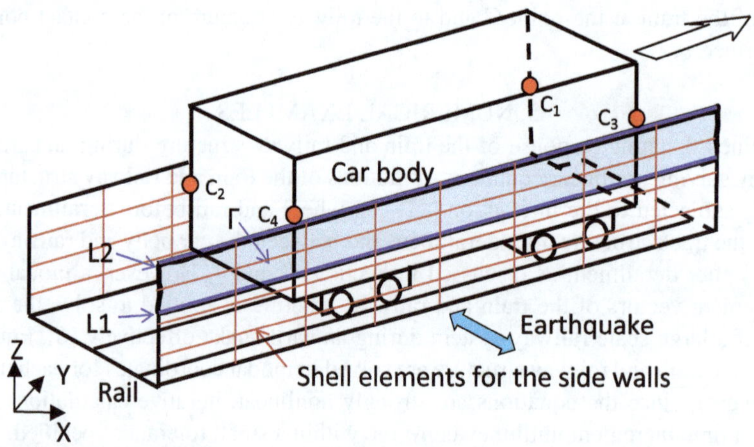

Figure 7: Contact between censor point C_i (i = 1,4) on the car body and contact surface Γ on the railway structure.

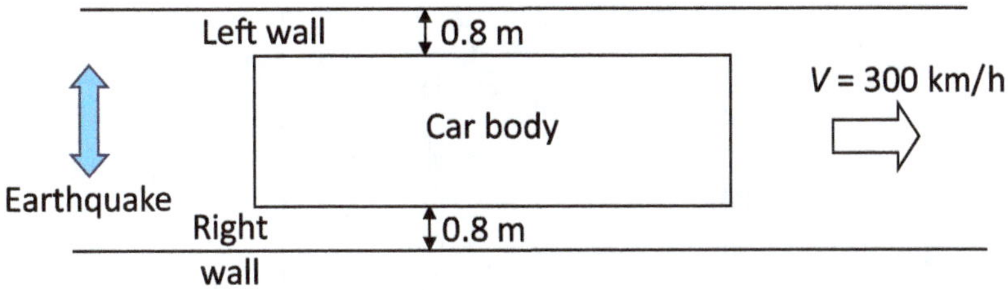

Figure 8: Car body and walls of railway structure.

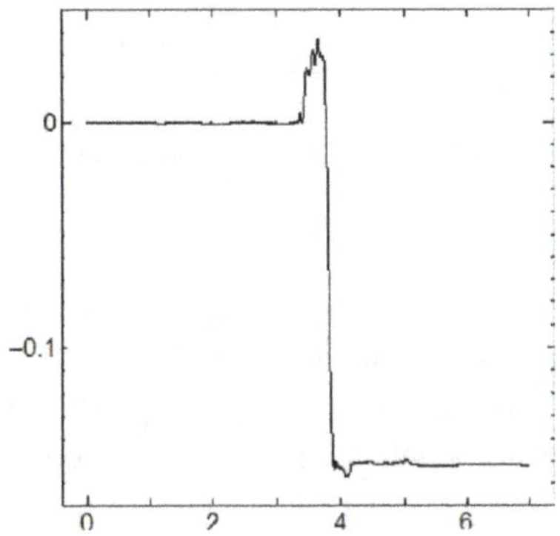

Figure 9: Vertical displacement of right wheel.

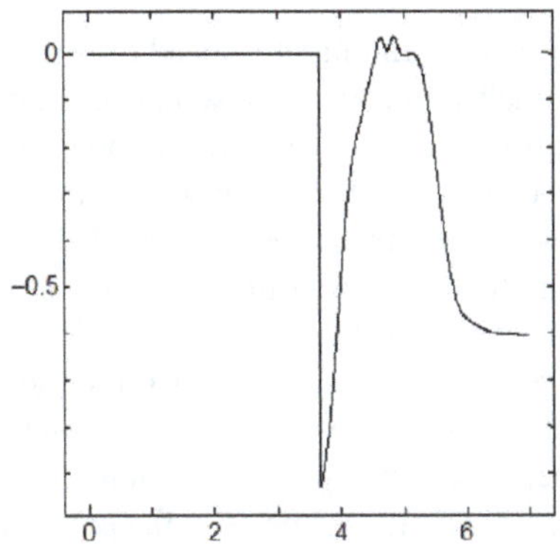

Figure 10: Relative displacement between right wheel and rail in the transverse direction.

contact force-displacement relationship per a half length of the car body given at points C_i (i = 1,4) was obtained from experiments when the car body contacted on the rigid wall [10]. Figure 8 shows the vertical displacement response of the right wheel at the 1st wheel-set. It is shown that the wheel runs onto the rail due to the impact between the wheel and rail in the transverse and vertical directions, lifts off, derails and touches down on the track at the time around $t = 3.66$ (s). Figure 9 shows the relative displacement response between the right wheel and rail in the transverse direction, d_y, at the 1st wheel-set. When d_y exceeds u_F (7 cm in this case; shown in Fig. 3(a)), the derailment to field side is initiated. Figure 10 shows the

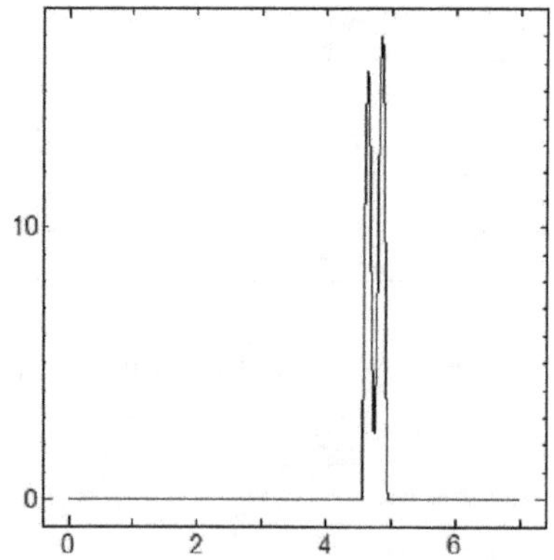

Figure 11: Contact force at $C_{3.}$

contact displacement, δ_C, at C_3. When δ_C is positive, it shows that there is a contact between the car body and the wall at the time around $t = 4.6$. Figure 11 shows the contact force at C_3. The maximum contact force between the car body at C_3 and the side wall of railway structure is shown to be about 16.9 tonf for a half-length of the car body.

5.2 Simulation of a Shinkansen car running on the railway structure with irregular side wall attached in the transverse direction during an earthquake

The simulation of a Shinkansen car running at a speed of 300 km/h on the railway structure with an irregular side wall attached during a sinusoidal seismic wave with the frequency of 1 Hz, the maximum amplitude of 3.8 m/s^2 and wave number of 5 given from the base in the transverse direction has been conducted. The railway structure is also assumed to be stiff and side walls of the railway structure are modelled with rigid shell elements. An irregularity is given on the right side wall in the transverse direction in the slope of 1/200 in the rail direction as shown in Fig. 12. It is expected that the body contacts and impacts on the side wall during the earthquake. Four censor points Ci (i = 1,4) at four corners of the car body are also given at the height of the centre of gravity of the car body as in the previous example (Fig. 6). The contact force-displacement relationship per a half length of the car body given between the points C_i (i = 1,4) and the rigid wall is same as in the previous example. Figure 13 shows the vertical displacement response of the left wheel at the 1st wheel-set. It is shown that the wheel derails and touches down on the track at the time around $t = 3.66$. Figure 14 shows the relative displacement response between the left wheel and rail in the transverse direction, d_y, at the 1st wheel-set. When d_y exceeds u_G (–7 cm in this case; shown in Fig. 3(b)), the derailment to the gauge side is initiated. Figure 15 shows the contact displacement, δ_C, at C_1. It is shown that the car body contacts on the right side wall at C_3 at the time around $t = 4.8$ and on the left side wall at C_1 at the time around $t = 6.1$. Figure 16 shows the contact force at C_1. The maximum contact force between the car body at C_1 and the left side wall is shown to be about 13.1 tonf.

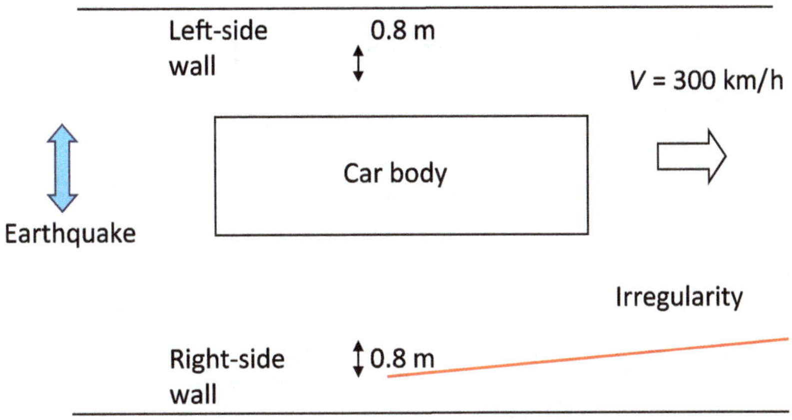

Figure 12: Car body and walls where an irregularity with the slope of 1/200 is given in the transverse direction on the right side wall.

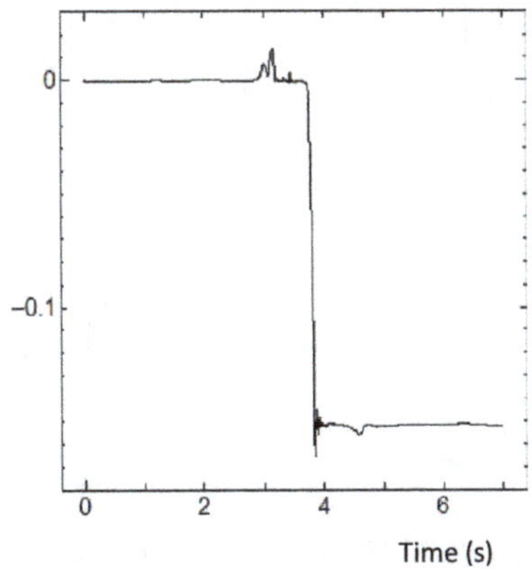

Figure 13: Vertical displacement of left wheel.

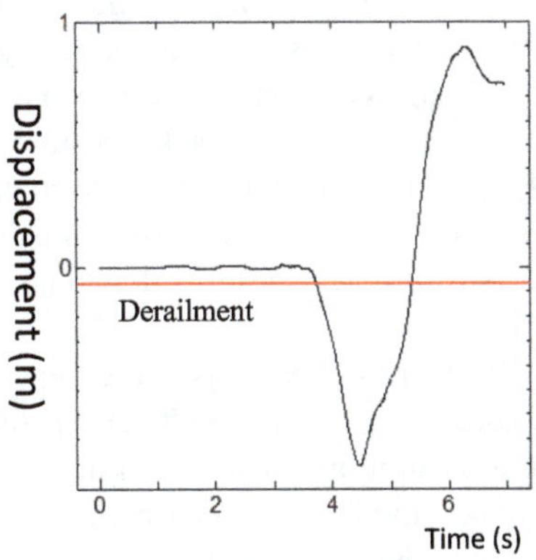

Figure 14: Relative displacement between left wheel and rail in the transverse direction.

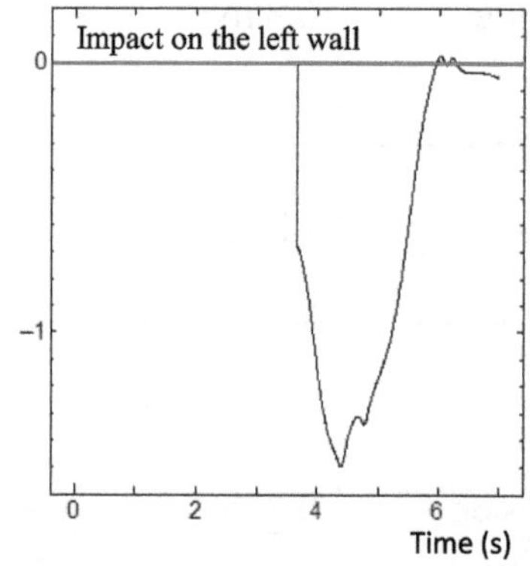

Figure15: Contact displacement at C_1.

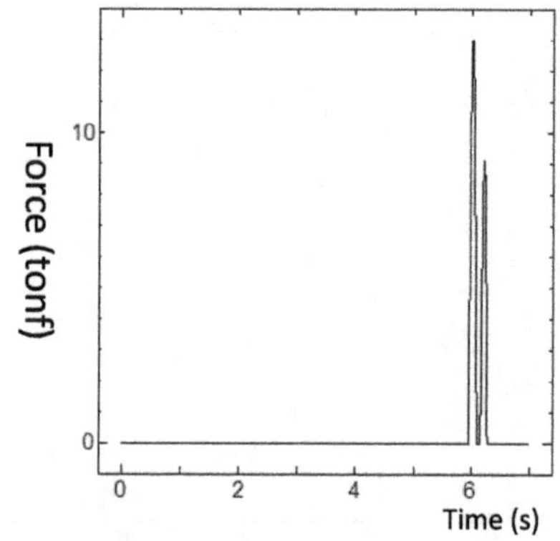

Figure 16: Contact force at C_1.

6 CONCLUSIONS

A computational method to solve for the dynamic interaction of a high-speed train and railway structure including derailment and post-derailment during an earthquake was given. An efficient mechanical model to express contact-impact behaviours between wheel and rail before derailment and between wheel and the track structure after derailment was given to solve the interaction during an earthquake effectively. A contact model to express contact-impact behaviour between the car body and railway structure during and after derailment was presented, where the contact stiffness was described as a nonlinear function of the relative displacement.

The motion of the train with nonlinear springs and dampers was modelled in multibody dynamics. The railway structure was modelled with various finite elements. A modal reduction was applied to equations of motions of the train and railway structure to solve them effectively. Based on the contact model presented here, the simulation program has been developed to solve the dynamic interaction of a Shinkansen high-speed train and railway structure including derailment and post-derailment behaviours during an earthquake.

Simulations of a Shinkansen car on the railway structure including radical dynamic behaviours of derailment and post derailment and also contact-impact of the car body on the side wall of railway structure during an earthquake were demonstrated.

The computational method developed here will be effective to design an earthquake-safe railway system since experiments of derailment and post-derailment of a high-speed train on the railway structure during an earthquake are not available.

REFERENCES

[1] Pombo, J., Ambrosio, J. & Silva, M., A new wheel-rail contact model for railway dynamics. *Vehicle System Dynamics*, **45(2)**, pp. 165–189, 2007. DOI: 10.1080/00423110600996017.

[2] Ju, S.H., A simple finite element for nonlinear wheel/rail contact and separation simulations. *Journal of Vibration and Control*, **20(3)**, pp. 330–338, 2014. DOI: 10.1177/1077546312463753.

[3] Montenegro, P.A., Neves, S.G.M., Calcada, R., Tanabe, M. & Sogabe, M., Wheel-rail contact formulation for analyzing the lateral train-structure dynamic interaction. *Computers & Structures*, **152**, pp. 200–214. 2015. DOI: 10.1016/j.compstruc.2015.01.004.

[4] Xiao, X., Ling, L. & Jin, X., A study of the derailment mechanism of a high speed train due to an earthquake. *Vehicle System Dynamics*, **50(3)**, pp. 449–470, 2012. DOI: 10.1080/00423114.2011.597508.

[5] Sunami, H., Termichi, Y. & Adachi, M., An analytical study of derailed vehicle motion from wheel-sleeper impacts, *Journal of Railway Technology*, **2(2)**, pp. 35–63, 2013. DOI: 10.4203/ijrt.2.2.2.

[6] Tanabe, M., Matsumoto, N., Wakui, H., Sogabe, M., Okuda, H. & Tanabe, Y., A simple and efficient numerical method for dynamic interaction analysis of a high-speed train and railway structure during an earthquake. *Journal of Computational and Nonlinear Dynamics*, **3(041002)**, ASME, 2008. DOI: 10.1115/1.2960482.

[7] Tanabe, M., Matsumoto, N., Wakui, H. & Sogabe, M., Simulation of a Shinkansen train on the railway structure during an earthquake. *Japan Journal of Industrial and Applied Mathematics,* **28**, pp. 223–236, 2011. DOI: 10.1007/s13160-011-0022-4.

[8] Kalker, J.J., *Three-Dimensional Elastic Bodies in Rolling Contact, Dordrecht, The Netherlands:* Kluwer, Academic Publishers 1990.

[9] Japan Society of Mechanical Engineers, *Dynamics of Railway Vehicle,* Train Research Company, 1994 (in Japanese). ISBN 4-88548-074-4.

[10] Okino, T. & Ujita, Y., Evaluation of car body strength against loads from the side. *RTRI Report*, **25(8)**, pp. 23–28, Railway Technical Research Institute, 2011 (in Japanese).

RISK ASSESSMENT OF RAILROAD FOR THE PRESENCE OF UNDERGROUND CAVITIES BASED ON A STATISTICAL APPROACH

J. OH[1], H. YOO[1], B. PARK[2], J. KIM[3] & J. OH[4]
[1]Department of Railroad Facility Engineering, Korea National University of Transportation, Republic of Korea.
[2]Department of Railroad Management and Logistics, Korea National University of Transportation, Republic of Korea.
[3]Department of Geotechnical Engineering, Korea Railroad Technical Corporation, Republic of Korea.
[4]Department of Research and Development, Korea Rail Network Authority, Republic of Korea.

ABSTRACT
Collapses due to underground cavities have been drastically increasing in urban areas of South Korea. This article establishes a statistical model to assess the risk potential of railroads with respect to underground cavities. The authors first identified the risk factors of the event based on case studies where the collapses of underground cavities took place. The database was then established, taking into account the risk factors, to come up with a statistical model that estimates the risk level. In this study, the maximum likelihood estimation (MLE) method was employed to estimate the parameters in a statistical model. Thorough the statistical analysis, the probability of underground cavity occurrences was found to be expressed in terms of the depth of alluvial layer, groundwater level, water and sewage utilities, and their age. Consequently, an attempt was made to generate a preliminary hazard map for a specific railway route by employing the statistical model.
Keywords: database, maximum likelihood estimation, railroad statistical model, underground cavities.

1 INTRODUCTION

The numbers of occurrences of underground cavities have been dramatically increased in South Korea, particularly in metropolitan areas. The sudden nature of this phenomenon can be very detrimental for the various types of infrastructures and for the residences. It should be noted that the sinkhole is different from underground cavities, dealt with in this article. Several researches have reported that sinkhole hazard is common in limestone karst, associated with subcircular surface depression or collapse structure formed by the collapse of small subterranean karst cavities [1, 2]. Therefore it is deemed more likely a natural phenomenon.

An extensive effort has been made to identify the causes of underground cavities by the Seoul metropolitan government in conjunction with engineers in academic and industry fields [3]. The findings of this research can be summarized as follows:

I. Since 2010, around 600 events per year have mainly taken place in pavements, sidewalks and vicinities of underground construction areas.
II. Around 40% of events occurred during summer season, which has frequent heavy rainfalls.

DOI: 10.2495/TDI-V1-N3-552-557

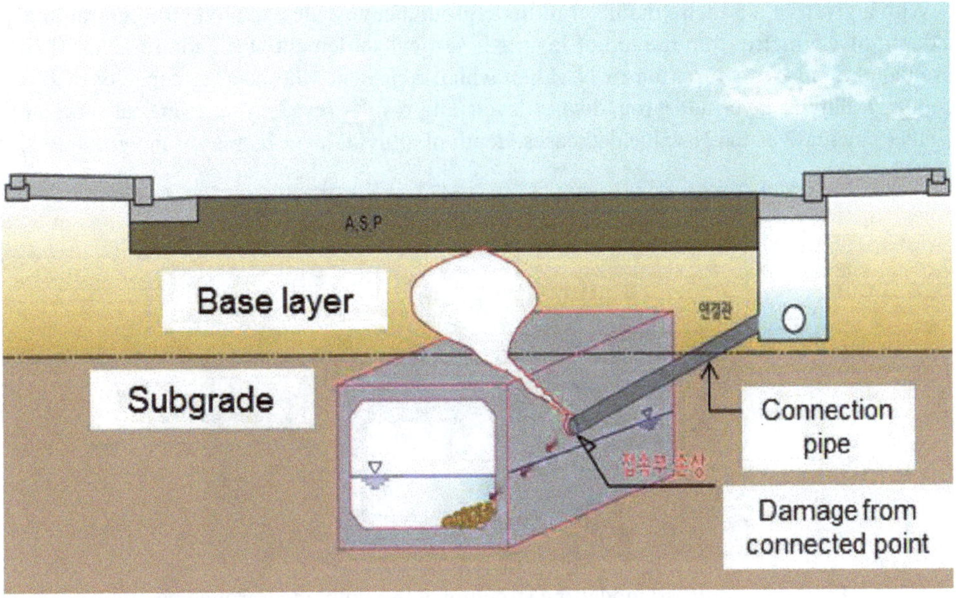

Figure 1: Illustration of the development of underground cavity.

III. The causes of the events have been categorized as:
   - Damage of water and sewage utilities, which allows surrounding soil loss through the holes made as illustrated in Fig. 1;
   - Inappropriate backfill compaction during excavation activities that include open-cut construction and installation of underground utilities;
   - Sudden drop of groundwater table due to pumping activities.

In this article, the author focused on conducting risk assessments of railroads that are being operated throughout metropolitan areas with respect to underground cavities in spite of such events rarely occurring in railroads of South Korea. In order to conduct the risk assessment, the following works have been performed consequently:

I. Identify the critical factors based on numerical analysis
II. Establish database of critical factors
III. Develop a statistical model to assess the risk of underground cavity
IV. Apply the model for a preliminary risk assessment

2 NUMERICAL ANALYSIS

Numerical analysis was employed to identify the critical factors associated with the development of underground cavities. Since this study focuses on railroad structure, the following numerical model was tested. With respect to loading condition, 50 kPa of distributed load was applied at the surface of layer #1 (Figs. 2 and 3).

Table 1 presents the material properties that were used for this analysis. Finite element modelling was utilized along with Mohr Coulomb's failure theory, which is widely adopted for analysing granular or soil material behaviour.

With a given of a 3 metre diameter of underground cavity located along the central line at a depth of 1.5 metre from the top of layer #3, vertical settlement at the top of layer #1 was evaluated along with variation of N-value, which represents the bearing capacity of layer, depth of alluvial layer and groundwater level. The results revealed that vertical settlement tends to increase as the N-value decreases, depth of alluvial layer increases and groundwater level lowers.

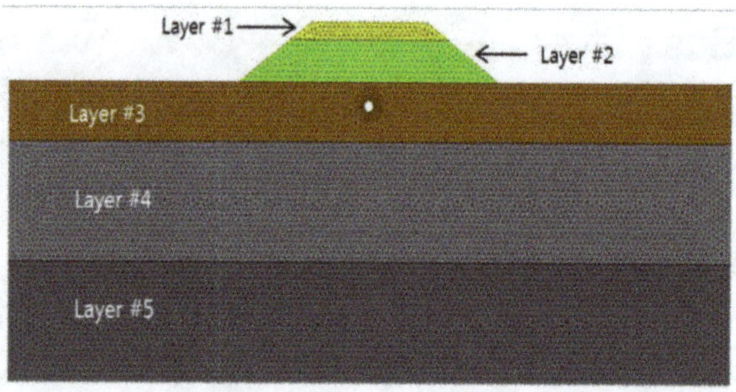

Figure 2: Railroad modelling structure along with underground cavity.

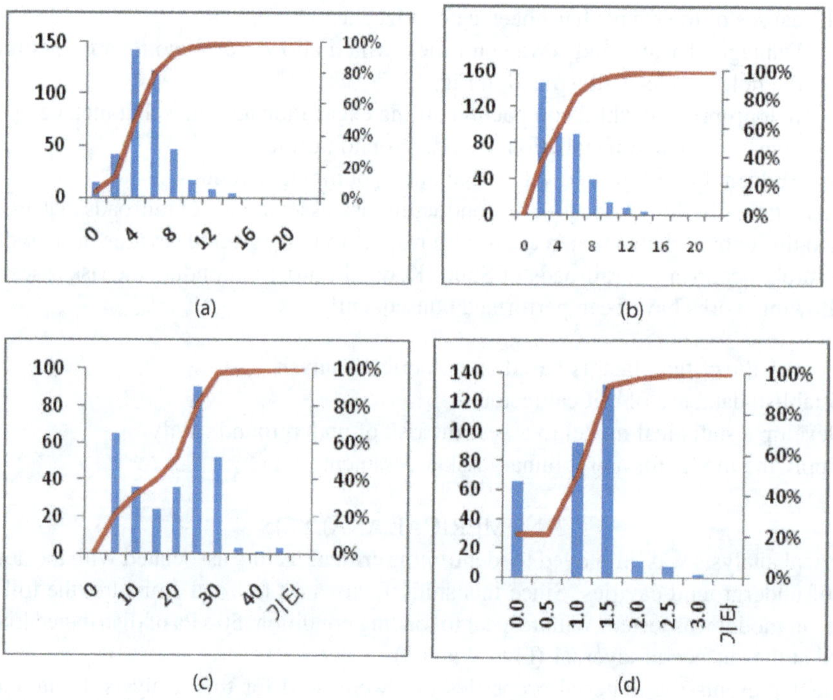

Figure 3: Histogram of each dataset: (a) groundwater table depth (m), (b) depth of alluvial layer (m), (c) age of water pipelines, (d) cover depth of water pipelines.

Table 1: Geometric and material properties used.

Property	Layer #1	Layer #2	Layer #3	Layer #4	Layer #5
Elastic modulus (MPa)	80	60	20	300	2000
Poisson's ratio	0.3	0.3	0.32	0.29	0.28
Unit weight (kN/m^3)	18	18	18	20	23
Cohesion (kPa)	15	15	1	50	150
Internal friction angle (°)	25	25	28	33	35
Thickness (m)	1.5	3.5	5.0	10.0	10.0

3 ESTABLISHMENT OF DATABASE

Based on previous research findings and numerical analysis results, the authors made an attempt to establish database in order to develop a statistical model that assesses the risk of underground cavity around railroads. First of all, all of the locations where the event took place were identified, and geotechnical and underground utility information at the vicinity of the site was extracted using existing database systems. Here are examples of database analysed as follows.

4 DEVELOP A STATISTICAL MODEL

The purpose of developing a statistical model is to generate a hazard map that shows the level of risks, taking into account various critical factors identified. Several researches have established sinkhole hazard models incorporating various data interpretation techniques [1, 4, 5]. The sinkhole hazard models, suggested from previous studies, highly depend on the degree of event occurrences. However, as stated earlier, since there have not been such cases where the underground cavities have been detected at the vicinity of railroads, a rational statistical model that can assess the risk of the event using only the critical factors needs to be developed for the generation of a hazard map.

4.1 Application of maximum likelihood estimation

There are two general methods of parameter estimation: least-squares estimation (LSE) and maximum likelihood estimation (MLE) [6]. While LSE has been widely applied to numerous statistical approaches such as linear regression, sum of squares error and root mean squared deviation, MLE is regarded as a standard method to parameter estimation and inference in statistics [6]. MLE is feasible to have many optimal properties in estimation compared to LSE. Due to the uncertainty of dependent variable, MLE seems to be more appropriate to apply in this study. Using a logistic regression model incorporating MLE, the following eqn (1) can now be written as follows:

$$p \equiv P\left(Y = 1\middle|\overline{X}\right) = \frac{e^{b_0 + \sum_{i=1}^{m} b_i \overline{X}_i}}{1 + e^{b_0 + \sum_{i=1}^{m} b_i \overline{X}_i}} \quad . \tag{1}$$

In this equation, p is the level of risk, Y is an indicator (when underground cavity occurs, Y=1, otherwise, Y=0), $\overline{X}$ is an average of each dataset (which is regarded as an individual critical factor), m is number of dataset, and b_0 and b_i are MLE coefficients.

Table 2: Results of statistical test.

Variable	b	S.E.	Walls	D.O.F	P-value	Exp (b)
Alluvial layer depth	1.536	0.734	4.379	1	0.036	4.644
Cover depth of sewage	–0.280	0.131	4.592	1	0.032	0.756
Cover depth of water pipelines	–7.134	2.762	6.671	1	0.010	0.001
Intercept	8.430	3.663	5.295	1	0.021	4581.263

For the development of an MLE-based statistical model, six variables were first considered: groundwater table depth, alluvial layer depth, age of sewage, cover depth of sewage, age of water pipelines and the cover depth of water pipelines. The statistical model accounting for all the variables was found to be ineffective due to multi-collinearity among independent variables. After the completion of statistical tests, as presented in Table 2, the following model is tentatively suggested in this study.

$$p = \frac{\exp(8.430 + 1.536x_1 - 0.280x_2 - 7.134x_3}{1 + \exp(8.430 + 1.536x_1 - 0.280x_2 - 7.134x_3}, \tag{2}$$

where x_1 is alluvial layer depth, x_2 is the cover depth of sewage and x_3 is the cover depth of water pipelines.

4.2 Preliminary risk assessment

A preliminary risk assessment was conducted using a statistical model proposed. The authors selected a railroad that travels 11.9 km as denoted by five arrows, representing the location of station. There are five stations, so four segments were evaluated using obtained database along with the statistical model. For the evaluation, the database was used along the railroad which covers 100 metres with respect to the centre of the route. Figure 4 shows the results of

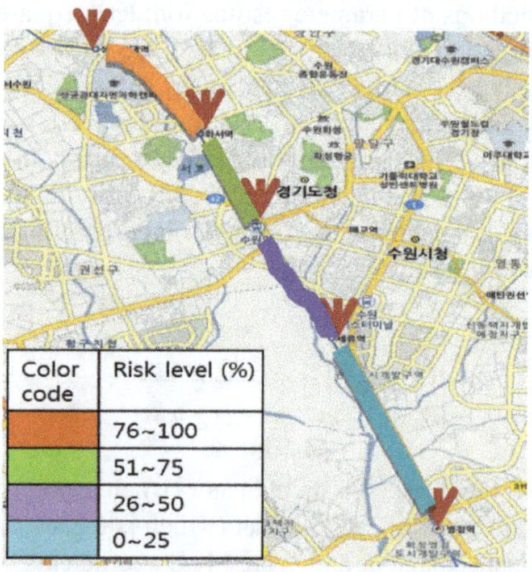

Figure 4: Results of preliminary risk assessment.

the preliminary risk assessment. Further efforts are being conducted to upgrade the statistical model, such as incorporating artificial neural network to improve its efficiency.

5 CONCLUSION

This study aimed to develop a statistical model to generate a hazard map with respect to underground cavities along the railroads in metropolitan areas of South Korea. Employing a numerical model was successfully conducted to evaluate the impact of critical factors on vertical settlements of the railroad structures. Extensive efforts are going on to establish a database of critical factors that entails geological properties and underground utility information at the vicinity of actual underground cavities that have taken place. Based on analysing database, a tentative statistical model incorporating the MLE method was applied to a preliminary risk assessment on a selected railroad segment. As a result, four different risk levels were identified by taking into account critical factors. Further efforts are being conducted to improve the statistical model by employing artificial neural network technique. In addition, field experimental programmes have been planned to verify the hazard map.

REFERENCES

[1] Al-Kouri, O., Al-Fugara, A., Al-Rawashdeh, S., Sadoun, B. & Pradhan, B., Geospatial modeling for sinkholes hazard map based on GIS&RS data. *Journal of Geographic Information System*, **5**, pp. 584–592, 2013.

[2] Klimchouk, A., Morphogenesis of hypogenic caves. *Geomorphology*, **106**(1–2), pp. 100–117, 2009.

[3] Seoul Metropolitan Government, Investigation of underground cavity mechanism, *Research Report*, 2015.

[4] Galve, J.P., Remondo, J. & Gutierrez, F., Improving sinkhole hazard models incorporating magnitude-frequency relationships and nearest neighbor analysis. *Geomorphology*, **134**, pp. 157–170, 2011.

[5] Galve, J.P., Gutierrez, F., Lucha, P., Bonachea, J., Remondo, J., Cendrero, A., Gutierrez, M., Gimeno, M.J., Pardo, G. & Sanchez, J.A., Sinkholes in the salt-bearing evaporate karst of the Ebro River valley upstream of Zaragoza city (NE Spain) – Geomorphological mapping and analysis as a basis for risk management. *Geomorphology*, **108**, pp. 145–158, 2009.

[6] Myung, I.J., Tutorial on maximum likelihood estimation. *Journal of Mathematical Psychology,* **47**, pp. 90–100, 2003.

A RAILWAY SIMULATION SUITE FOR MODELLING ADVANCED RAILWAY CONTROL SYSTEMS

P. STANKAITIS, A. ILIASOV & A. ROMANOVSKY
Centre for Software Reliability, Newcastle University, UK.

ABSTRACT

All large railway networks use a mixture of outdated, modern and emerging signalling and train operation principles. There is a need to develop novel modelling and verification mechanisms to support mixed traffic scenarios, including, for example, mixing different types of signalling and driving. In our previous work we introduced the Unified Train Driving Policy (UTDP) formal modelling language for uniformly capturing diverse signalling principles and mixing, in a demonstratively safe manner, at the node and/or network-level novel and legacy signalling principles. This paper describes our work on making UTDP practical and useful for the engineers.

Keywords: advanced railway control systems, simulation, formal methods, automatic train operation (ATO), Unified Train Driving Policy.

1 INTRODUCTION

1.1 Background and literature review

The transport sector has been seriously affected by the economical progress in many countries (Abril et al. [1]). The traffic congestion and high negative environmental impact in the metropolitan cities are the key reasons behind the global initiative to improve railway and make it more competitive. A number of research projects (e.g. [2, 3]) were launched to investigate the current problems of the railway industry and look into possible solutions.

One of the EU-funded railway projects [2] identifies the effective management and optimization of railway networks as a grand challenge for the future of railway. There is a number of issues making it difficult to address. First of all, there is a correlation between capacity and service reliability. For example, for the heavily congested metropolitan city's rail networks, which are run close to the maximum capacity, ensuring service reliability becomes critical. Paper [1] shows a clear correlation between capacity and service reliability, where adding extra trains exponentially increases average delays in the network. Furthermore, ensuring the efficient use of energy in the railway network is crucial too. For example, González-Gil et al. [4] show this for maintaining urban rail as a solution for transport problems.

Significant network improvements can be achieved by upgrading the current railway infrastructure. The European Rail Traffic Management System (ERTMS) is a leading initiative to overcome critical railway problems. A modern train control system and continuous radio-based communication are the two key components of the new management systems. Optimization of capacity will be achieved by using systems capabilities to reduce the block lengths, or even by switching in yet to be implemented a moving block signalling.

DOI: 10.2495/TDI-V1-N3-558-567

Further improvements in capacity, energy and stability can be achieved by fitting trains with the Automatic Train Operation (ATO) system, which allows optimization of train speed profiles. Even though these modifications would dramatically optimize railway network use of resources the deployment of adjustments can take a long time and be very costly.

In the short term, a more cost-effective solution is to increase the capacity of the existing railway by utilizing computer-based simulation techniques [1]. The railway companies use proprietary, or externally developed, simulation suites like OpenTrack [5] and RailSyS [6] in a timetable and infrastructure optimization. A visual and user-friendly aspects of simulation platforms offer an intuitive method to analyse railway network without requiring a tedious simulator configuration process.

As discussed by Stankaitis et al. [7] any type of improvement of railway systems has to meet the paramount requirement of safety. Railway certification standards require applying semi-formal or formal methods for description of the system and even correctness validation. The industry has been successfully applying formal techniques for reasoning about safety for some time now with many developed systems currently in operation. Perhaps the best example was the use of the B-method Abrial [8] in major railway projects including a Paris metro line and a more complex New York Canarsie line (Essame et al. [9]). However, simulation techniques are still extensively used by the railway industry to ensure that safety standards are preserved within system. And even though simulation tools are progressively becoming more powerful the main drawback of this technique is a small state-space coverage. In contrary to the simulation approach formal safety verification of railway networks is a complicated process, which requires expertise in the mathematical modelling; however, it provides techniques to guarantee safety for a complete system state-space.

A significant effort has been put by academic research projects like SafeCap [10] into hiding mathematical complexity to make it easier for railway engineers to use formal methods. The SafeCap project was started with an objective to develop techniques for overcoming railway capacity problems without weakening safety constraints. The key outcome of the first stage of the project was a SafeCap tooling environment, which allowed railway engineers not only to design and analyse railway junctions/stations, but also to formally and automatically verify their safety.

A major challenge of optimizing the use of railway network resources like capacity and energy without undermining safety constraints or minimizing service reliability for large-scale industrial projects requires state-of-the-art techniques and tools. The user-friendly railway simulation platforms currently used by the industry are typically rigid. Advanced functionality and expressiveness of simulation suites are sacrificed for its simplicity. Railway academic community has successfully analysed and applied a variety of analytical, iterative methods and even genetic algorithms to optimize railway networks. However, our analysis of a number of academic research projects shows a lack of adequate and multi-purpose simulation tooling for the railway domain. A large number of projects (e.g. [11, 12]) have used popular and powerful suites like OpenTrack or RealSyS to design or evaluate novel algorithms or optimization methods. Yet in many instances researchers had to conduct a manual experiment reconfiguration of the tools to test the newly developed methods. Some other projects developed single-purpose railway simulators for their experiments.

There has been few attempts to create a more expressive and flexible simulators. For example, Grube et al. [13] developed a dynamic event-driven railway simulator for metro systems. The developed railway simulator enabled users adjusting a number of parameters in order to model complex metro systems. Furthermore, using a widely applied software package, Matlab users were able to experiment and design novel control and optimization algorithms. Paolucci and Pesenti [14] attempted to exploit object-oriented paradigm and design a discrete-event simulator for underground railway systems. The ability to define a network of relationships between physical and decisional objects allowed users to design a whole system. The base of a code and simulation code were automatically generated by the simulation suite. However, the research group was yet to make the tool more usable by implementing a functionality to modify system objects and processes.

The application programming interface (API) enables some tools, like Open-Track, to be connected to other third-party tools to add extensions, implementing, analysing or developing train control systems or dispatch algorithms. But configuring the software may sometimes be tricky, or tools can be simply incompatible. In the next section we present the SafeCap project approach to allowing optimization of railway networks by developing an expressive simulator.

1.2 Modelling advanced railway systems

The SafeCap team at Newcastle University has been working on railway modelling and formal verification for nearly six years. One of the topics we have been looking at recently is the definition, simulation and validation of novel, hybrid or simply exotic signalling principles. In our previous work Iliasov et al. [15] introduced the Unified Train Driving Policy (UTDP) (Fig. 1), a generic approach that can capture both existing and novel signalling as well as scenarios involving a mixture of signalling solutions and transient arrangements like virtual trains and platoons. The primary subject of this formal language is the definition of laws of train movement that result in a safe railway operation, achieving optimum capacity. The reasoning about capacity at the same level as safety means that one is able to formulate a range of QoS criteria such as average train speed, minimum speed at a given location, maximum separation between trains and so on.

In UTDP, all stateful parts of railway are known as actors. A train and a point (a track switching device) are two possible examples of an actor. New actor types may be defined in a model (i.e. a station controller) and all actors are permitted to interact with each other

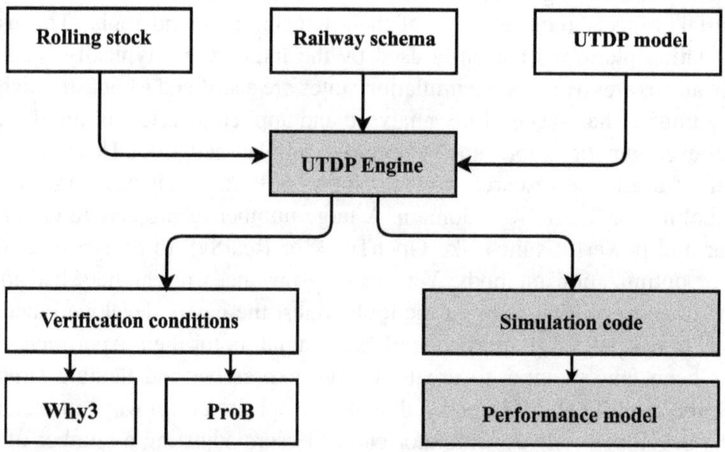

Figure 1: The UTDP approach with highlighted executional side.

through some shared state. An actor lifecycle is split into three stages: continuous evolution (e.g. a train moving in accordance with the laws of physics), mutation (a step-change due to a logical trigger) and hibernation (inability to evolve or mutate). An evolution of a train is described by a continuous or piece-wise continuous function of the effective longitudinal force. It defines a smooth progression from one position to another and accounts for such effects as air drag, gradient, rolling resistance, variable tractive effort and others. A mutation is a change of law governing an actor evolution; it may be dictated, for instance, by signalling logic or speed regime. An actor goes into hibernation when it is unable to evolve (a train waiting for a proceed signal).

At the core of UTDP is a formal notation used to express both static and dynamic properties of railway. The static part is based on a formal domain-specific language and captures railway topology, route boundaries and track-side equipment. The dynamic part describes train properties and the way trains move over the track.

In Fig. 2, a railway schema example and an excerpt of its signalling in UTDP are given. The layout is a simple junction with a fixed block signalling. In fixed block signalling, a simple programme (called a control table) helps a driver to maintain safe speed. The programme observes the infrastructure state (points and train detection circuits) and interacts with a driver via track-side signals that communicate commands to a driver via colour-coded messages. If a driver obeys the commands then, assuming the control table is correct, it may be shown that a train is never within a dangerous proximity to another train. UTDP is not specific to fixed block signalling; one may express various forms of the moving block principle.

In this paper, we focus on executional semantics of UTDP shown in Fig. 1, which is exactly the simulation of railway operation. In our research, we propose to develop a new simulator on making UTDP practical and useful for the engineers.

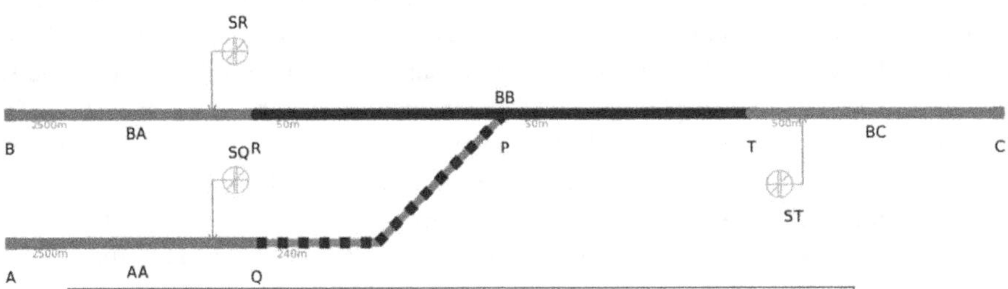

track flyout
behavior
    stop on entering(B_C.B_SR) **when** occ(BA) ∨ reverse(P)
    stop on entering(B_C.SR_C) **when** occ(BB, BC)
    stop on entering(A_C.A_SQ) **when** occ(AA)
    stop on entering(A_C.SQ_C) **when** occ(BB, BC) ∨ normal(P)
    stop on entering(C_A.C_ST) **when** occ(BC)
    stop on entering(C_A.ST_A) **when** occ(BB, AA) ∨ normal(P)
    stop on entering(C_B.C_ST) **when** occ(BC)
    stop on entering(C_B.ST_B) **when** occ(BB, BA) ∨ reverse(P)
    ...

Figure 2: In the layout, the valid paths through the junction are A_C, B_C, C_A and C_B; signals SR, SQ and ST define routes B_SR, SR_C, ...; AA, BB, ... are the train detection circuits that report the presence or absence of a train; point P is located on train detection circuit BB; finally, R, Q and T are the labels of train detection circuits boundaries.

2 SIMULATOR PROGRAMMING AND IMPLEMENTATION

The traditional approach to railway simulation consists in time or distance slicing where within small time or distance extents train movement laws are fairly straightforward and determined only by the boundary conditions. A more advanced version could use an adaptive step size to maintain a degree of modelling fidelity.

Still such technique has a number of drawbacks. One must take care that errors do not accumulate over time (since in machine arithmetic $3 * x$ is not necessarily the same as $x + x + x$, especially for small values of x) and important but short-lived phenomena are accounted for (e.g. a delay in train occupation confirmation). According to the Shannons theorem, the time delta of a time-slicing simulator must be at least two times smaller than the extent of the shortest event and this means time delta of around 0.01 second.

The original SafeCap simulator [16] took a different approach of event-based simulation where the temporal extent of a simulation step is determined by the amount of interference between various entities of a simulated railway: trains, signals, points and virtual entities such as route reservation logic. Such a technique does not have fidelity limitations and has a far greater potential performance than a naive time-slicing simulator. It also is far better compartmentalized in actor-like entities that dictate operational rules. For instance, the determination of the most restrictive target of train is done by collating and processing a list of actors affecting train speed. Actors may be added and removed, giving a compositional approach to train behaviour modelling.

Also the original simulator is flexible and extensible; the changes still have to be implemented at the level of source code of the simulator. To test some speculative scenarios it was deemed necessary to have an even more general approach to simulation without disrupting the code base of the existing simulator.

Traditionally, driver behaviour is simulated by driving a train to some percentage of its overall acceleration with braking capability that is always within the safety conditions of signalling and interlocking. For out-of-the-box scenarios, signalling rules stand in the way; so to enable the freedom of experimentation we decided to scale back the new simulator to the very bare bones and to use a more basic, though more expressive, time-slicing approach.

Let us consider what a train is. A train driver has but one essential form of control over train: acceleration and braking (assuming both are never together). From a high-level perspective, it is a function of many variables including the capability, position and speed of the controlled and nearby trains, infrastructure state, weather, staff rotation and so on. For any such set of variables this function yields just one number: the effective acceleration (a negative value representing braking). Somewhat simplistically we could try to summarize the statement as a set of simultaneous equations defining train acceleration control TC and infrastructure control IC (the time domain t is implied, $\mathbf{T}$ is a set of all train states, $\mathbf{C}$ is infrastructure state, a subscript is a train name and superscript is a discrete time step):

$$\begin{cases} a_i = TC_i(T, C^j, \ldots), \\ c^{j+1} = IC(T, C^j, \ldots) \end{cases} \quad (1)$$

The starting point of every simulation scenario is then just a layout made of tracks, points and, optionally, detection circuits, with some trains appearing on the boundary nodes according to some pattern. Such a starting point is clearly unsafe: there is nothing preventing train collision or other safety violations. However, the driver behaviour may be specified to an unlimited level of detail and a variety of actors (trains, points, signals or some new kind of actors) are free to share information and exchange messages. The goal is to programme a driver who would, at least for a variety of simulation scenarios, move a train safely across a layout. The driver's logic is defined simply as a function of the form

```
double acceleration(World world) {
    ...
}
```

The function computes acceleration or braking in the context of the current world, a shared data structure describing both current and other trains position, acceleration as well as the state of other actors, i.e. points. A simple yet unsafe driving function is

```
double acceleration(World world) {
    return 0.5;
}
```

It prescribes a constant acceleration of 0.5 m/s^2 and disregards any obstacles or hazards. A more realistic one would have to assess the situation on the train horizon, in particular state of the points, trains occupying same paths or appearing from converging paths, as well as to comply to scheduled stops and fixed speed limits. The conceptual difference is that a safety envelope is taken away. The logical entity of a driver is responsible both moving a train forward (liveness) and ensuring absence of collisions and derailments (safety).

A train is but one kind of actor. There are predefined actors for a signal, point and train detection circuit. Generic actors may be defined to cater for and structure new kinds of signalling logic. One example is an actor commanding route reservation and point locking on the basis of a timetable. For a system to function as a whole, all the constituent actors need to coordinate with each other. This is achieved either through accessing shared state directly or via message passing.

3 METHODOLOGY EVALUATION

In this section, we present an application example we used to assess a time-slicing simulator. The primary objective was to design a novel train control mechanism and evaluate its performance with respect to the original simulator with a route-based signalling using a simple bridge layout shown in Fig. 3.

3.1 Experiment description

To demonstrate the control system's adaptability to various service patterns as part of the new simulator suite we designed a scenario generator. The tool in a non-deterministic order and frequency would construct a train actor, which would appear on boundaries of the layout. The generated train actor would randomly inherit several parameters including a train class, which describes train physical properties and a route.

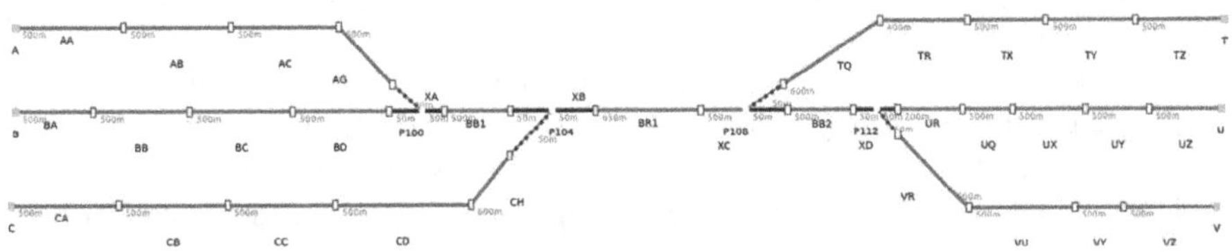

Figure 3: A bridge layout used to evaluate train control algorithm.

In order to demonstrate the flexibility of the designed control system several scenarios were simulated. We evaluated the application example by performing identical train service patterns in the original simulator, which used a route-based signalling, and in the newly developed train control system. The performance was compared using the capacity measurement together with the analysis of train speed profiles.

In the following sections we provide a more detailed description of the control algorithm, capacity results and tool evaluation.

3.2 Train control algorithm

To design a more dynamic and optimal train control mechanism one cannot rely on physical signals any longer. Such signalling system would require a precise and continuous knowledge of the railway network status including the rolling stock and infrastructure parameters. In our design a control system has a full autonomy to manage both trains and the infrastructure. A similar way of thinking is used in ERTMS Level 3, where a centralized unit called Radio-Block Centre receives a continuous information about the network and issues a movement authority.

To understand our algorithm let us reconsider eqn (1). The first technical obstacle is that any meaningful approximation of *TC* yields non-linear differential equations while infrastructure control by its nature is decidedly discrete. The historical perspective on the problem is to replace the computation of train acceleration with a simpler task of the determination of a safe upper boundary

$$a_i \leq SA_i(T, C^j, \ldots). \tag{2}$$

This style of thinking is used, for instance, in ERTMS Level 3 to maintain train acceleration within some sensible braking curve. Computing such *SA* while maintaining confidence in its safety properties is quite challenging and a meaningful discretization requires a high fidelity proximity or position sensing. When such means are unavailable one has to rely on a cruder technique where the control variables are discretized in temporal or spacial domain. Direct acceleration control then becomes too finely grained a tool for the task and bounds are normally given in terms of train velocity

$$v_i^{k+1} \leq SV_i(T^k, C^j, \ldots). \tag{3}$$

The safe velocity bound used in our control system was based on a moving block signalling principle. A control system would continuously compute speed, V_i, to maintain a safe distance between trains and avoid train derailment over unset point or unsafe curve speed. Further system's performance optimization can be achieved by integrating a dynamic speed adjustment function. A typical example is over-speeding trains, which usually results in situation, where train is required to stop and wait before crossing the point. As our system is able to access any information about the network, train speed can be adjusted with respect to other trains more dynamically.

The second essential part of the control algorithm is the train scheduling sub-system. Scheduling algorithm is mainly responsible for solving the use of shared resources (e.g. points) issue. A variety of aspects can be considered in the scheduling function. For example, a simple yet widely used approach in queuing algorithms in various domains is a First in, First out method.

The bridge is a topologically limited layout with a heavily used single line. In order to obtain a better performance one would like to schedule trains, so the fastest trains would have a priority over sections XB, BR1 and XC in Figure 3. So, we considered a physical distance aspect and used it to improve the system. Since we constructed an experiment with randomly appearing trains on the boundaries no pre-computation of optimal schedule is done. Nonetheless, the scheduling system attempts to optimize the performance, though; if we were to plot scheduling algorithm output, one would notice oscillations for a short period before steady-state is reached.

3.3 Capacity and speed assessment

To evaluate the design of the developed train control system we compared its capacity performance against the fixed block signalling system. The definition of railway network capacity is a highly debatable subject. In our experiment, a *used capacity* metric described in [1] was applied. Often, a *used capacity* measurement results in a slightly lower value from the practical capacity; however, it does not require a rigorous computation.

In order to clearly show our systems advantage we needed to demonstrate performance improvements for a large number of scenarios. Therefore, we the tool generated multiple service patterns and applied them to both systems. The experiment results are shown in Fig. 4, where in an individual experiment *used capacity* number is plotted against the capacity for both signalling systems. The performance of the novel train control system is represented by red points, whereas the fixed block system by blue.

Results clearly demonstrate that a developed control system, which was based on a moving block signalling principle, outperforms a traditional fixed block system. Noticeably, there are some improvement deviations; for example experiment 2 has a relatively lower improvement coefficient. Throughout the study we noticed that greatest improvements are seen when a large number of trains run on a frequent service.

Furthermore, we wanted to investigate the effect of dynamic speed adjustment functionality we integrated in the speed calculation function. Previously, we discussed an over-speeding train scenario, where a train is required to stop and allow another train to pass. This was captured in one of the simulation runs with a fixed block route-based signalling system; the speed profile is shown in Fig. 5 (red plot). The identical scenario simulated in the system with the newly developed control system produced a blue speed profile. The latter train was

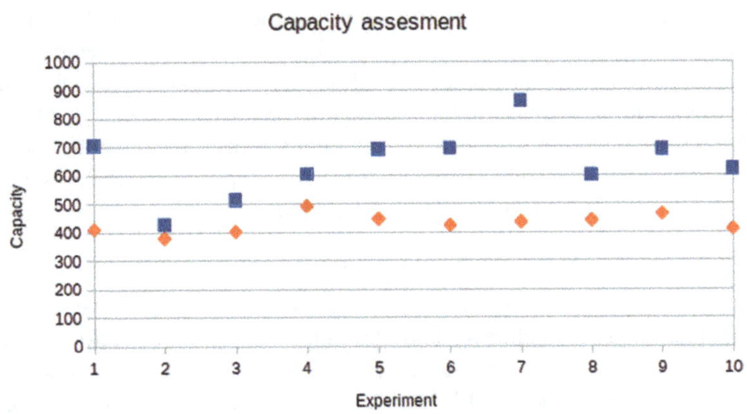

Figure 4: A capacity evaluation of two train control systems.

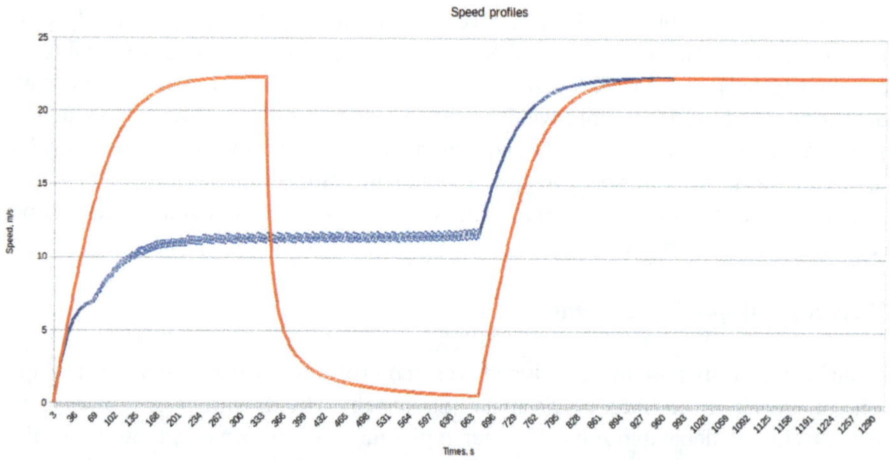

Figure 5: A speed profile comparison of control systems.

running at lower velocity before crossing the point and was able to enter a new section from a running state, and thus, reduce a route completion time.

One of the ongoing work packages in the SafeCap project investigates the energy aspect of the railway. In the future, we plan to integrate energy consumption plug-in, which could analyse speed profiles like the one shown in Fig. 5, and provide essential information for railway engineer.

All these measurements were produced using a new simulator, which is now integrated into the SafeCap platform. The platform, also has a replay functionality, which can be used to further investigate the model. A further use analysis of the simulator is given in the following section.

3.4 Simulation tool evaluation

A lack of adequate railway simulation tools restrains our progress for improving development of railway networks. In our research, we designed a vigorous simulation tool in attempt to make UTDP approach for modelling complex railway networks practical.

We demonstrated capabilities of the new simulation suite by designing a novel train control system. However, to fairly assess the simulation platform the effort factor required to develop a new control system has to be considered. To simplify the development phase, tool provides a number of functions for computing a variety of railway network parameters, which can be easily called in the development framework. Moreover, a user-friendly graphical interface can provide user with important information about the capacity or speed profiles, as well as can visually replay a scenario. Although, the simplicity aspect of simulation tool utilized in many commercials tools like OpenTrack and RailSyS is reduced, as, the present use of the simulator requires some basic programming knowledge.

4 CONCLUSIONS

An expressive time-slicing simulator was developed in an effort to make the formal approach [15] for modelling novel signalling systems feasible. A simple application example of the simulation suite was demonstrated by developing an exotic train control system, performance of which was compared to the fixed block signalling system. As authors pointed out, the simulation platform is suitable for a variety of railway development and optimization purposes.

However, the current implementation of the simulation tool requires a basic programming knowledge. In order not to undermine expressiveness of the tool, yet make it more usable for railway engineers, two future work directions were identified. A signalling library with existing and novel train control mechanism will be developed to allow engineers to design more complex railway networks with heterogeneous signalling. The second direction is to link UTDP modelling language and object-oriented programming language currently used by the simulation platform. Finally, in the future we intend to focus on safety verification aspect of the approach. However, in order to be able to verify advanced train control systems formal semantics have to be revisited.

ACKNOWLEDGEMENTS

This work is supported by the RSSB SafeCap+ project. We are grateful to our colleagues from Siemens Rail Automation for invaluable feedback. We also thank Roberto Palacin and Luke Martin for their suggestions and comments.

REFERENCES

[1] Abril, M., Barber, F., Ingolotti, L., Salido, M.A., Tormos, P. Lova, A., An assessment of railway capacity. *Transportation Research Part E: Logistics and Transportation Review*, 44, pp. 774–806, 2008.

[2] IMPROVERAIL project. *Deliverable D10 of the project*. 2003.

[3] European Commission. *EU Policy and Its Impact on the Rail System*. 2006.

[4] A. González-Gil, R. Palacin, P. Batty, J.P. Powell., A systems approach to reduce urban rail energy consumption. *Energy Conversion and Management*, 80, pp. 509–524. 2014.

[5] OpenTrack Simulation Platform, www.opentrack.ch

[6] RailSyS Simulation Platform, www.rmcon.de

[7] Stankaitis, P., Iliasov, A., Safety Verification of Heterogeneous Railway Networks. *International Conference: Reliability, Safety and Security of Railway Systems*. 2016.

[8] Abrial, J.R., *The B-Book*. Cambridge University Press. New York, NY, 1996.

[9] Essame, D. Dolle, D., B in Large-Scale Projects: The Canarsie Line CBTC Experience. *LNCS*, vol. 4355. Springer, pp. 252–254, 2007.

[10] Iliasov, A., Romanovsky, A., SafeCap domain language for reasoning about safety and capacity. *Workshop on Dependable Transportation Systems at the Pacific-Rim Dependable Computing Conference*. 2012.

[11] Luethi, M., Nash, A., Weidmann, U., Laube, F., Wuest, R., Increasing Railway Capacity and Reliability through Integrated Real-Time Rescheduling. *Institute for Transport Planning and Systems*, ETH Zurich. 2007.

[12] Li, Wei., Tang, Tao., The impact of signalling system heterogeneity on urban transit capacity. *IEEE International Conference on Intelligent Rail Transportation (ICIRT)*, pp. 123–127, 2013.

[13] Grube, P., Núñez, F., Cipriano, A., An event-driven simulator for multi-line metro systems and its application to Santiago de Chile metropolitan rail network. *Simulation Modelling Practice and Theory*, 19, pp. 393–405, 2011.

[14] Paolucci, M., Pesenti, R., An Object-Oriented Approach to Discrete-Event Simulation Applied to Underground Railway Systems. *Transactions of the Society for Modelling and Simulation*, 72(6) pp. 372–383, 1999.

[15] Iliasov, A., Lopatkin, I., Romanovsky, A., Unified Train Driving Policy. *In Formal Methods Applied to Complex Systems*, eds. J.-L. Boulanger, Wiley, pp. 447–473, 2014.

[16] SafeCap Simulation Platform, www.safecap.co.uk

LED RAILWAY SIGNAL VS FULL COMPLIANCE WITH COLORIMETRIC SPECIFICATIONS

G. SCHIRRIPA SPAGNOLO[1], D. PAPALILLO[1], C. MALTA[2] & S. VINZANI[2]
[1]Department of Mathematics and Physics, University 'Roma Tre', Italy.
[2]Rete Ferroviaria Italiana (RFI), Roma, Italy.

ABSTRACT
In recent times, the transportation industry has generated a number of developments involving new technology in signalling. Important developments have involved the production of light by means of light-emitting diode (LED). Unfortunately, the maximum light output, dominant wavelength, reliability and the lifetime of LEDs are all closely related to the junction temperature. Accordingly, LED performance largely depends on the ambient temperature of its operating environment. Consequently, adequate heat-sinking is required to maintain a long lifetime and compliance. Today again, designing an LED railway signal that is able to meet compliance over wide range of temperatures is problematic. In particular, for yellow aspects compliance is difficult to achieve at extreme ambient temperature because the shift in the colour of yellow LEDs over the specified temperature range can be greater than the magnitude of colour specification. Use of phase change materials (PCMs) can provide practical solution of these problems. PCMs can store and release thermal energy during the process of melting and freezing (changing from one phase to another). When such a material freezes, it releases large amounts of energy in the form of latent heat of fusion or energy of crystallization. Conversely, when the material is melted, an equal amount of energy is absorbed from the environment as it changes from solid to liquid. This paper presents studies that enable the design of railway signals to meet, in a rigorous way, colorimetric compliances on a large range of temperature change. Furthermore, the proposed system is applicable, in a simple way and without operational changes, to the existing systems. It is sufficient to insert inside of the signals a suitable amount of a specific PCM.
Keywords: light-emitting diode (LED), phase change materials (PCM), railway signals.

1 INTRODUCTION

Light-emitting diodes (LEDs) are one of the fastest growing technologies in the lighting industry today. LEDs are solid-state semiconductor devices that directly convert electrical energy into light. Power, energy efficiency, maintenance savings, impact resistance and durability make LEDs suitable for safety signals for railway, road traffic and other applications in stationary transportation [1].

However, thermal management is still one of the major issues to be improved for implementing LEDs into lighting fixtures. In general, any electrical components generate heat whilst running; if these components are housed in a confined space, the internal temperature of enclosure can rise well above the ambient temperature, resulting degraded performance or burnout for the system. This is exacerbated if the enclosure is exposed to solar radiation. In fact, the reliability of electronic components is strongly influenced by the operating temperature. For the United States Air Force, more than 55% of failures in electronics are caused by high temperature [2]. Therefore, a good thermal management system is required to maintain specifications and have a long lifetime. Furthermore, when

DOI: 10.2495/TDI-V1-N3-568-577

the LEDs are used as light source in the railway signals, the optical proprieties of these have to maintain rigorous colour-intensity specifications on ample temperature range (typically −25 to +70°C). Again, one drawback with using LEDs in stationary transportation signal is that optical proprieties of LEDs degrade with increasing temperature. In other words, the intensity and the colorimetric coordinates of LED radiation depend on junction temperature [3]. In a typical railway signal, the heat generated inside the junction is dissipated into the ambient; changing the ambient temperature modifies the junction temperature and hence the characteristics of emitted light. The ambient temperature affects both the intensity and the colour of the radiation emitted by the LED systems; the intensity of the radiation decreases with higher ambient temperature and dominant wavelength increases. Therefore, in designing a signal system that makes use of LEDs, a good thermal management system is required to maintain a long lifetime and compliance. Hence, designing a LED railway signal that is able to meet compliance over wide range of temperatures is very problematic. In particular for yellow aspects, compliance is difficult to achieve at extreme ambient temperature because the shift in the colour of yellow LEDs over a specified temperature range can be greater than the magnitude of colour specification. Without a mechanism to compensate this phenomenon, unacceptable situations can occur on stationary transportation signals carried out with LEDs, especially when these signals are installed in climates where solar radiation can produce high temperatures inside their signal enclosure. To solve this problem, a number of conventional temperature compensation circuits have been designed which use sensors and opportune feedback techniques [4–9]. All of these systems work well. Unfortunately, they increase the complexity of the system. Consequently, with the same reliability and safety, we have higher production costs. The problem of incomplete conformity of the signals within a wide temperature range is very important. For this reason, the most important companies producing LED (e.g. NICHIA Corporation) have designed and manufactured LEDs for application in stationary transportation signals. These LEDs are devices made by combining blue LEDs and special phosphors. Consequently, it is possible to obtain LEDs with stable photometric characteristics [10]. However, while on one hand blue LED chips can have high reliability, on the other hand several reports have demonstrated that the phosphors can severely degrade during device lifetime. Mainly this is due to the high temperature levels reached in the proximity of the LED chips during operation [11–13]. For all these reasons, in this paper we propose a passive cooling that is able to accumulate energy during the day and release it during the night. In this way, we can minimize thermal excursions. In other words, we can eliminate the temperature peaks and get smaller thermal oscillation. The temperature fluctuations within a railway signal can be attenuated with adding thermal mass. Thermal mass acts as a passive energy store to attenuate temperature fluctuations inside signals [14]. In particular, adding phase change material (PCM) to thermal mass increases the effective heat capacity during the phase transition. This can anchor the temperature of the mass in a narrow band around the melting point of the PCM, reducing the temperature swings of the LEDs used in the signal. This paper presents the use of PCMs for thermal management solutions of railway signals. With use of the phase transition materials it is possible to design railway signals that are able to meet, in a rigorous way, colorimetric compliances on a large range of temperature. Furthermore, the proposed approach is applicable, in a simple way and without changes, on the signals in the existing systems. Simply insert a suitable quantity of specific PCM inside the railway light housings. The amount and the melting point of the PCM must be chosen in function of the environmental temperature variation.

2 OVERVIEW OF PCM

PCMs are materials which undergo a change in physical phase when a sufficient amount of heat is supplied. An example is transition from solid into liquid (melting). During the time when the material actively melts, additional heat, applied to material, does not change the temperature of the same. This phenomenon continues until the entirety of material melts. A typical temperature profile of a PCM is shown in Fig. 1. Like it shows, if heat is added to a solid PCM below its melt temperature or a liquid PCM above its melt temperature, the energy is stored as sensible heat and increases the temperature of solid or liquid. When phase change occurs PCMs operate by storing energy at a constant temperature, like transition from solid to liquid as illustrated in the centre of Fig. 1. In particular, in this phase transition, heat added to the material does not rise the temperature, and indeed it drives the phase change of material to a higher energy phase. In fact, liquid phase has a kinetic energy of the motion of atoms that is not present in the solid, so its energy is higher than that. In particular, the energy difference between the liquid and solid state defines latent heat – heat used for phase transition. Of course, when the solid is fully transformed into liquid, adding energy reverts to going into sensible heat and raising the temperature of the liquid [15].

Therefore, as liquid freezes and returns to a solid, it releases energy (heat) to the surrounding environment, which is stored in the kinetic energy of the motion of atoms. When a PCM freezes, it releases a large amount of energy in the form of latent heat at a relatively constant temperature. Conversely, when such material melts, it absorbs a large amount of heat from the environment. The property of storing and releasing thermal energy during the process of melting and freezing (transition from one phase to another) makes PCMs the ideal products for thermal management solutions. Like physical mass, PCMs offer the potential to reduce fluctuations in temperature and shift cooling loads to off-peak periods. However, in contrast to physical mass, whose energy storage capability is restricted by sensible heat, the ability of a PCM to store energy is characterized by its latent heat of fusion. So if latent heat of fusion increases, the material's capacity to store heat also increases. PCMs can be used for a variety of everyday applications that require temperature control. To meet better the various types

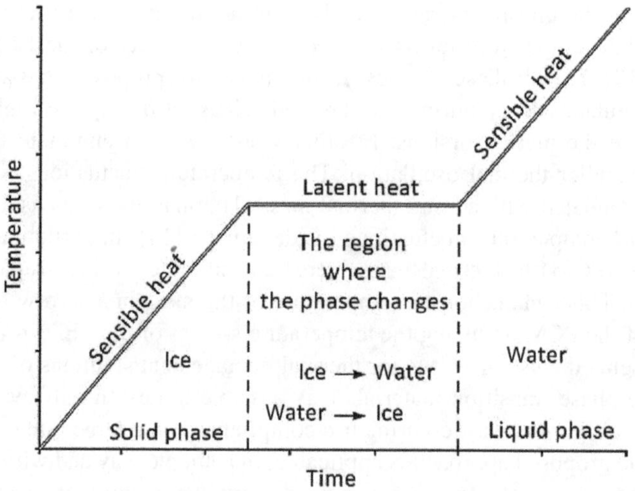

Figure 1: Temperature profile of a PCM. In the region where latent heat is effective, the temperature keeps either constant or in a narrow range. The phase of the material turns from one to another and both phases appear in the medium.

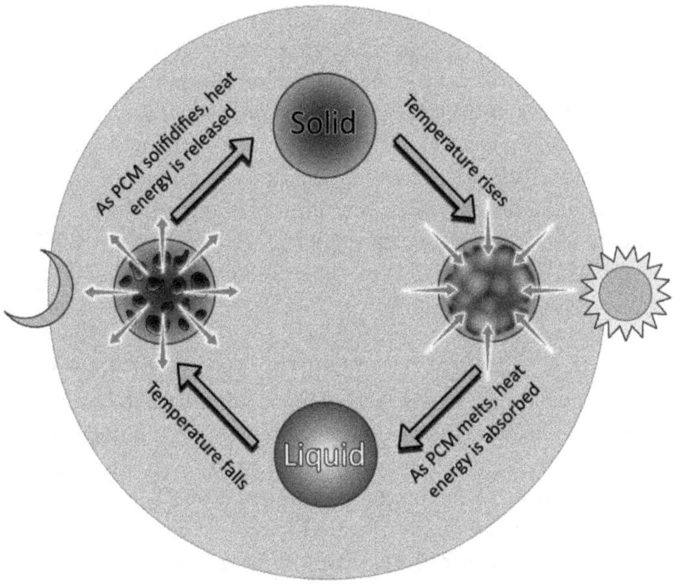

Figure 2: Description of how a PCM works.

of uses, like all materials that change phase at specific temperatures, PCMs are engineered to change phase at a temperature appropriate for their use and to store an unusual amount of thermal energy. PCMs are able to store 5–14 times more thermal energy per unit volume than conventional thermal storage materials. Figure 2 depicts schematically how PCM works. As a cycle, when temperature rises PCM melts and absorbs heat, and when temperature decreases PCM solidifies and provides a warming effect.

Heat is absorbed and released almost isothermally so that PCM reduces thermal peak loads. In fact, Fig. 3 shows the objective of PCM. As can be seen, PCM limits excessive temperatures by storing the excess heat during the day and releasing it during the night. This mode mimics the effect of thermal mass.

The use of PCMs has been proposed as a passive cooling technique of electronic systems [17–21]. Every electrical component generates heat during operation; if these components are housed in a confined space, the internal temperature of the housing can rise well above

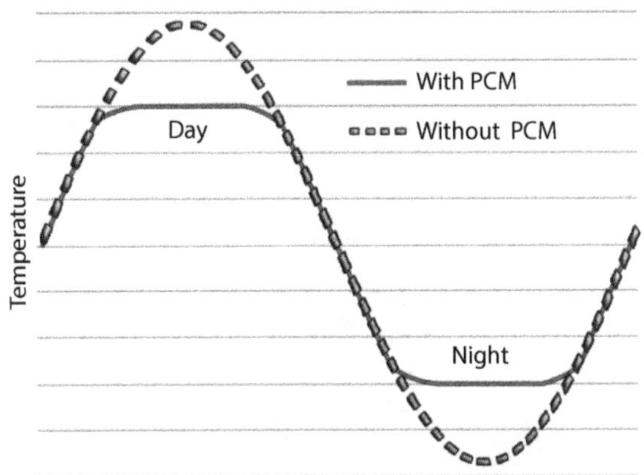

Figure 3: Objective of PCM; PCM limits excessive temperature variations.

the ambient temperature, resulting degradation of the performance. This is exacerbated if the housing is exposed to the sun. So, in many case, use of PCMs is a better alternative than classic cooling techniques. In fact, PCM heat sinks have been recognized, by the NASA, as an important tool in optimizing thermal control systems for space exploration vehicles and habitats that must deal with widely varying thermal loads and environments. Naturally, it is difficult to get a PCM ideal for a particular application. Certain amount of trade-off between different properties is therefore essential. Selection of a suitable PCM, determination of its quantity requirements based on heat-balance calculations, packaging of PCM, design of heat exchanging and heat distribution surfaces are other important steps in developing an effective PCM-based device or product [22].

3 SIGNAL CHARACTERIZATION: THERMAL EFFECT

In general, a railway signal is consisted of a housing inside which the actual signal is inserted. The thermal performance of signal system depends on a large number of factors. A block diagram that shows various factors affecting the heat balance of a signal system is presented in Fig. 4. The influence of these factors on performance of a signal system can be studied using an appropriate heat transfer model.

In our study, for the total heat rate (enthalpy flow rate at constant pressure), we have considered [23, 24]

$$C\frac{\partial T}{\partial t} = A \cdot K \cdot \Delta T + Q_{\text{internal}}, \tag{1}$$

where

K = global heat transfer coefficient associated to a transfer area A

$C = \sum_j m_j \cdot c_{pj}$, with m_i element mass and c_{pi} the corresponding specific heat (the subscript 'j' represents the signal system elements).

Q_{internal} = heat produced by the signal during its normal condition

ΔT = average temperature jump between the system and the surroundings.

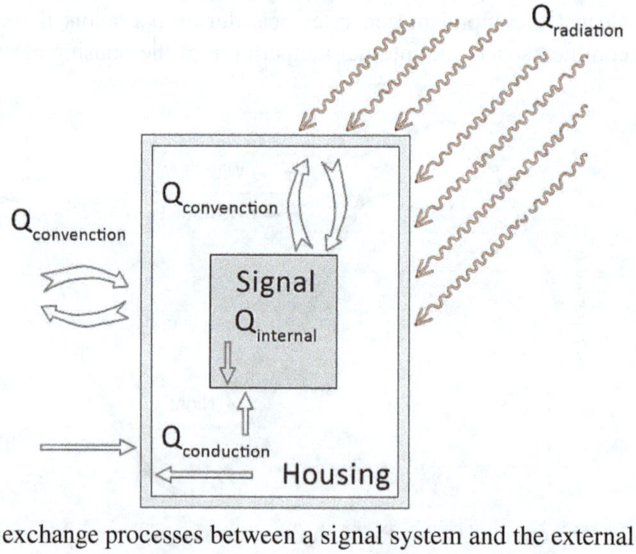

Figure 4: Heat exchange processes between a signal system and the external environment.

If the surface is exposed (as it is normally done for signals positioned along the railway line), we can consider $\Delta T = T_{so} - T_i$, where T_i is the signal housings temperature and T_{so} is the sol-air temperature, calculated using the expression

$$T_{so} = T_o + \frac{\alpha S_T}{h_o} - \frac{\varepsilon \Delta R}{h_o}, \tag{2}$$

where

T_o = daily average value of hourly ambient temperature
α = absorbance of the surface for solar radiation
S_T = daily average value of hourly solar radiation incident on the surface
h_o = outside heat transfer coefficient
ε = emissivity of the signal housings surface
ΔR = difference between the long wavelength radiation incident on the housings surface from the sky and the surroundings, and the radiation emitted by a black body at ambient temperature.

Determining all parameters of interest is extremely difficult. For this reason, after a first estimate, the model was calibrated by experimental measurements. Therefore, in the signal system some temperature sensors were positioned. Furthermore, temperature, wind direction and speed, and solar radiation were measured outside the housing. Figure 5 shows experimental set-up used. With this experimental apparatus we measured the temperature variations inside the signal vs the variability of environmental conditions. These measurements have allowed us to fully characterize our heat rate model. Figure 6 shows an example of the measurements used to calibrate the model.

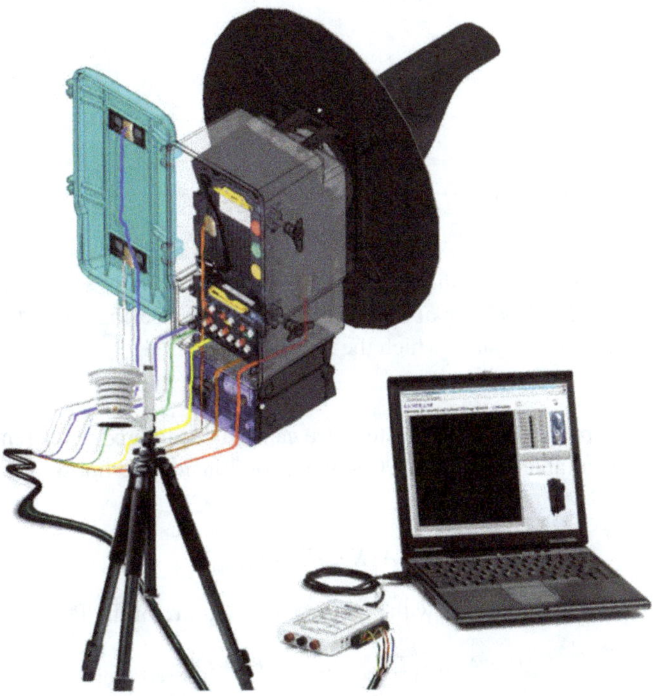

Figure 5: Experimental apparatus used to calibrate the total heat rate model.

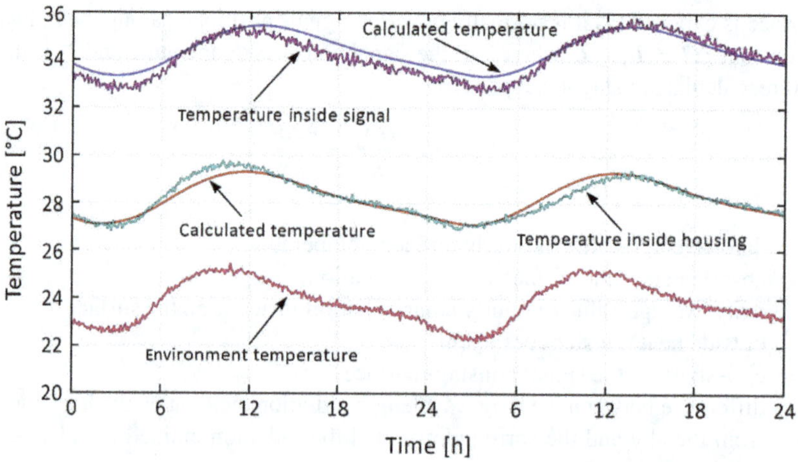

Figure 6: Example of measures made to calibrate the model.

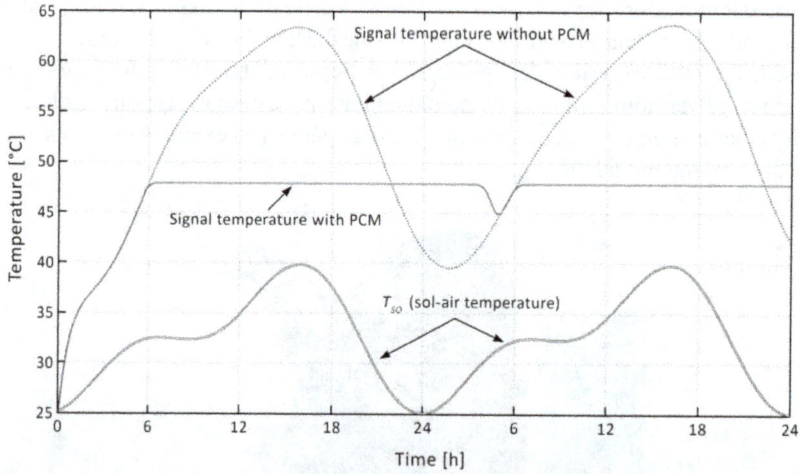

Figure 7: Numerical simulation showing the thermal behaviour of signal system with and without PCM. Looking at the figure is easy to see that the PCM reduces the temperature variation to which the signal is subjected.

Obtaining a realistic model for signal system, we can use this to predict the temperature trend in the case when a suitable amount of PCM is inserted inside the housing. In this case, the model becomes

$$C\frac{\partial T}{\partial t} = A \cdot K \cdot \Delta T + Q_{\text{internal}}, \tag{3}$$

which is valid for temperatures below the melting temperature of the PCM:

$$0 = A \cdot K \cdot \left(T_{\text{melting-PCM}} - T_{\text{i}}\right) + Q_{\text{internal}}, \tag{4}$$

in other cases, with the restriction that the PC is not completely melted.

Therefore, we have hypothesized to place inside the housing 6 kg (about 8 dm^3) of PCMRT-50Rubitherm®GmbH(latentheat168kJ/kgandmeltingpoint48°C).Wealsoconsidered that the signal is subjected to a sol-air temperature variation between 25 and 40°C. Finally, we made the assumption that every day, and at night, the PCM solidifies completely. Figure 7 shows the numerical simulation obtained with our model.

Figure 7 shows that the use of PCMs reduces the temperature range to which the signal is subjected. Reduced thermal excursions, in particular the high temperature, improve the performance of the LEDs and increase the reliability of the system. Obviously, the type and characteristics of the used PCM depend on the thermal characteristics of the signal and the average environmental conditions present in the installation area.

4 CONCLUSIONS

The objective of this paper has been to investigate the feasibility of using PCMs for application in thermal control of railway signals. In particular, we have shown that the use of the PCM can reduce the temperature fluctuations to which a rail signal is subjected. In particular, for yellow aspects compliance is difficult to achieve at extreme ambient temperature because the shift in the colour of yellow LEDs over the specified temperature range can be greater than the magnitude of colour specification. Use of PCMs can provide practical solution of these problems. In this paper, we introduced an innovative technical to allow the use of PCMs for thermal management solutions of railway signals. With the use of PCMs can be designed railway signals to maintain rigorous colour-intensity specifications on ample temperature range. Furthermore, since the use of the PCM decreases the likelihood that the equipment works at relatively high temperatures, the reliability and the lifetime of the signals increases. On the other hand, although the use of PCM has been proposed to improve the performance of railway signals, the same methodology can be used to improve reliability and decrease the probability of becoming out of order, for all those electronic systems subject to considerable temperature fluctuations. In fact PCM-based electronic cooling devices in aerospace and defence application are being used successfully by government agencies like NASA or defence industry. Unfortunately, these systems are unavailable off the shelf commercially and design information is a closely guarded trade secret. Anyway, successful systems with thermal management obtained by means of PCMs depend on thermal reliability and stability of PCM used. Many international companies like BASF, Climator, Cristopia, EPS Ltd., Mitsubishi Chemical Corporation, Rubitherm GmbH, TEAP, Witco., etc., are marketing PCMs, but it is still important to study PCMs for industrial applications and develop and test new materials.

REFERENCES

[1] Suozzo, N., Borg, D., Frering, M., Lucas, P. Vrabel, LED traffic lights: Signaling a global transformation. *Proceedings American Council for an Energy Efficient Economy Summer Study on Energy Efficiency in Buildings (ACEEE)*, Panel 6: Market Transformation, pp. 6.391–6.402, 2000, available at http://www.eceee.org/library/conference_proceedings/_ buildings/2000/Panel_6/p6_34 (accessed 22 February 2016).

[2] Yeh, L., Review of heat transfer technologies in electronic equipment. *Journal of Electronic Packaging,* **117(4)**, pp. 333–339, 1995. DOI: 10.1115/1.2792113.

[3] Zheludev, N., The life and times of the LED – a 100-year history. *Nature Photonics*, **1(4)**, pp. 189–192, 2007. DOI: 10.1038/nphoton.2007.34.

[4] Grossman, H. & Lambertville, N.J., LED driving circuitry with light intensity feedback to control output light intensity of an LED. *United States Patent* n. 6,153,985, 28

November 2000, available at https://www.google.com/patents/US6153985 (accessed 22 February 2016).

[5] Wu, C.H. & Chuang, H.J., LED power supply with temperature compensation, *United States Patent* n. 6,111,739, 29 August 2000, available at https://www.google.com/patents/US6111739 (accessed 22 February 2016).

[6] Schirripa Spagnolo, G., Papalillo, D. & Martocchia, A., LED applications in railway signals: wavelength and intensity vs temperature variation. *Journal of Transportation Technologies*, **2**, pp. 78–83, 2012. DOI: 10.4236/jtts.2012.21010

[7] Papalillo, D., Del Vecchio, P. & Schirripa Spagnolo, G., LED applications in road and railway signals: is it possible to fit specifications? *Proc. SPIE 8306, Photonics, Devices, and Systems V*, 83061N, 11 October 2011. DOI: 10.1117/12.912384.

[8] Schirripa Spagnolo, G., Papalillo, D. & Martocchia, A., Light emitting diode in stationary transportation applications: wavelength response to varying temperature. *Proceedings SPIE 8278, Light-Emitting Diodes: Materials, Devices, and Applications for Solid State Lighting XVI*, 82781P, 9 February 2012. DOI: 10.1117/12.909490.

[9] Schirripa Spagnolo, G., Papalillo, D., Martocchia, A. & Makary, G., Application of LEDs to traffic signal, *Environment and Electrical, 11th International Conference on Engineering (EEEIC)*, pp. 864–868 (Venice, Italy, 18–25 May 2012. DOI: 10.1109/EEEIC.2012.6221497.

[10] Mills, A., Phosphors development for LED lighting. *III-Vs Review*, **18(3)**, pp. 32–34, 2005. DOI: 10.1016/S0961-1290(05)01052-5.

[11] Meneghini, M., Trevisanello, L.R., Meneghesso, G. & Zanoni, E., A review on the reliability of Gan-based LEDs. *IEEE Transactions on Device and Materials Reliability*, **8**(**2**), pp. 323–331, 2008. DOI: 10.1109/TDMR.2008.921527.

[12] Meneghesso, G., Meneghini, M. & Zanoni, E., Recent results on the degradation of white LEDs for lighting, 19 August 2010. *Journal of Physics D: Applied Physics*, **43**(**35**), p. 354007, 2010. DOI: 10.1088/0022-3727/43/35/354007.

[13] Meneghini, M., Dal Lago, M., Trivellin, N. & Meneghesso, G., Thermally activated degradation of remote phosphors for application in LED lighting, *IEEE Transactions on Device and Materials Reliability*, **13**(**1**), pp. 316–318, 2013. DOI: 10.1109/TDMR.2012.2214780.

[14] Richardson, M.J. & Woods, A.W., An analysis of phase change material as thermal mass. *Proceedings of the Royal Society A*, pp. 1029–1056, 2008. DOI: 10.1098/rspa.2007.0312.

[15] Al-Hallaj, S. & Kizilel, R., Applications of phase change materials for sustainable energy, OpenStax-CNX module: m41734, version 1.4, Mar 1, 2012, available at http://cnx.org/content/m41734/1.4/ (accessed 22 February 2016).

[16] Hale, D.V., Hoover, M.J. & O'Neill, M.J., Phase change materials handbook. NASA CR-61363, Marshal Space Flight Centre. Alabama. 1971, available at http://ntrs.nasa.gov/archive/nasa/casi.ntrs.nasa.gov/19720012306.pdf (accessed 22 February 2016).

[17] Kumar, R., Misra, K.M., Kumar, R., Gupta, D., Sharma, P.K., Tak, B.B. & Meena, S.R., Phase change materials: technology status and potential defence applications. *Defence Science Journal*, **61**(**6**), pp. 576–582, 2011. DOI: 10.14429/dsj.61.363.

[18] Kandasamy, R. & Wang, X.Q., Application of phase change materials in thermal management of electronics. *Applied Thermal Engineering,* 27(17–18), pp. 2822–2832, 2007. DOI: 10.1016/j.applthermaleng.2006.12.013.

[19] Pal, D. & Joshi, Y.K., Melting in a side heated tall enclosure by a uniformly dissipating heat source. *International Journal of Heat and Mass Transfer*, **44**, pp. 375–387, 2001. DOI: 10.1016/S0017-9310(00)00116-2.

[20] Tan, F.L. & Tso, C.P., Cooling of mobile electronic devices using phase change materials. *Applied Thermal Engineering*, **24**, pp. 159–169, 2004. DOI: http://dx.doi.org/10.1016/j.applthermaleng.2003.09.005

[21] Lu T.J., Thermal management of high power electronics with phase change cooling. *International Journal of Heat and Mass Transfer*, **43(13)** pp. 2245–2256, 2000. DOI: 10.1016/S0017-9310(99)00318-X.

[22] Mehling, H., & Cabeza, L.F., *Heat and Cold Storage with PCM. Hand Book*, Springer: Berlin, Germany, 2008. ISBN: 978-3540685562

[23] Kreith, F., Manglik, R.M. & Bohn, M.S., *Principles of Heat Transfer*, 7th ed., CL Engineering: Stamford, CT, 2010. ISBN-13: 978-0495667704.

[24] Dutila, Y., Roussea, D.R., Salahb, N.B., Lassuec, S. & Zalewskic, L., A review on phase-change materials: Mathematical modeling and simulations, *Renewable and Sustainable Energy Reviews*, **15**(**1**), pp. 112–130, 2011 DOI: 10.1016/j.rser.2010.06.011.

TRAVELLING VIRTUAL BALISE FOR ETCS

A. FILIP
Faculty of Electrical Engineering and Informatics, University of Pardubice, Czech Republic.

ABSTRACT

The railway industry has taken a great effort and is currently focused on exploitation of global navigation satellite system (GNSS) for the European train control system (ETCS). It has been assessed that replacement of track balises, used for safe train location determination, with virtual balises (VBs) detected by means of GNSS will significantly reduce the track-side infrastructure and operational costs. However, this innovated ETCS can be put into operations only in the case when detection of VBs by means of GNSS will achieve the same safety integrity level (SIL 4) and availability as it is required for physical balise groups (BGs).

This paper describes a novel travelling virtual balise (TVB) concept, which was proposed to meet the demanding ETCS safety requirements for GNSS using the existing European Geostationary Navigation Overlay Service (EGNOS) safety-of-life (SoL) service. The TVB concept profits from the basic feature of GNSS – i.e. the ability of abundant train position determination in GNSS service volume, which cannot be realized by current track balise groups (BGs) with a spacing of hundreds of metres or more. The frequent GNSS train positions are utilized for (1) fast diagnostics of on-board location determination system (LDS), (2) introduction of reactive fail-safety into LDS and (3) derivation and justification of the ETCS safety requirements for EGNOS.

The TVB concept brings one significant advantage to ETCS in contrast to the static VBs – i.e. the safety requirement for LDS doesn't depend longer on the distance between successive VBs. It means that the existing spacing between physical BGs (up to 2.5 km) can be also preserved in case of TVBs. It can significantly improve the availability of LDS. Further it was found that a less demanding tolerable hazard rate (THR) requirement for GNSS of about 1e-7/1 h or more still enables to meet the ETCS THR requirement for VB determination, i.e. THR_{VB} of 0.67e-9/1 h. Thus the ETCS TVB concept opens the door for efficient use of the EGNOS SoL service, originally developed for aviation.

Keywords: EGNOS, ETCS, Galileo, GPS, high-safety integrity, LDS, railway signalling, reactive fail-safety, safety-related systems, SIL 4.

1 MOTIVATION

Safe train position, velocity and time (PVT), among other basic quantities, are required for railway signalling. Nowadays all these quantities can be effectively provided by global navigation satellite system (GNSS). The very idea of the exploitation of global positioning system (GPS) satellites for signalling and train control is not new. The first experiments and field tests focused on train position determination using GPS and differential GPS (DGPS) were mainly performed in the United States and Europe more than 25 years ago. At that time the United States started work on the positive train control (PTC) system, which is based on GPS, and Europe chose the European train control system (ETCS), where track balises and on-board balise transmission module (BTM) were used for safe train position determination. Both PTC and ETCS have been successfully implemented on thousands of kilometres of tracks.

DOI: 10.2495/TDI-V1-N3-578-588

Besides Europe, ETCS has also become successful in Asia – mainly in China, Turkey, Taiwan and South Korea. However, additional installations of ETCS with 'classical' track balises on heavy haul and other lines may not be in some countries efficient (Australia, Russia, China, etc.) – mainly due to high investment and maintenance cost of ETCS balises. The European Commission was aware of this potential ETCS limitation and therefore gradually initiated several R&D projects oriented on the use of GNSS for signalling and ETCS in the 1990s.

There are two important dates regarding the use of GNSS for ETCS. Firstly, on 8 May 1998, a mixed train position determination solution by means of ETCS track balises and virtual GPS balises was described by Sterner [1]. Secondly, on 2 March 2011, the European Geostationary Navigation Overlay Service (EGNOS) based on GPS with its safety-of-life (SoL) service was officially declared available for safety operations in aviation. EGNOS belongs to the family of wide-area satellite-based augmentation systems (SBAS), which is similar to the following: US WAAS (RTCA DO-229 [2]), Japanese MSAT/QZSS and Russian SDCM. In spite of the fact that SBAS was originally developed and certified for safety operations in aviation, it also represents a strategic infrastructure for safety-related systems in land transport, including rail (Pullen et al. [3], Neri et al. [4]). It should also be mentioned that the European railway community also had a lot of expectations from nascent Galileo and its SoL service in the first decade of the new century. But finally the stand-alone Galileo SoL service was cancelled and it was decided to include Galileo in the multi-constellation and multi-frequency EGNOS V3, which should be operational around 2022.

Safe exploitation of aeronautical GNSS SoL service in railway signalling primarily means that the SoL service must be employed according to CENELEC standards (EN 50126 [5], EN 50128 [6], EN 50129 [7]). Excepting this, the GNSS SoL service must also be utilized effectively according to a realistic and generally acceptable strategy. The strategy says that GNSS should only replace the train location determination function, which is currently performed by means of track ETCS BGs and on-board BTM, a balise reader. Other ETCS subsystems, such as safe odometry (SIL 4), should remain untouched. And finally, there is a strong will in Europe to preferably employ the current well-working single-constellation EGNOS V2 (based on GPS) for ETCS and not to wait until 2022 for EGNOS V3. It is often said that 'what is sufficient for airplanes from viewpoint of safety, it must be also applicable for trains'. Later on Galileo or other constellations (GLONASS, BeiDou) could be used for availability of safety integrity improvement. In order to be able to think upon the EGNOS V2 exploitation for ETCS, it is necessary to make evidence first that EGNOS V2 is able to meet safety requirements for virtual balise (VB) detection. Due to this reason the travelling virtual balise (TVB) concept was proposed. It is demonstrated in this paper that the existing EGNOS SoL service, originally developed for safety operations in aviation, is using the TVB concept to meet the ETCS THR requirements for VB detection.

Differences in train position determination by means of physical BGs and VBs are outlined in Section 2. Section 3 briefly describes the allocation of ETCS tolerable hazard rate requirement to the VB and thus also for the entire LDS based on GNSS. Basic fail-safe principles applicable in railway safety-related systems compliant with SIL 3 and SIL 4 and their relation to the mono/multi-constellation LDS solutions are described in Section 4. Finally, the applicability of existing EGNOS V2 for the ETCS LDS with the reactive fail-safety architecture is justified by means of the so-called TVB and rapid LDS diagnosis in Section 5. The rapid and independent GNSS diagnosis undoubtedly represents the key to the LDS solution.

2 PHYSICAL AND VIRTUAL BALISES FOR TRAIN POSITION REPORTING

In order to demonstrate the positive impact of the GNSS diagnosis on the ETCS safety integrity, a train position determination function within the ETCS using classical track BGs and VBs is outlined first.

The classical ETCS track BG, also called information point (IP), which shall be compliant with SIL 4 and λ_{IP} of 1e-9/1 h (SUBSET-088 [8]), determines together with the on-board BTM the absolute position of a train. The ETCS odometry (SIL 4) provides the instant speed of a train and the relative distance from the last relevant balise group (LRBG) including its confidence interval (CI). The train position, velocity and other data are reported via GSM-R to the track-side Radio Block Centre (RBC). One of the important odometry functions is called linking of balises via relative distance measurement. It is in fact an independent diagnosis of balises and on-board unit (ONB) because it enables detection of a deleted (missing) balise, incorrectly inserted balise or ONB fault.

In case of the VB concept the absolute position of a train is determined using the LDS based on GNSS. The instant position of the train is compared with the position of VBs whose coordinates are stored in the on-board European vital computer (EVC) and in RBC. If the actual GNSS train position together with the relevant CI match with a VB stored in the database, then the VB is considered as the last relevant virtual balise (LRVB) (see Fig. 1). The odometry together with the track database perform two following functions: (1) diagnosis of the consecutive VBs by linking with its direct positive impact on the desirable reduction of the safety integrity requirement for the GNSS LDS – i.e. GNSS THR increasing, and also 2) provision of the relative train position from the LRVB for ordinary train position reporting to the RBC. The relative train position is also provided when a VB is missed due to temporal GNSS signal-in-space (SIS) service outages or SIS shadowing in a harsh railway environment. The above described VB concept copies in fact the ETCS train position determination function using track BGs.

Now it should be said how ETCS can profit from GNSS. Let's compare for this purpose the diagnosis of train position determination in case of ETCS with track balises and diagnosis in the ETCS VB concept.

The ETCS ONB is able to perform fault diagnosis of physical BGs and also its own diagnosis only in locations of BGs. It is possible, thanks to BG linking, because position of next BG with respect to the LRBG position is known to the ONB and the correct BG detection can

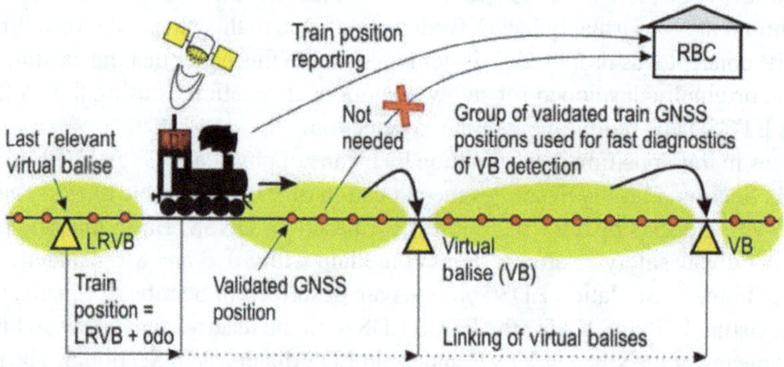

Figure 1: Exploitation of EGNOS SoL service within ETCS VB concept.

be validated using the so-called expectation window (ExW). The ExW includes all potential uncertainties due to odometry and BG position errors. As it is depicted in Fig. 1, GNSS LDS is able to perform its fault diagnosis also in the vicinity of VBs or on the whole track section between VBs, depending on SIS visibility. It is important for LDS initialization (system start-up) in staff responsible (SR) mode or during entry into an ETCS area from an unfitted area. At this moment it is considered that the position of a train on parallel tracks can be determined using existing track elements such as track circuits, axle counters or balises. But in the future the safe parallel track discrimination function will also be performed by GNSS (Neri et al. [9]).

Abundant GNSS train positions outside the VB vicinity are not in fact needed under normal operation (after LDS initialization) for the train position reporting to RBC because it is provided by means of the relative distance measurement from the LRVB. However it is evident that these abundant GNSS positions together with the odometry data can be effectively used for the GNSS diagnosis and it can finally lead to reduction of safety requirements for the LDS based on GNSS.

3 SAFETY REQUIREMENTS FOR VIRTUAL BALISE AND LDS

The ETCS THR requirements for VB and GNSS LDS were derived by means of the ETCS core THR allocation in SUBSET-088 [8] and SUBSET-091 [10]. The target ETCS core THR of 2e-9/1 h/train is equally allocated to ETCS ONB (1e-9/1 h) and all ETCS track-side equipment. Then THR related to balise transmission hazard THR_{BTX} of 0.67e-9/1 h was determined. THR_{BTX} was further sub-allocated to different track IP failure modes, such as balise deletion ($THR_{BTX\ Deletion}$ < 3.3e-10/1 h), balise insertion ($THR_{BTX\ Insertion}$ < 3.3e-10/1 h) and balise corruption ($THR_{BTX\ Corruption}$ < 1e-11/1 h). Since GNSS position is determined on board of a train, then only two following failure modes for VB were analysed: VB deletion and VB insertion (Filip and Rispoli [11], Filip [12]). These two VB failure modes can be described as follows:

- *VB Deletion* – means an event, when a VB (i.e. virtual IP) was not detected by means of on-board GNSS LDS. It can happen due to (1) excessive latent LDS error (wrong position) or (2) absence of train position in the GNSS LDS output. In both cases no VB is detected within the ExW provided by the odometry.
- *VB Insertion* – means an event when a wrong VB is determined due to wrong GNSS LDS position.

Since both VB failure modes are caused by a wrong GNSS LDS position (i.e. incorrect or no position), and diagnosis for both failure modes is provided by rapid and independent diagnosis in GNSS service volume, then the total THR_{BTX} of 0.67e-9/1 h was taken as THR for VB, i.e. THR_{VB} = 0.67e-9/1 h. THR_{VB} determines in fact THR for the entire GNSS-based LDS, i.e. $THR_{GNSS\ LDS}$, consisting of GNSS, odometry, etc. Tolerable hazard rate for single GNSS, i.e. THR_{GNSS} (THR_{SBAS}), will be then derived for a given LDS architecture and for the most demanding LDS operational mode within ETCS. If we omit temporally speed restriction (TSR) mode, then the most demanding operational mode from viewpoint of safety for baseline ETCS is start of mission. It is because linking (diagnosis) of BGs cannot be used during this mode. Therefore start of mission will be used for derivation of THR_{GNSS} – provided that safe parallel track discrimination can be solved using classical means, by two-tier GNSS augmentation (Neri et al. [9]), or by EGNOS V3 after 2022.

4 SINGLE- VERSUS MULTI-CONSTELLATION EGNOS FOR LDS

Railway safety-related systems to be compliant with SIL 3 or SIL 4 must ensure that they will remain safe in the event of any kind of single random HW fault. This principle is known as fail-safety and can be achieved by means of the following techniques: inherent fail-safety, composite fail-safety or reactive fail-safety (EN 50129 [7] and Lovric and Gülker [13]). It is evident that implementation of these techniques determines not only the level of LDS safety achieved, but also how efficiently GNSS will be used within the LDS.

The inherent fail-safety technique allows a safety-related function to be performed by a single channel, provided that all the credible failure modes of the channel are not hazardous. It would be very difficult or impossible to make such evidence in case of complex SBAS and therefore inherent fail-safety is not further considered for the SBAS (EGNOS)-based LDS.

The composite fail-safety technique allows a safety-related function to be performed by at least two independent channels. Hazardous fault in one channel shall be detected and negated sufficiently quickly to meet the required THR. The fault is detected by the comparison of the output values of these two or more channels, or also by means of an additional independent diagnosis. This technique has been already employed in case of the dual-constellation EGNOS-R interface (see Fig. 2) (Filip [12], Filip and Rispoli [14]). The EGNOS-R was mainly proposed with the intention to meet THR of 1e-8/1 h (SIL 4) for SIS and simplify the required safety case and certification according to railway CENELEC safety standards. It was found that introduction of EGNOS-R also enables to reduce the EGNOS CI magnitude in the position domain.

Finally, the reactive fail-safety technique allows a safety-related function to be performed by a single channel, provided its safe operation is assured by fast detection and negation of any dangerous fault. The existing single-constellation EGNOS V2 itself can be considered as a system with reactive fail-safety, because the safety function is performed by the GPS, and its correctness is checked by the EGNOS infrastructure (see Fig. 3(a)). Nevertheless, the stand-alone EGNOS V2 is not yet able to meet the ETCS SIL 4 requirement for train position determination. It is because the position determination function must also meet the required

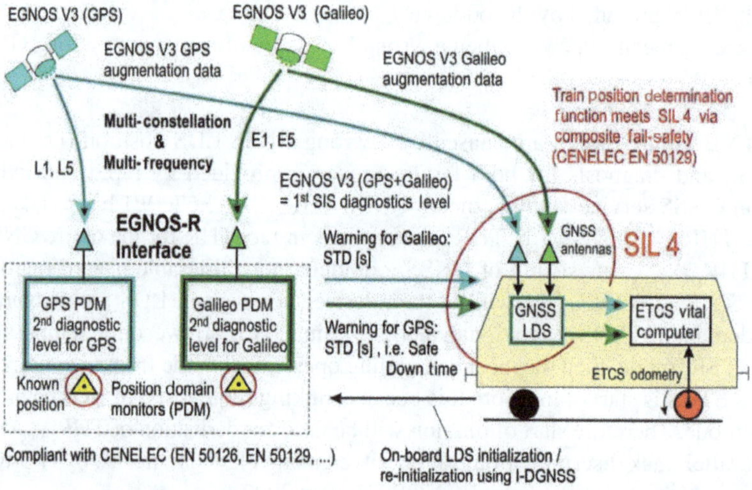

Figure 2: LDS based on dual-constellation EGNOS V3 and composite fail-safety.

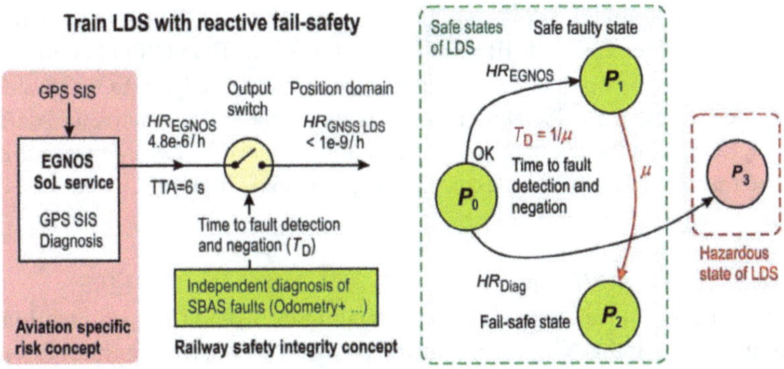

(a) (b)

Figure 3: LDS based on single-constellation EGNOS V2 with reactive fail-safety: (a) High-level architecture. (b) Markov model.

integrity level in case of local effects, such as multipath, EMI, spoofing, etc., against which SBAS does not protect. That's why the SBAS fault diagnosis must be completed with an additional independent fault diagnosis realized, e.g. using the safe ETCS odometry (SIL 4), 3-dimensional track database (SIL 4) and other relevant techniques.

Markov model of the LDS based on single-constellation EGNOS and reactive fail-safety is depicted in Fig. 3(b). The following four system states are defined for the model:

- P_0 – Fully functional LDS state: both SBAS and independent SBAS diagnosis work well according to the specifications;
- P_1 – Safe faulty LDS state: SBAS is faulty and rapid diagnosis is functional;
- P_2 – Fail-safe state of the LDS: SBAS fault was detected and negated;
- P_3 – Hazardous LDS state: independent diagnosis of SBAS is faulty. Note: Although SBAS is functioning properly according to the specifications, the LDS is in a dangerous state.

The corresponding time-dependent LDS state probability $P_1(t)$ can be derived from the model as follows:

$$P_1(t) = -\frac{HR_{\text{SBAS}}[e^{-(HR_{\text{SBAS}} + HR_{\text{Diag}})\cdot t} - e^{-\mu\cdot t}]}{HR_{\text{SBAS}} + \text{HR}_{\text{Diag}} - \mu}, \tag{1}$$

where HR_{SBAS} – hazard rate of SBAS per 1 h, HR_{Diag} – hazard rate of SBAS independent diagnosis, μ – rate of diagnosis and fault negation. $P_1(t)$ is the safe faulty state probability in case of SBAS fault. Since $(HR_{\text{SBAS}} + HR_{\text{Diag}})$ is much smaller than μ, eqn (1) can be simplified as follows:

$$P_1(t) \approx -\frac{HR_{\text{SBAS}} \cdot [1-0]}{-\mu} = HR_{\text{SBAS}} \cdot T_D, \tag{2}$$

where T_D is time to fault detection and negation, which is also sometimes called safe down time (SDT) (see (EN 50129 [7])). It is evident from eqn (2) that $P_1(t)$ depends on T_D (i.e. on $1/\mu$) and is no longer dependent on the time t. The corresponding LDS hazard rate during 1 h long mission can be expressed as $HR_{GNSS\ LDS} = HR_{SBAS} \times T_D \times 1\ h^{-1}$. This relation will be further used for derivation of the ETCS THR requirement for SBAS/EGNOS in Section 5.

The conclusion: while composite fail-safety was mainly used for EGNOS safety integrity improvement to meet ETCS safety requirements for the LDS, reactive fail-safety implemented in the LDS is on the contrary indented for reduction of ETCS safety requirements for EGNOS. The reactive LDS solution exploiting the existing aviation EGNOS V2 SoL service is described below.

5 FROM STATIC TO TRAVELLING VIRTUAL BALISE

The baseline ETCS requires both track balises and ONB for safe train position determination. On the other hand GNSS estimates the position on board of a train. Let us assume that λ_{ONB} is the rate of occurrence of ONB being unable to detect a correctly working ETCS IP. If linking of IPs is active, then the duration of ONB failure corresponds to the time interval T_L between two successive IPs marked as linked. Further if the average speed of a train is v and the linking distance D_L, then the probability of ONB failure causing the IP deletion is

$$P_{f,\mathrm{ONB}} = \lambda_{\mathrm{ONB}} \times T_L = \lambda_{\mathrm{ONB}} \times (D_L / v). \tag{3}$$

There is no safety requirement in respect of not being able to detect an IP when IP linking is active (see (SUBSET-088 [8])). As lately as two expected consecutive IPs announced by linking are not detected by ONB in the ExW, measured from the LRBG, the on-board vital computer shall consider the linking command of the second IP as a command to apply the service brake. Then the hazardous failure rate of ONB corresponding to the deletion of any IP met during one-hour-long mission is

$$HR_{\mathrm{ONB}} = \lambda_{\mathrm{ONB}} \times (2 \times T_L) \times 1\mathrm{h}^{-1}. \tag{4}$$

It should be mentioned here that balise insertion due to a wrong train position could have much more serious consequences; it could e.g. happen when system is initiated. In order to check the ONB functionality even before the detection of a regular and properly working BG by the ONB, an additional hypothetical 'testing' BG can be placed on the track ahead of the regular BG in the direction of movement from the LRBG (see Fig. 4). A much shorter ONB failure duration T_D is achieved in this case. Equation (3) can be then modified as

$$P_{f,\mathrm{ONB}} = \lambda_{\mathrm{ONB}} \times T_D, \tag{5}$$

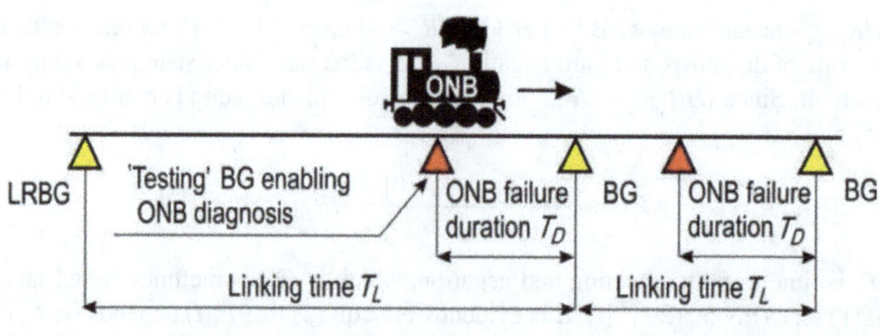

Figure 4: Diagnosis of ETCS on-board unit using 'testing' balise group.

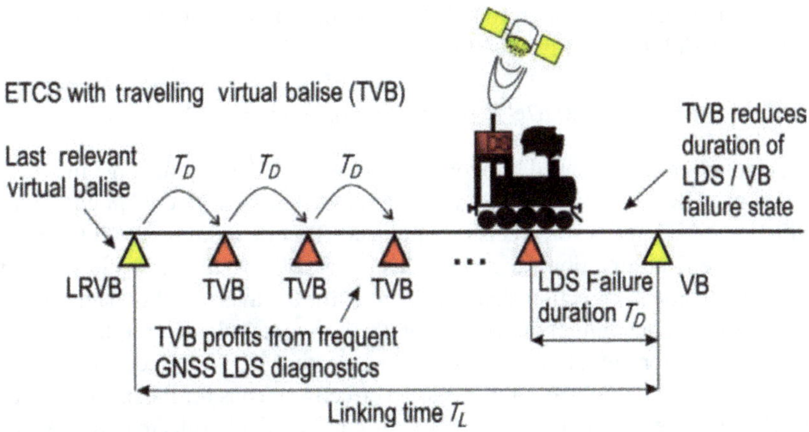

Figure 5: Travelling virtual balise introduced for justification of ETCS LDS architecture with reactive fail-safety.

and the corresponding ONB hazardous failure rate per mission (1 h) is

$$HR_{\mathrm{ONB}} = \lambda_{\mathrm{ONB}} \times T_D \times 1\ \mathrm{h}^{-1}. \tag{6}$$

Hazardous ONB failure rate related to IP deletion according to eqn (4) can be thus reduced with respect to HR in eqn (6) significantly. However it is evident that installation of the additional physical 'testing' BGs on a track would be very inefficient. Nevertheless, this reactive fail-safety principle can be easily implemented in case of the GNSS LDS. The 'testing' BG is simply replaced by the so-called TVB, as it is depicted in Fig. 5.

The TVB is equivalent to LRBG or LRVB from viewpoint of safety integrity because it is a validated GNSS train position using the independent diagnosis. The TVB arises from the LRVB as the validated subsequent GNSS train position generated just after the LRVB is detected and further *travels* to the next VB location in a given direction of movement. The TVB can also originate on a track section between two VBs during LDS initialization.

The detection function of the presence of an IP by ETCS on-board unit is a critical function and this function is the most critical when IPs are employed in such scenarios where linking is not used. It is e.g. during ONB initialization in SR mode or during entry into an ETCS area from an unfitted area when wrong IP can be inserted or IP can be deleted (missed). The ETCS THR requirement for GNSS must be derived using these scenarios considering that VB insertion could cause a more dangerous situation than VB deletion.

It is evident that the *travelling* VB can be utilized for the LDS diagnosis related to next VB deletion or insertion in the same manner as the hypothetical *static* 'testing' BG is used for the ONB diagnosis in Fig. 4. The ETCS THR requirement for GNSS, i.e. $\mathrm{THR}_{\mathrm{GNSS}}$, can be determined for the LDS start-up from the THR requirement for VB deletion or insertion per mission, i.e. $\mathrm{THR}_{\mathrm{GNSS\ VB}}$ of 0.67e-9 h^{-1} (see Sections 3 and 4), as follows:

$$0.67\mathrm{e}\text{-}9\ \mathrm{h}^{-1} = \mathrm{THR}_{\mathrm{GNSS}} * T_D * \mathrm{h}^{-1}, \tag{7}$$

where T_D is the duration of GNSS hazardous failure defined as the time interval between the two consecutive linked TVBs or linked TVB and next VB. Let's assume e.g. $\mathrm{HR}_{\mathrm{GNSS}}$ of 1e-7/h which corresponds to the SBAS integrity risk (IR) requirement for the aviation non-precision approach (NPA or en route). Then according to eqn (7) the acceptable hazard duration T_D due to VB deletion/insertion is

$$T_D = 0.67\text{e-}9/\ 1\text{e-}7 * 1\ \text{h} = 6.7\text{e-}3\ \text{h} = 24.12\ \text{s}.$$

It should be noted that the allowed horizontal alert limit (HAL) is quite large in this case, i.e. 0.3 nmi (556 m). HR of on-board GNSS receiver subsystem is neglected in this estimation.

The SIS IR of 2e-7/150 s for precision approach (PA) including LPV-200 operations is required in the vertical direction. Excepting this the SIS IR of 1e-9/150 s in the horizontal/lateral (one-dimensional) direction shall also be met for the aviation PA operations. It seems that the integrity (i.e. guarantee) of accuracy in the horizontal plane or in the track direction would be sufficient for signalling in case of the reactive LDS architecture. Nevertheless, three-dimensional (3-D) track map appears as an effective means for the independent diagnosis of SBAS, and therefore the IR of 2e-7/150 was conservatively selected for signalling. The SBAS SIS hazard rate approximately corresponds to 4.8e-6/1 h. Then the allowed duration of SBAS failure can be estimated as

$$T_D = 0.67\text{e-}9/\ 4.8\text{e-}6 * 1\ \text{h} = 1.36\text{e-}4\ \text{h} = 0.50\ \text{s}$$

The HAL of 40 m and VAL (vertical AL) of 35 m are required for LPV-200 operations, where the pilot's decision height is 200 feet (60 m) above the runway. The actual WAAS/EGNOS accuracies (95%) in horizontal/lateral and vertical directions are typically better than 1.1 m and 1.5 m, respectively. The real SBAS performance in terms of SIS integrity is better than it is required by aviation. Let's consider the real EGNOS IR for of 6e-8/150 s for LPV I operations. Then the corresponding EGNOS SIS hazard rate is 1.44e-6/1 h and the acceptable duration of failure T_D of 1.44 s can be estimated according to eqn (7). If SBAS receiver with an output rate of 10 Hz will be used, then all the above calculated values of T_D are realistic – however under the condition that all SBAS failure modes will be properly detected and negated.

In spite of the fact that the TVB is not a priori defined by means of its geographic coordinates, it can efficiently substitute a number of static VBs with known positions and small spacing between them, which might be otherwise introduced for reduction safety requirements for GNSS, but might be also hardly detectable.

6 CONCLUSION

The travelling virtual balise was introduced into the ETCS VB concept with the intention to justify the exploitation of the existing aviation EGNOS V2 SoL service for the train LDS to be compliant with SIL 4 at a system level. The TVB concept fully supports the harmonization of the aviation and railway safety requirements for efficient use of the single EGNOS V2 SoL service for both aviation *safety-critical systems* and railway *safety-related systems*. Further the TVB ensures the continuity in the ETCS balise concept evolution oriented from the baseline ETCS platform with physical track balises to more efficient virtual ones stored in the ONB and track-side RBC.

The term 'TVB' has been proposed to reflect the analogy between the ETCS static 'testing' track BG and 'testing' VB intended for rapid fault diagnosis of the entire ETCS LDS. The adjective 'travelling' means that geographic coordinates of the TVB are not a priori assumption. The TVB propagates on the track section between two subsequent VBs, which can have the same spacing as the track BGs. The abundant GNSS train positions together with the odometry data and combined with other diagnostic methods, e.g. pseudo-range validation using a 3-D track database, are used for fast TVB validation for the required safety integrity.

The TVB is a validated train GNSS position that meets the THR requirement for VB deletion or insertion, i.e. THR_{VB} of 0.67e-9/1 h. The diagnosis of both LDS ONB and GNSS SIS mainly relies on TVB/VB linking. The TVB concept is fully consistent with reactive fail-safety where the main channel itself, i.e. GNSS, may not meet the safety requirements, but the diagnostic channel must detect all possible dangerous failures so quickly that the overall safety target is met. The TVB brings the following additional benefits:

- TVB enables to preserve or even enlarge the VB spacing with respect to the maximum allowed ETCS BG spacing (2,500 m) without any impact on the entire LDS safety;
- Temporal TVB unavailability doesn't influence safety of train position determination because the safe ETCS odometry (SIL 4) is used for train position reporting in respect to the LRVB when it is required;
- Additional GNSS constellations such as Galileo, GLONASS or BeiDou can be used for LDS availability of integrity improvement;
- TVB efficiently substitutes a number of static VBs with small VB spacing;
- Introduction of the TVB doesn't influence the ETCS safety concept and the backward compatibility with the baseline ETCS is preserved.

ACKNOWLEDGEMENT

This work was supported from the European H2020 research and innovation programme within the RHINOS project (2016–2017).

REFERENCES

[1] Sterner, B.J., On the method of combining GPS and ETCS for localization purposes. *The European Railway Research Institute (ERRI)*, 6pp., Draft of 8th May 1998.

[2] *RTCA DO-229D – Minimum operational performance standards for GPS WAAS Airborne Equipment.* RTCA Inc.: Washington DC, 2006.

[3] Pullen, S., Walter, T. & Enge, P., Integrity for non-aviation users. *GPS World*, pp. 28–36, July 2011.

[4] Neri, A., Sabina, S., Rispoli, F. & Mascia, U., GNSS and odometry fusion for high integrity and high availability train control systems. *ION GNSS+ 2015*, Tampa, 11 pp., 14–18 September 2015.

[5] *CSN EN 50126 Railway Applications: The Specification and Demonstration of Dependability Reliability, Availability, Maintainability and Safety (RAMS).* CENELEC European Standard, 2002.

[6] *CSN EN 50128 Railway Applications: Communications, Signalling And Processing Systems – Software for Railway Control and Protection Systems.* CENELEC European Standard, 2003.

[7] *CSN EN 50129 Railway Applications: Safety related electronic systems for signalling.* CENELEC European Standard, 2003.

[8] *ETCS/ERTMS – Class 1, ETCS Application Levels 1 & 2 – Safety Analysis, Part 3 – THR Apportionment.* SUBSET-088 Part 3, ISSUE: 2.3.0, 91pp., 2 April 2008.

[9] Neri, A., Capua, R., Salvatori, P. & Palma, V., *ERSAT EAV High Integrity Two-tiers Augmentation Systems*, manuscript, 2016.

[10] *ETCS/ERTMS Safety Requirements for the Technical Interoperability of ETCS in Levels 1 & 2.* SUBSET-091, ISSUE: 3.3.0, 51pp., 8 May 2014.

[11] Filip, A. & Rispoli, F., Safety concept of GNSS based train location determination system SIL 4 compliant for ERTMS/ETCS. *Proceedings of ENC GNSS 2014*, Rotterdam, 15pp., 15–17 April 2014.

[12] Filip, A., Multi-constellation railway SBAS interface: A common platform for advanced signalling compliant with SIL 4 world-wide. *Proceedings of the International Heavy Haul Association 2015 Conference (IHHA)*, Perth, Australia, 10pp., 21–24 June 2015.

[13] Lovric, T. & Gülker, J., Singe channel ATP architectures, a new trend in Europe? *Proceedings of the World Congress on Railway Research (WCRR) 2001*. Köln, 9pp., 25–29 November 2001. http://www.uic.org/cdrom/2001/wcrr2001/pdf/sessions/3_5/040.pdf

[14] Filip, A. & Rispoli, F., SIL 4 Compliant train location determination system based on dual-constellation EGNOS-R for ERTMS/ETCS. *Proceedings of the International Symposium on Certification of GNSS System (CERGAL 2014)*, Dresden, Germany, pp. 109–114, 8–9 July 2014.

AN ANALYTICAL METHODOLOGY FOR EXTENDING PASSENGER COUNTS IN A METRO SYSTEM

R. DI MAURO, M. BOTTE & L. D'ACIERNO
Department of Civil, Architectural and Environmental Engineering, Federico II University of Naples, Italy

ABSTRACT

The planning of a rail system requires the definition of travel demand in terms of passenger (or freight) flows for sizing physical and technological elements (such as number of trains, signalling system type, length and width of platforms). Moreover, once a system has been set up and functional elements have been acquired, system management in terms of services and related timetables requires knowledge of travel demand flows. Much has been written about the methods and techniques for estimating travel demand by means of analytical models (calibrated by surveys), statistical processing of survey data and/or correcting model results by using properly collected traffic counts. However, whatever the adopted approach, it is necessary to proceed with survey campaigns to acquire experimental data. Obviously, the greater the number of detected data (and related acquisition costs and times), the greater the accuracy of travel demand estimations. Hence, in real cases, a fair compromise between survey costs and estimation accuracy has to be struck.

In this context, we propose an analytical methodology for identifying space–time relations between passenger counts to reduce the amount of data to be surveyed without affecting estimation accuracy. In particular, our proposal is based on defining analytical functions to provide boarding and alighting flows depending on the station (space component) and the time period (time component) in question. Finally, in order to show the feasibility of the proposed methodology and related improvements with respect to traditional approaches, we applied our proposal to the case of a real metro line in Naples (Italy) by comparing different levels of detail in passenger surveys.

Keywords: OD matrix estimation, public transport management, traffic count accuracy, travel demand estimation.

1 INTRODUCTION

The promotion of public transport may be viewed as one of many useful strategies to reduce negative externalities on road networks (see, for instance, European Commission [1]). Indeed, although numerous policies have been applied to improve road system conditions in terms of safety (as shown by Dell'Acqua and Russo [2] and Dell'Acqua et al. [3]) or analysis of driver behaviour (as proposed by Bifulco et al. [4] and Pariota et al. [5]), the most successful strategies are based on the variation in user modal choices which allows sharp reductions in accidents, energy consumption, traffic congestion and air and noise pollution. In this context, adoption of interventions aimed at increasing the attractiveness of public transport (see Ca-scet-ta et al. [6, 7]) or reducing mass-transit operator costs and passenger disutilities (D'Acierno et al. [8], Gallo et al. [9] and D'Acierno et al. [10]) may represent a useful way for achieving the intended purposes.

However, whatever the adopted approach, it is necessary to estimate travel demand in terms of potential or expected passengers (or freights) with related features (i.e. starting and

DOI: 10.2495/TDI-V1-N3-589-600

arrival stations, adopted time slot, trip duration, etc.). Indeed, such information is useful both for planning and management phases of transportation systems. This has generated an extensive literature on the estimation and forecasting methodologies of travel demand, which is summarized below.

In general, estimation of current and future demand can be performed by (Cascetta [11]) direct estimations, disaggregated estimations and aggregated estimations. The first approach, indicated in the literature (see, for instance, Smith [12], Brog & Ampt [13] and Ortuzar & Willumsen [14]) as *direct estimation*, can be adopted to determine only 'present' travel demand. It is based on the application of sampling theory in the case of mobility choices. The main limits of this methodology consist in the huge number of information to be collected and the inability to predict future developments due to transportation network or socio-economic variations.

The second approach, known as *disaggregated estimation* (see, for instance, Domencich & McFadden [15], Horowitz [16], Manski & McFadden [17] and Ben-Akiva & Lerman [18]), consists in specifying (i.e. providing the functional form and related variables), calibrating (i.e. determining numerical values of model parameters) and validating (i.e. verifying the ability of the model to reproduce original data) a model by means of proper data. These data express disaggregate information related to a sample of individuals, where the size and the sample characteristics generally differ from those used in the first approach. This methodology allows mobility choices to be simulated in current conditions (based on the ability to reproduce sampling data) and in the case of future conditions (based on the ability to simulate user reactions to transportation network or socio-economic variations). The above disaggregated approach is referred to in the literature as *the revealed preference* approach (Cascetta [11]) since it is based on the use of data related to real behaviour of travellers. Recently (see, for instance, Ben-Akiva & Morikawa [19] and Ortuzar [20]), the *stated preference* approach has been developed, based on the statements of travellers about their appropriately described and designed preferences in hypothetical scenarios. With the use of this second approach the prediction abilities of the calibrated demand models can be improved.

Finally, the last approach (see, for instance, Lo & Chan [21], Cascetta et al. [22] and Lu et al. [23]), known as *aggregated estimation*, is based on modifying demand model results after correcting them by means of traffic counts (i.e. vehicular or passenger flows). The aim of this approach is to identify an origin-destination (OD) matrix which is closest to its estimation by model and, once it is assigned to the network, generates flows closest to the counting data.

The brief analysis of methods for estimating travel demand shows that it is always necessary to conduct survey campaigns to acquire experimental data. In particular, the greater the number of surveyed or detected data (and related acquisition costs and times), the greater the accuracy of travel demand estimations. Hence, in real cases, a fair compromise between survey costs and estimation accuracy should be achieved.

In this context, our proposal is to provide an analytical procedure for identifying some space–time functions in order to reduce the number of data to collect without significantly affecting estimation accuracy. In the particular case of a metro system, we aim to identify analytical relations which express boarding and alighting flows of passengers depending on the station (space component) and the time period considered (time component).

The article is organized as follows: Section 2 provides general features of the proposed approach; Section 3 applies the methodology in the case of a real metro line; finally, conclusions and research prospects are summarized in Section 4.

2 DEFINITION OF SPACE–TIME RELATIONS

The proposed methodology is based on the assumption that the spatial correlation (i.e. the correlation among different stations) and temporal correlation (i.e. the correlation among different time periods) of passenger flows is generally not null. The reason lies in the framework of travel demand and its temporal distribution.

In the literature, the definition of functions for determining travel demand can be addressed through two main approaches (Cascetta [11]): descriptive and behavioural. The former (see, for instance, Oi & Shuldiner [24] and Wilson [25]) provides relations among variables without any assumption on users' behaviour; the latter (see, for instance, Ortuzar & Willumsen [14], Domencich & McFadden [15] and Ben-Akiva & Lerman [18]) is based on explicit assumptions of users' choice behaviours. Our proposal is based on adopting the first approach.

In particular, in order to investigate whether it is possible to identify one or more space–time functions to calculate passenger flows with a descriptive approach, we propose the following five-step procedure:

- design and execute a survey campaign in order to acquire a sufficient amount of data for applying the following statistical analyses;
- define a partial set of data, obtained by hiding (i.e. assuming not detected) some data with a predetermined sampling rate;
- perform a mono-dimensional statistical analysis on the partial data set in order to identify the optimal functional form;
- perform a multi-dimensional statistical analysis on the partial set of data, in order to specify, calibrate and validate one or more space–time functions;
- validate the methodology by comparing the simulation of the metro system by using data of the whole set (assumed as '*absolute truth*') and those in the case of data obtained by space–time functions identified.

The first step consists in analysing the main features of the context such as the number and location of stations, the layout in terms of platforms and accesses of any station, operating hours and timetables, and rolling stock characteristics. This information allows a proper survey plan to be designed, indicating 'when' and 'where' passenger flows must be counted. Obviously, this phase includes the execution of surveys, whose appropriately noted results can be shown as in Fig. 1.

Since the aim of the procedure is to provide a tool for reducing the amount of data to be surveyed, the second phase consists in determining a partial data set to be analysed. In particular, we simulate the adoption of a prefixed sampling rate during surveys, for instance 50%, by

		Station					
		st_1	st_2	...	st_i	...	st_m
Time period	tp_1	$f_{1\text{-}1}$	$f_{1\text{-}2}$	...	$f_{1\text{-}i}$	...	$f_{1\text{-}m}$
	tp_2	$f_{2\text{-}1}$	$f_{2\text{-}2}$	...	$f_{2\text{-}i}$	...	$f_{2\text{-}m}$
	...	...	...	...	...	...	...
	tp_n	$f_{n\text{-}1}$	$f_{n\text{-}2}$	...	$f_{n\text{-}i}$	...	$f_{n\text{-}m}$

Figure 1: Surveyed data (i.e. real surveyed data).

hiding some detected data. In this way, we apply the proposed methodology in the case of a limited subset of surveyed data (see Fig. 2).

With the information collected and properly processed, the first kind of statistical analysis can then be implemented, that is the mono-dimensional approach. It consists in checking the class of functions which best describes the simulated survey data. This procedure is indicated as mono-dimensional because the analysed functions are defined in an R^2 space, where the abscissa is the sequence of stations or the time periods and the ordinate is the surveyed flow (as shown in Fig. 3).

Hence, for each class of functions, it is necessary to analyse $(n_{st} \times 2) \times n_{tp}$ set of data, where n_{st} is the number of the stations (multiplier 2 is required to consider outgoing and return trips separately) and n_{tp} is the number of time periods considered. Since there are two flow types (i.e. boarding and alighting flows) to be considered, assuming n_{fc} as the number of function classes to be analysed, the second phase requires the calibration and the validation of n_f mono-dimensional functions, where

$$n_f = n_{fc} \cdot 2 \cdot (n_{st} \cdot 2) \cdot n_{tp}. \tag{1}$$

Obviously, for each class of function, suitable validation tests have to be performed.

Having defined the best class of mono-dimensional functions, the fourth phase consists in providing the specification, calibration and validation of four surfaces, where the independent parameters are the stations and the time periods, while the surface expresses the value of passenger flows. The number of functions to be defined, which is equal to four, is related to the fact that we have to consider two kinds of passenger flows (boarding and alighting flows) and two kinds of trips (outgoing and return trips). Obviously, also in this case the calibration set consists in the simulated (i.e. partial) surveyed data, shown in Fig. 2.

		Station					
		st_1	st_2	…	st_i	…	st_m
Time period	tp_1	$f_{1\text{-}1}$	missing	…	$f_{1\text{-}i}$	…	missing
	tp_2	missing	$f_{2\text{-}2}$	…	missing	…	$f_{2\text{-}m}$
	…	…	…	…	…	…	…
	tp_n	$f_{n\text{-}1}$	missing	…	$f_{n\text{-}i}$	…	missing

Figure 2: Partial set of surveyed data (i.e. simulated surveyed data).

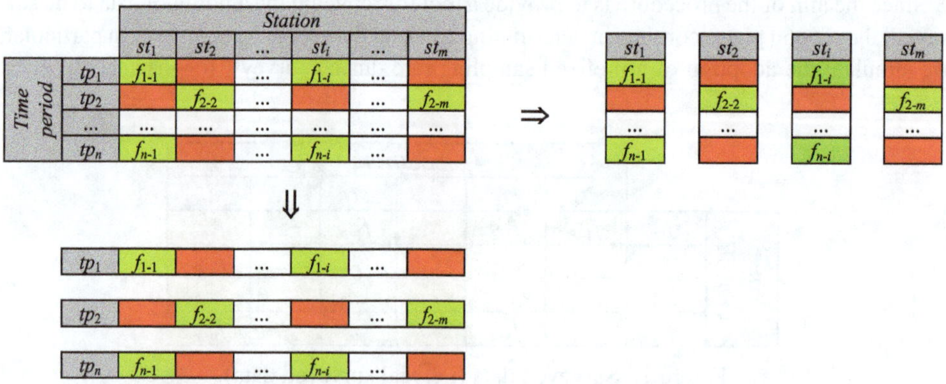

Figure 3: Organization of data for mono-dimensional analyses.

		Station					
		st_1	st_2	...	st_i	...	st_m
Time period	tp_1	$f_{1\text{-}1}$	*function data*	...	$f_{1\text{-}i}$	...	*function data*
	tp_2	*function data*	$f_{2\text{-}2}$	...	*function data*	...	$f_{2\text{-}m}$
	...	...	...	...	...	...	...
	tp_n	$f_{n\text{-}1}$	*function data*	...	$f_{n\text{-}i}$	...	*function data*

Figure 4: Subset extension by means of function data.

		Station					
		st_1	st_2	...	st_i	...	st_m
Time period	tp_1	*function data*	*function data*	...	*function data*	...	*function data*
	tp_2	*function data*	*function data*	...	*function data*	...	*function data*
	...	...	...	...	...	...	...
	tp_n	*function data*	*function data*	...	*function data*	...	*function data*

Figure 5: Function data for all values.

The last phase consists in comparing application results obtained by using the whole set of the survey data and those using the data of calibrated functions properly combined with the data of calibration subsets. Three different combinations of sets may be obtained for comparison: only the calibration subset (Fig. 2), the calibration subset extended by replacing missing data with function data (Fig. 4) and only function data for all values (Fig. 5).

3 APPLICATION OF THE PROPOSED METHODOLOGY

In order to show the feasibility of the proposed methodology, we applied it in the case of Metro Line 1 in Naples (Italy) which is about 18 km long and consists of 18 stations (further details can be found in Botte et al. [26]) from the suburbs (Piscinola) to city centre (Garibaldi).

Survey activities were implemented in July 2015 to collect data related to daily flows on an average working day in summer. It is worth noting that investigations were organized to detect flows for each single access (gate, stair, elevator, etc.) which were subsequently grouped according to platforms and travel directions. Finally, data were organized by considering three time periods and 18 stations; therefore we obtained four matrices of dimensions 3×18, whose framework is shown in Fig. 1.

3.1 Calibration of analytical functions

As described in the previous section, the calibration phase was divided into two sub-phases: a mono-dimensional phase and a multi-dimensional phase. Obviously, it was first necessary to define the simulated surveyed rate which was assumed equal to 50%.

The first analysis consisted in dividing each of the four survey matrices into 54 (i.e. 3 ×× 18) vectors and testing the goodness of fit (i.e. the discrepancy between surveyed data and function data). The following classes of functions were tested: linear, quadratic, cubic, fourth-degree polynomial, fifth-degree polynomial, power, logarithmic and exponential functions. However, due to the scarcity of data along the matrix columns (i.e. there were at most two data), linear functions were adopted only in row analyses.

The goodness of fit of each class of function was estimated by means of the coefficient of determination $\Re^2$, calculated as follows:

$$\Re^2 = \left(\sum_i (f_i - \bar{\phi})^2\right) \Big/ \left(\sum_i (\phi_i - \bar{\phi})^2\right), \tag{2}$$

with

$$\bar{\phi} = \sum_i \phi_i \Big/ n \quad, \tag{3}$$

where φ_i is the i-th simulated survey data (i.e. known value of Fig. 3), n is the number of simulated survey data φ_i, $\bar{\varphi}$ is the mean of data φ_i and f_i is the i-th value assumed by the calibrated function.

The class which provided the best R^2 values were polynomial functions. Hence, we implemented the multi-dimensional phase by adopting this category of functions, obtaining the following formulations:

$$f_1^k(x,y) = a_1^k x^2 y^3 + a_2^k x\ y^4 + a_3^k x\ y^3 + a_4^k y^4 + a_5^k y^3, \tag{4}$$

$$\begin{aligned} f_2^k(x,y) = {} & b_1^k x^4 + b_2^k x^3 y + b_3^k x^2 y^2 + b_4^k x\ y^3 + b_5^k y^4 + b_6^k x^3 + b_7^k x^2 y + \\ & + b_8^k x\ y^2 + b_9^k y^3 + b_{10}^k x^2 + b_{11}^k x\ y + b_{12}^k y^2 + b_{13}^k x + b_{14}^k y + b_{15}^k, \end{aligned} \tag{5}$$

with

$$k \in \{AG; BG; AP; BP\}, \tag{6}$$

where x represents the time period; y represents the sequence of stations ($y = 1$ in the case of Piscinola and $y = 18$ in the case of Garibaldi); AG represents alighting flows (A) in the case of the outgoing trip (i.e. Garibaldi direction); BG represents boarding flows (B) in the case of the outgoing trip (i.e. Garibaldi direction); AP represents alighting flows (A) in the case of the return trip (i.e. Piscinola direction); BP represents boarding flows (B) in the case of the return trip (i.e. Piscinola direction).

It is worth noting that eqn. (4) is defined in the case of $k \in \{AG; AP; BP\}$, while eqn. (5) is defined only in the case of $k = BG$. Therefore, values f_i are calculated by means of eqn. (4) or eqn. (5) according to the values assumed by parameter k.

However, we adopted as global statistical tests: $\Re^2$ (formulated as eqn. (2)); adjusted $\Re^2$ (indicated as $\bar{\Re}^2$) and F-test (indicated as F), formulated as follows:

$$\bar{\Re}^2 = \Re^2 - \left(1 - \Re^2\right) \cdot p/(n-p-1), \tag{7}$$

$$F = \Re^2 \cdot (n-p-1) \Big/ \left(\left(1 - \Re^2\right) \cdot p\right), \tag{8}$$

where p expresses the number of function parameters, which is equal to 5 in the case of function (4) and equal to 15 in the case of function (5). Moreover, we also performed the t-student (indicated as t) test of coefficients, calculated as

$$t = \left|a_i^k\right| \Big/ \sqrt{\mathrm{Var}\left(a_i^k\right)} \quad \text{or} \quad t = \left|b_i^k\right| \Big/ \sqrt{\mathrm{Var}\left(b_i^k\right)}, \tag{9}$$

where Var(), or equivalently Var(a_i^k), is the i-th element of the main diagonal of variance–covariance matrix S, obtained as

$$\Sigma = -\left[\partial^2 \varepsilon / \left(\partial a_i^k \ \partial a_j^k\right)\right]^{-1} \text{ or } \Sigma = -\left[\partial^2 \varepsilon / \left(\partial b_i^k \ \partial b_j^k\right)\right]^{-1}, \tag{10}$$

with

$$\varepsilon = \sum_i (f_i - \phi_i)^2 . \tag{11}$$

Table 1 reports the results of the global statistical tests; Tables 2–5 describe the results of statistical tests of coefficients.

Table 1: Global statistical tests.

Function type (k)	$\Re^2$	$\bar{\Re}^2$	*F*-test		
			F value	Threshold	Confidence level (%)
AG	0.764	0.701	12.273	8.018	99.90
BG	0.829	0.572	3.220	3.190	94.00
AP	0.621	0.521	6.218	5.967	99.50
BP	0.514	0.386	4.023	4.016	97.00

Table 2: Coefficient statistical tests of coefficients: *AG* condition.

Parameter	a_1^k	a_2^k	a_3^k	a_4^k	a_5^k
Value	0.0624	0.0184	–0.5150	–0.0272	0.7285
t-Value	489.62	588.89	706.461	398.376	588.973
Threshold	5.077	5.077	5.077	5.077	5.077
Confidence level (%)	99.99	99.99	99.99	99.99	99.99

Table 3: Coefficient statistical tests: *AP* condition.

Parameter	a_1^k	a_2^k	a_3^k	a_4^k	a_5^k
Value	–0.0552	0.0096	0.0677	–0.0585	0.7942
t-Value	375.497	184.120	69.562	525.132	468.483
Threshold	5.077	5.077	5.077	5.077	5.077
Confidence level (%)	99.99	99.99	99.99	99.99	99.99

Table 4: Coefficient statistical tests: *BG* condition.

Parameter	b_1^k	b_2^k	b_3^k	b_4^k	b_5^k
Value	136.82	–71.15	3.82	1.05	0.16
t-Value	$2.381{\cdot}10^6$	908.94	367.44	$2.892{\cdot}10^6$	$1.845{\cdot}10^8$
Threshold	6.412	6.412	6.412	6.412	6.412
Confidence level (%)	99.99	99.99	99.99	99.99	99.99
Parameter	b_6^k	b_7^k	b_8^k	b_9^k	b_{10}^k
Value	–557.32	362.08	–48.62	–8.59	129.16
t-Value	7,285.76	958.53	1,220.60	5,718.63	3,016.26
Threshold	6.412	6.412	6.412	6.412	6.412
Confidence level (%)	99.99	99.99	99.99	99.99	99.99
Parameter	b_{11}^k	b_{12}^k	b_{13}^k	b_{14}^k	b_{15}^k
Value	–201.01	172.87	380.50	–970.94	2,035.15
t-Value	342.91	3,513.05	465.52	$1.040{\cdot}10^8$	$1.063{\cdot}10^5$
Threshold	6.412	6.412	6.412	6.412	6.412
Confidence level (%)	99.99	99.99	99.99	99.99	99.99

Table 5: Coefficient statistical tests: *BP* condition.

Parameter	a_1^k	a_2^k	a_3^k	a_4^k	a_5^k
Value	–0.0400	–0.0006	0.1864	–0.0142	0.1726
t-Value	332.221	14.512	234.541	165.639	124.745
Threshold	5.077	5.077	5.077	5.077	5.077
Confidence level (%)	99.99	99.99	99.99	99.99	99.99

3.2 Validation of the proposed procedure

In order to validate the proposed procedure, the daily travel demand, estimated by authors for the previous studies (see D'Acierno et al. [27] and Ercolani et al. [28]) in the case of an average working day in winter, was adapted in the case of new surveyed data by adopting an aggregated estimation approach. Details on the *a priori* travel demand estimation and related improvement by means of surveyed data can be found in Botte et al. [29].

The travel demand estimated by means of the whole set and assumed as '*absolute truth*' was compared with OD matrices obtained by adopting three different counting sets: the partial survey set (Fig. 2), the set obtained by replacing missing data with function outputs (Fig. 4) and the set obtained by using only function outputs (Fig. 5). The comparisons were implemented by identifying the optimal intervention strategies in the case of metro system failures. Details on the considered breakdown and related intervention strategies can be found in D'Acierno et al. [10].

Table 6: Objective function values for each different calibration set.

Intervention strategy	Complete survey data	Partial survey set	Replaced missing data	Function data
0	€ 231,034.54	€ 253,518.18	€ 208,026.81	€ 168,733.07
1	€ 245,446.51	€ 268,782.33	€ 243,390.79	€ 229,185.58
2	€ 245,446.51	€ 268,782.33	€ 243,390.79	€ 229,185.58
3	€ 244,296.97	€ 267,618.70	€ 241,918.38	€ 228,309.20
4	€ 245,378.78	€ 268,671.10	€ 243,344.24	€ 229,131.18
5	€ 245,378.78	€ 268,671.10	€ 243,344.24	€ 229,131.18
6	€ 244,362.09	€ 267,724.45	€ 241,952.16	€ 228,324.68
7	€ 243,316.37	€ 268,612.45	€ 243,414.95	€ 229,163.03
8	€ 245,107.97	€ 268,431.45	€ 242,980.34	€ 228,863.49
9	€ 245,031.45	€ 268,338.34	€ 242,614.05	€ 228,665.49
10	€ 244,295.08	€ 267,631.59	€ 241,938.04	€ 228,309.45
11	€ 226,969.63	€ 248,628.12	€ 203,790.44	€ 165,513.24
12	€ 226,969.63	€ 248,628.12	€ 203,790.44	€ 165,513.24
13	€ 225,819.36	€ 247,460.93	€ 202,974.86	€ 164,773.21
14	€ 226,901.90	€ 248,516.89	€ 203,685.02	€ 165,427.05
15	€ 226,901.90	€ 248,516.89	€ 203,685.02	€ 165,427.02
16	€ 225,885.21	€ 247,570.24	€ 203,024.01	€ 164,822.67
17	€ 226,839.49	€ 248,458.24	€ 203,741.70	€ 165,473.23
18	€ 226,631.09	€ 248,277.24	€ 203,599.50	€ 165,342.21
19	€ 226,554.57	€ 248,184.13	€ 203,490.71	€ 165,242.52
20	€ 225,818.20	€ 247,477.38	€ 202,967.43	€ 164,769.11

Table 7: Objective function accuracy for each different calibration set.

Intervention strategy	Partial surveyed set (%)	Replaced missing data (%)	Function data (%)
0	9.73	9.96	26.97
1	9.51	0.84	6.63
2	9.51	0.84	6.63
3	9.55	0.97	6.54
4	9.49	0.83	6.62
5	9.49	0.83	6.62
6	9.56	0.99	6.56
7	10.40	0.04	5.82
8	9.52	0.87	6.63
9	9.51	0.99	6.68
10	9.55	0.96	6.54
11	9.54	10.21	27.08
12	9.54	10.21	27.08
13	9.58	10.12	27.03
14	9.53	10.23	27.09

(Continued)

Table 7: (Continued).

Intervention strategy	Partial surveyed set (%)	Replaced missing data (%)	Function data (%)
15	9.53	10.23	27.09
16	9.60	10.12	27.03
17	9.53	10.18	27.05
18	9.55	10.16	27.04
19	9.55	10.18	27.06
20	9.59	10.12	27.03
Average	9.59	5.71	17.28
Median	10.40	10.23	27.09
Minimum	9.49	0.04	5.82
Maximum	9.54	9.96	26.97

Table 6 provides the objective function calculation for each different calibration set (the value in dark gray identifies the first best solution, light gray shows the second and third best strategies). Table 7 shows variation in the objective function with respect to the '*absolute truth*'.

4 CONCLUSIONS AND RESEARCH PROSPECTS

By following the proposed methodology the amount of data to be collected can be reduced without significantly compromising estimation accuracy. The space–time functions allow missing data to be replaced, providing an accuracy reduction of less than 6%.

In terms of future research, we propose to apply the proposed methodology in other contexts both for different time periods (for instance by collecting winter data) and different networks. A further research aim would be to verify the performance of different spatial reference systems such as curvilinear abscissa or polar coordinates.

REFERENCES

[1] European Commission. *Climate Action: Reducing emissions from transport*, http://ec.europa.eu/clima/policies/transport/index_en.htm (accessed March 2016).

[2] Dell'Acqua, G. & Russo, F., Safety performance functions for low-volume roads. *The Baltic Journal of Road and Bridge Engineering*, **6(4)**, pp. 225–234, 2011. DOI: 10.3846/bjrbe.2011.29.

[3] Dell'Acqua, G., Russo, F. & Biancardo S.A., Risk-type density diagrams by crash type on two-lane rural roads. *Journal of Risk Research*, **16(10)**, pp. 1297–1314, 2013. DOI: 10.1080/13669877.2013.788547.

[4] Bifulco, G.N., Pariota, L., Brackstione, M. & McDonald, M., Driving behaviour models enabling the simulation of advanced driving assistance systems: Revisiting the action point paradigm. *Transportation Research Part C*, **36**, pp. 352–366, 2013. DOI: 10.1016/j.trc.2013.09.009.

[5] Pariota, L., Bifulco, G.N., Galante, F., Montella, A. & Brackstone, M., Longitudinal control behaviour: Analysis and modelling based on experimental surveys in Italy and the UK. *Accident Analysis & Prevention*, **89**, pp. 74–87, 2016. DOI: 10.1016/j.aap.2016.01.007.

[6] Cascetta, E., Cartenì, A., Pagliara, F. & Montanino, M., A new look at planning and designing transportation systems: A decision-making model based on cognitive rationality, stakeholder engagement and quantitative methods. *Transport Policy*, **38**, pp. 27–39, 2015. DOI: 10.1016/j.tranpol.2014.11.005.

[7] Cascetta, E., Cartenì, A. & Montanino, M., A behavioral model of accessibility based on the number of available opportunities. *Journal of Transport Geography*, **51**, pp. 45–58, 2016.

[8] D'Acierno, L., Gallo, M., Montella, B. & Placido, A., The definition of a model framework for managing rail systems in the case of breakdowns. *Proceedings of the 16th IEEE Conference on Intelligent Transportation Systems (ITSC)*, The Hague, The Netherlands, pp. 1059–1064, 2013. DOI: 10.1109/ITSC.2013.6728372.

[9] Gallo, M., Simonelli, F., De Luca, G. & De Martinis, V., Estimating the effects of energy-efficient driving profiles on railway consumption. *Proceedings of the 15th IEEE International Conference on Environment and Electrical Engineering (EEEIC)*, Rome, Italy, pp. 1059–1064, 2015. DOI: 10.1109/EEEIC.2015.7165269.

[10] D'Acierno, L., Placido, A., Botte, M. & Montella, B., A methodological approach for managing rail disruptions with different perspectives. *International Journal of Mathematical Models and Methods in Applied Sciences*, **10**, pp. 80–86, 2016.

[11] Cascetta, E., *Transportation Systems Analysis: Models and Applications*. Springer: New York, 2009.

[12] Smith M.J., The existence, uniqueness and stability of traffic equilibria. *Transportation Research Part B*, **13(4)**, pp. 295–304, 1979. DOI: 10.1016/0191-2615(79)90022-5.

[13] Brog, W. & Ampt, E., State of the art in the collection of travel behaviour data. *Travel Behaviour for the 1980s*, Special Report 201, National Research Council: Washington, DC, 1982.

[14] Ortuzar, J.de D. & Willumsen, L.G., *Modelling Transport*, 4th ed., John Wiley and Sons: Chichester, UK, 2011.

[15] Domencich, T.A. & McFadden, D., *Urban Travel Demand: A Behavioural Analysis*. Elsevier: New York, 1975.

[16] Horowitz, J., Identification and diagnosis of specification errors in the multinomial logit model. *Transportation Research Part B*, **15(5)**, pp. 345–360, 1981. DOI: 10.1016/0191-2615(81)90019-9.

[17] Manski, C.F. & McFadden, D., *Structural Analysis of Discrete Data with Econometric Applications*. The MIT Press: Cambridge, MA, 1981.

[18] Ben-Akiva, M. & Lerman, S.R., *Discrete Choice Analysis: Theory and Application to Travel Demand*. The MIT Press: Cambridge, MA, 1985. DOI: 10.1016/0191-2607(90)90037-7.

[19] Ben-Akiva, M. & Morikawa, T., Estimation of switching models from revealed preference and stated intention. *Transportation Research Part A*, **24(6)**, pp. 485–495, 1990.

[20] Ortuzar, J. de D., Stated preference in travel demand modelling. *Proceedings of the 6th World Conference on Transportation Research (WCTR)*, Lyon, France, 1992.

[21] Lo, H.-P. & Chan, C.-P., Simultaneous estimation of an origin–destination matrix and link choice proportions using traffic counts. *Transportation Research Part A*, **37(9)**, pp. 771–788, 2003. DOI: 10.1016/S0965-8564(03)00048-X.

[22] Cascetta, E., Papola, A., Marzano, V., Simonelli, F. and Vitiello, I., Quasi-dynamic estimation of o–d flows from traffic counts: Formulation, statistical validation and perfor-

mance analysis on real data. *Transportation Research Part B*, **55**, pp. 171–187, 2013. DOI: 10.1016/j.trb.2013.06.007.

[23] Lu, C.-C., Zhou, X. & Zhang, K., Dynamic origin–destination demand flow estimation under congested traffic conditions. *Transportation Research Part C*, **34**, pp. 16–37, 2013. DOI: 10.1016/j.trc.2013.05.006.

[24] Oi, W.Y. & Shuldiner, P.W., *An Analysis of Urban Travel Demands*. Northwestern University Press: Evanston, IL, 1962.

[25] Wilson, A.G., *Urban and Regional Models in Geography and Planning*. John Wiley & Sons: London, 1974.

[26] Botte, M., Di Salvo, C., Placido, A., Montella, B. & D'Acierno, L., A Neighbourhood Search Algorithm (NSA) for determining optimal intervention strategies in the case of metro system failures. *International Journal of Transport Development and Integration*, **1**(**1**), pp. 63–73, 2016.

[27] D'Acierno, L., Gallo, M., Montella, B. & Placido, A., Analysis of the interaction between travel demand and rail capacity constraints. *WIT Transactions on the Built Environment*, **128**, pp. 197–207, 2012. DOI: 10.2495/UT120181.

[28] Ercolani, M., Placido, A., D'Acierno, L. & Montella, B., The use of microsimulation models for the planning and management of metro systems. *WIT Transactions on the Built Environment*, **135**, pp. 509–521, 2014. DOI: 10.2495/CR140421.

[29] Botte, M., Di Salvo, C., Caropreso, C., Montella, B. & D'Acierno, L., Defining economic and environmental feasibility thresholds in the case of rail signalling systems based on satellite technology. *Proceedings of the 16th IEEE Conference on Environment and Electrical Engineering (EEEIC)*, Naples, Italy, 2016. DOI: 10.1109/AEIT.2015.7415258.

THE IMPACT OF PLATFORM SCREEN DOORS ON RAIL CAPACITY

O. LINDFELDT
Sweco AB, Stockholm, Sweden.

ABSTRACT

A new railway line for commuter trains through central Stockholm, the 'City line', is under construction and will open in 2017. The line will have two new underground stations and these will be equipped with platform screen doors in order to enhance air quality and prevent suicides. Several stations along the line will have only two platform tracks and this will limit capacity.

Neither trains nor platform doors will be fully automatized and thus the drivers will have to stop the trains in exact position to match the platform doors. Consequently, adding extended times for opening and closing doors will prolong the dwell time by approximately 12 s/stop.

An extensive capacity evaluation has been carried out as a RailSys simulation in order to find out whether a capacity of 24 trains/hour/ direction can be reached. The core part of the study was to prepare dwell time distributions for all stations along the line. These distributions include several terms, namely additional braking time due to a more careful and precise braking, time to correct stopping position (stochastic), door opening time, passenger exchange time (stochastic), door closing time, door obstruction time (stochastic) and driver's reaction time before departure.

The simulation results indicate that the extended dwell times caused by the platform screen doors limit capacity to 22 trains/hour in one direction and 24 trains/hour in the other direction. This difference comes from the fact that the station with the longest dwell times is located adjacent to a station with four platform tracks where trains can recover delays. Thus, the delay level differs significantly between the two traffic directions. This delay difference influences the useful capacity in such a way that the most punctual direction has a higher capacity.
Keywords: dwell time, platform screen doors, railway capacity, simulation.

1 INTRODUCTION

A new railway line through central Stockholm, called 'City line', is under construction and will open in 2017. The line will be used as a commuter line. As a result the commuter traffic through central Stockholm will be completely separated from the regional and long-distance traffic. The traffic on the City line will be completely homogenous.

The line has two new underground stations, which are considered to be very attractive and frequently used by commuters in the future. These stations will be equipped with platform screen doors in order to enhance air quality and prevent suicides.

The Stockholm commuter network is X-shaped and consists of two branches to the north and two branches to the south (Fig.1). The City line will be a common, connecting link in the middle of the network. The overreaching goal is to operate all four branches with a regular timetable with five minutes between trains, which gives 24 trains/hour/direction through the City line.

However, since the train operation is not to be automatized, all trains will still be driven by human drivers. Moreover, the platform screen doors cannot be fully adjusted to the commuter

DOI: 10.2495/TDI-V1-N3-601-610

Figure 1: Stockholm's commuter train network.

train vehicles, since the City line ought to be used also as redundancy for other train categories. This implies that the screen doors will prolong the dwell time and affect capacity negatively.

In order to find out whether the capacity of 24 trains/hour/direction can be reached, an extensive capacity evaluation has been carried out as a RailSys simulation.

A crucial part of the work was to prepare distributions for the future dwell times at the stations. To do this, a special model for the dwell times was developed. Important parts of the model were extended braking time due to careful and precise stopping, door opening times, passenger alighting and boarding times, door closing times, extra dwell time caused by door obstructions, and driver's reaction time before departure. Several of these parts are stochastic and probability methods were applied in order to combine several stochastic contributions to the overall dwell time distributions derived. These distributions were used as input in the simulation.

Simulations were performed for different levels of traffic load, i.e. number of trains per hour. Results were compiled for mean delay and punctuality and from these a level of useful capacity could be found for each traffic direction.

This paper describes the procedure used to compile the dwell time distributions, the successive simulation as well as the final analysis, results and capacity recommendations.

2 LIMITATIONS

The study is a detailed simulation study of the mid-section of the commuter system in Stockholm. It is delimited in the following three important ways:

- geographically
- through delay modelling
- in time.

It is suitable to delimit the analysis to the section between the junctions where the two lines merge north and south of Stockholm. In order to make the simulation of the merging flows in the junctions realistic, the studied area also included the first station beyond each junction.

The simulated system is shown schematically in Fig. 2. Northbound trains are initiated in Älvsjö, whereas southbound trains are initiated in Sundbyberg and Solna. This makes it possible to model the critical merging procedures north of Älvsjö and south of Sundbyberg/Solna. The rest of the network is modelled indirectly through the use of distributions for the initial delays in Sundbyberg, Solna and Älvsjö, as will be discussed later in this paper.

Disturbances will also cause delays to trains within the simulated area. These can be divided into two categories: dwell time extensions and run-time extensions between scheduled stops. The former are in focus in this study, whereas the latter are not taken into account. This means that the effect of infrastructure, vehicle malfunctions, etc., is not further analysed and modelled.

The system will be fully utilized twice a day: during the morning peak period and during the afternoon peak period. For this reason only a time period of three hours, between 06:00 a.m. and 09:00 a.m., is evaluated in the simulation. This peak period is preceded and followed by a period with off-peak traffic in order to warm up the simulation model. Only trains operated during the peak traffic period are evaluated.

Figure 2 shows that the line has four stations between the junctions. Three of these stations have only two platform tracks, whereas the Stockholm City station has four tracks. Due to this configuration, the two-track stations will limit capacity and the dwell time in these stations will be critical for overall line capacity.

Platform screen doors will be installed in two underground stations: Stockholm City and Stockholm Odenplan. These doors require the trains to stop in a specific position in order to get a match between the train doors and the doors on the platform. Three alternative procedures for door closing and departure procedure are considered.

1. Platform screen doors are closed separately by the conductor.
2. Platform screen doors are closed automatically when train doors are closing. The conductor monitors the closing procedure while standing on the platform.
3. It is similar to alternative 2, but the conductor monitors the closing procedure while standing in his cabin on board of the train.

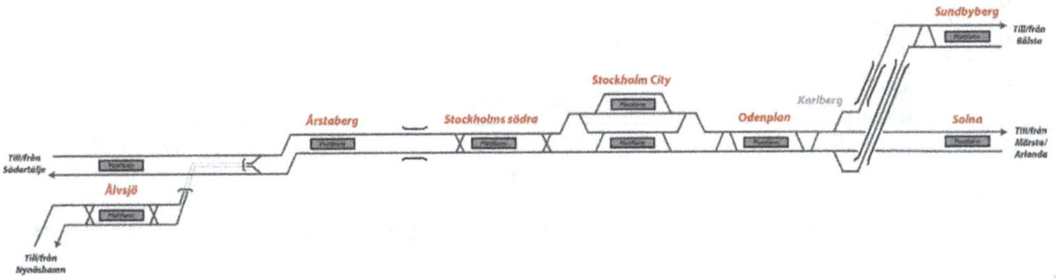

Figure 2: The analysed line section.

The three closing procedures are different in the sense that the total time – from initiation of the door closing procedure until clearance to drive is announced to the driver by the conductor – is estimated to be 16, 17 and 15 s, respectively.

3 TRAFFIC, TIMETABLES AND INITIAL DELAYS

A micro-simulation analysis of railway traffic requires a timetable to be designed and simulated (see Lindfeldt [1] and Sipilä [2]). Four alternative timetables, representing different capacity utilization levels, have been designed and simulated in order to find out how the platform screen doors impact the overall line capacity. Since traffic is completely homogenous it is easy to spread the trains evenly over the hour. The analysed timetables contained 20, 22, 24 and 26 trains/hour/direction. These traffic loads correspond to a time separation of 180, 164, 150 and 138 s between adjacent trains, respectively.

A fifth timetable, with only 2 trains/hour/direction, was evaluated as a reference. At this low-traffic density the knock-on delays are negligible, due to the large time separation, which means that the results will only contain primary delays and the robustness will be maximal.

All timetables were constructed so that every second train arrives from each of the two connecting branches in each junction.

Two important types of delay were modelled: initial delays and dwell time extensions. These are important inputs to the simulation model and have to be derived and compiled carefully (see Pachl [3] and Yuan and Hansen [4]).

Initial delays were assigned to the trains in Älvsjö, Sundbyberg and Solna through distributions that were compiled from historical delay data (Swedish Transport Administration [5]). Altogether four different distributions were compiled from the data for approximately 200 days of operation (Fig. 3).

These delays reflect the impact of the surrounding area, and they result in stochastic conflict situations that consume capacity in the City line. The cost, in the form of added delay, will determine the useful capacity, i.e. the number of trains that can be scheduled through the line.

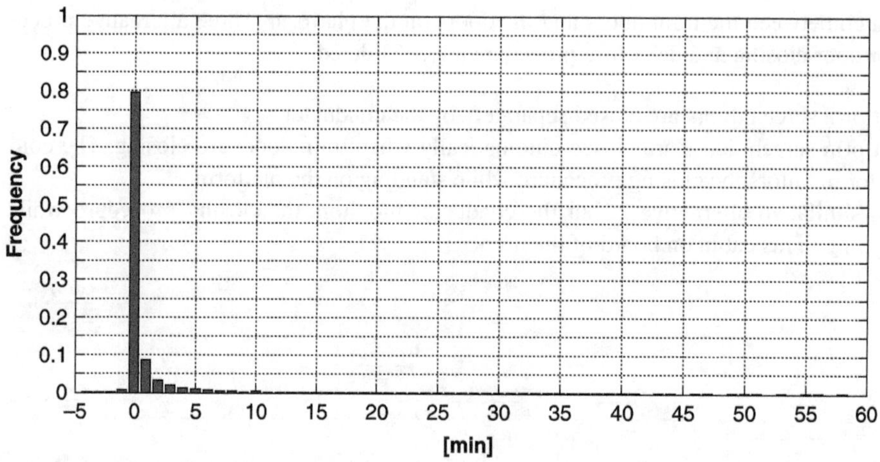

Figure 3: Initial delay distribution. This example is for trains from Älvsjö, bound for Solna.

4 DWELL TIME MODELLING

The dwell times are core inputs in the analysis, since they are directly affecting the capacity. A strong correlation is expected between capacity and the existence of platform screen doors, since these doors will affect dwell times and hereby also capacity.

Each station and traffic direction will have an individual distribution for the dwell time and it is essential to estimate these correctly to get reliable results from the simulation analysis (Yuan and Hansen [4]). The modelled area has four stations where delay distributions are needed as input in the simulation model:

- Årstaberg and Stockholms Södra
  These are existing stations that will not (initially) be equipped with platform screen doors. The number of passengers using these stations is not expected to increase significantly compared to the current situation.
- Stockholm City and Stockholm Odenplan
  These are newly constructed underground stations and the number of passengers is expected to increase, compared to the corresponding existing stations. Platform screen doors will be installed in these stations.

Dwell time distributions for the two existing stations, Årstaberg and Stockholms Södra, are rather easy to obtain through field time measurements. Such measurements were carried out on the platforms, analysed and compiled to one distribution, since the differences between the two stations and the two directions were small.

Figure 4 shows the distribution. The standard deviation is rather low and a scheduled dwell time of 45 s would be enough to cover most train stops without increase in delay. The mean dwell time is 38 s.

Finding the corresponding dwell time distributions for the two stations that will be equipped with platform screen doors is not this straightforward. These stations are non-existing and the size of the passenger flows is unknown. These flows affect the time for passenger exchange,

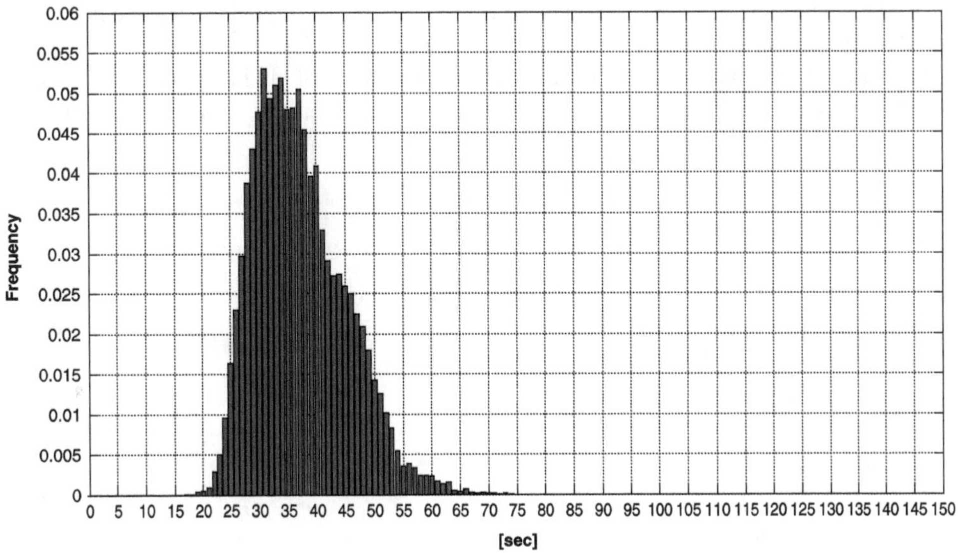

Figure 4: Dwell time distribution for Årstaberg and Stockholms Södra.

i.e. alighting and boarding. Moreover, the platform screen doors will affect the dwell time in several ways.

A first assumption, which might have to be corrected in a later study, is to make use of measured time distributions for alighting/boarding at existing stations. Assuming non-affected times for passenger exchange would imply that the City line will not increase the number of passengers. This assumption might hold as long as the increased passenger numbers are compensated with additional trains, i.e. as long as the number of alighting/boarding passengers per train is approximately the same as in the existing system.

A special model was set up in order to include the effect of the platform screen doors on the dwell time distributions (see Fig. 5). The model contains the following seven terms, correlated to the arrival, dwell or departure of a train:

- additional braking time
- time to correct stopping position (reverse the train a few metres)
- door opening time
- passenger exchange time
- door closing time
- door obstruction time
- driver's reaction time.

The platform screen doors impose the drivers to brake more carefully in order to stop the train within a stopping zone of approximately 0.5 m. This braking phase will take place on the platform track circuit, which means that the additional time consumed by the more carful braking will increase occupation time of the block section that is critical for capacity. This factor is assumed to be a constant that is equal to 4 s/arriving train.

From time to time, drivers will miss the stopping zone and have to reverse the train a couple of metres. Such a reverse procedure has been measured to take about 14 s. The occurrence of this situation is estimated to be 1/30. The stochastic feature of this factor is marked as dashed in the figure.

Door opening time has been measured to be 6.5 s. This time includes a short delay, compared to a situation without any platform screen doors.

The passenger exchange time depends only on the number of alighting and boarding passengers and their distribution on the train doors. As discussed earlier, these time distributions

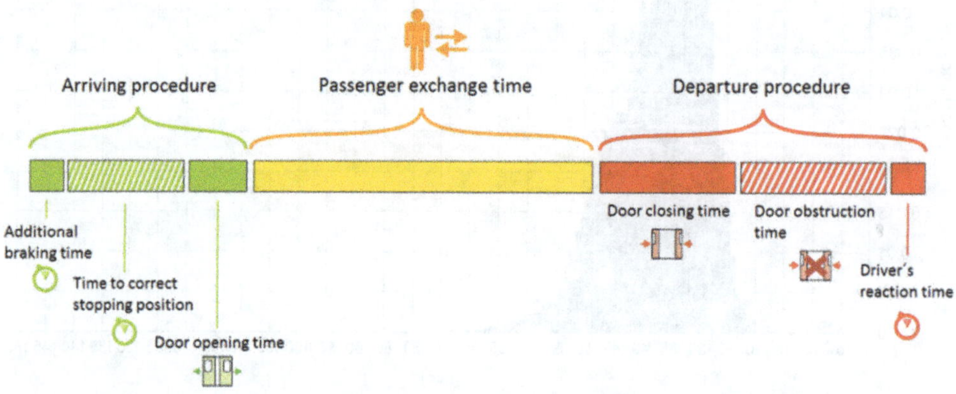

Figure 5: Factors causing dwell times in stations with platform screen doors.

have been assumed to be the same as in existing stations. Thus, Stockholm City is considered to correspond to the current Stockholm Central station and Stockholm Odenplan is considered to correspond to Årstaberg/Stockholms Södra.

The door closing procedure will follow one of the three alternatives. The time for these procedures has been measured to be 16, 17 and 15 s, respectively, as discussed earlier.

From time to time a train departure is delayed due to passengers who obstruct the doors. Interviewed train conductors estimate that obstructions occur as often as in 60% of the stops at Stockholm Central during rush hours.

Two types of obstructions occur: short and long, depending on whether the obstruction is short enough to let the doors close successively automatically or the conductor has to repeat the closing procedure for the obstructed door. Measured values of dwell time extension due to obstruction and conductors' estimates of frequencies are shown in Table 1.

The last factor, driver's reaction time, is measured to be 4 s and corresponds to the time for the conductor to give 'clear to drive' announcement and the driver to react on this clearance message.

Four of the terms are assumed to be constants, whereas the other three are known to be stochastic with a high degree of variance: time to correct stopping position, passenger exchange time and door obstruction time. The Monte Carlo approach was used in order to convolute these three distributions to one total dwell time distribution for each of the two stations: Stockholm City and Stockholm Odenplan. The four constants were added to these distributions, resulting in the total distributions as shown in Figs. 6 and 7.

Table 1: Data for door obstructions.

	Dwell time extension (s)	Estimated frequency	
		Stockholm Odenplan	Stockholm City
Short obstruction	5	17%	50%
Long obstruction	18	2%	10%

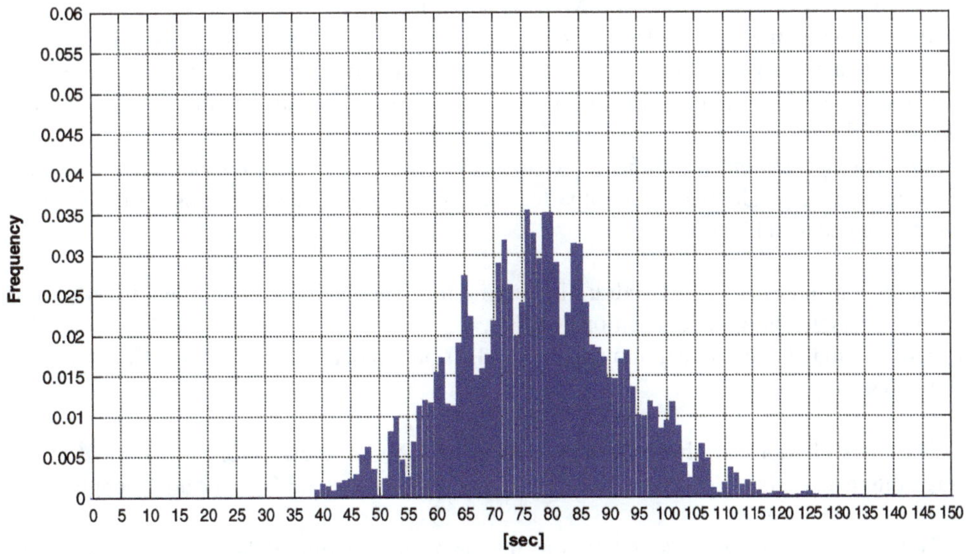

Figure 6: Dwell time distribution for Stockholm City.

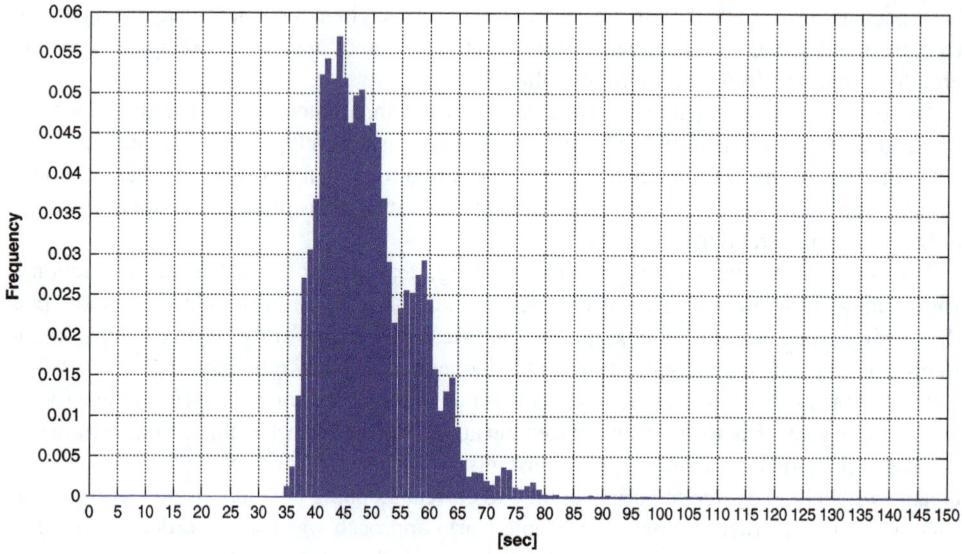

Figure 7: Dwell time distribution for Stockholm Odenplan.

The mean dwell time in Stockholm City is 78 s and in Stockholm Odenplan 50 s. This fact gives some very interesting results: the dwell time in Stockholm Odenplan is about 12 s longer than that in Årstaberg and Stockholms Södra, although the time for passenger exchange is the same. This is the expected capacity cost for the platform screen doors. Since Stockholm Odenplan is a two-track station, it is reasonable to believe that this station will become the bottleneck of the line.

On the other hand, Stockholm City proves to be a station where trains will recover delays, since the scheduled dwell time, 180 s, is well above even the longest dwells in the distribution.

5 SIMULATION RESULTS

A detailed infrastructure model for the system Älvsjö–Sundbyberg/Solna was built up in RailSys (Radtke [6]) and timetables for the five levels of capacity utilization were constructed. Each timetable included four traffic patterns, corresponding to the four branches in the network.

The four distributions for initial delays were assigned in Älvsjö for northbound traffic (two patterns) and in Sundbyberg and Solna for southbound traffic. Distributions for dwell times were assigned as follows:

- Årstaberg and Stockholms Södra: distribution for total dwell time compiled directly from measurements in Årstaberg station (see Fig. 4).
- Stockholm City and Stockholm Odenplan: distribution derived from Monte Carlo simulation (see Figs. 6 and 7).

These distributions for dwell time were adjusted in three levels to reflect the effect of the three alternative door closing procedures in Stockholm City and Stockholm Odenplan that were to be evaluated.

Figures 8 and 9 show the simulation results as mean departure delay for all stations. The mean arrival delays at Stockholm City are also included in the figures, since the scheduled dwell time in this station contains time supplements for delay recovery. Mean delays for the existing system are also shown for comparison.

Northbound traffic shows the expected effects of the traffic separation that is the main aim of the City line. The commuter trains can pass through Stockholm without interaction with regional and long-distance trains. This is shown in significantly lower delays. The useful capacity in this direction would be at least 24 trains/hour, given that all assumptions hold. Stockholm Odenplan is a bottleneck, which is reflected in increase of delay, whereas the long-scheduled dwell time in Stockholm City gives a good recovery effect. It is a considerable advantage for northbound traffic that the recovery is located just before the most capacity-restrictive passage of Stockholm Odenplan.

Southbound traffic shows less useful capacity, if mean delay is to be decreased compared to the current system with mixed traffic. This is due to the fact that the bottleneck station, Stockholm Odenplan, is located before the station where trains recover delays.

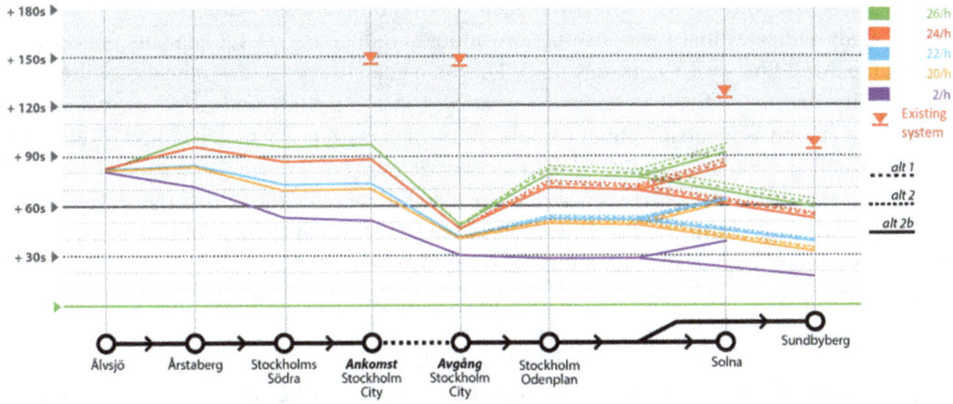

Figure 8: Simulation results for northbound traffic. The diagram indicates mean departure delays during rush hours for all stations. For Stockholm City both mean arrival and departure delays are presented.

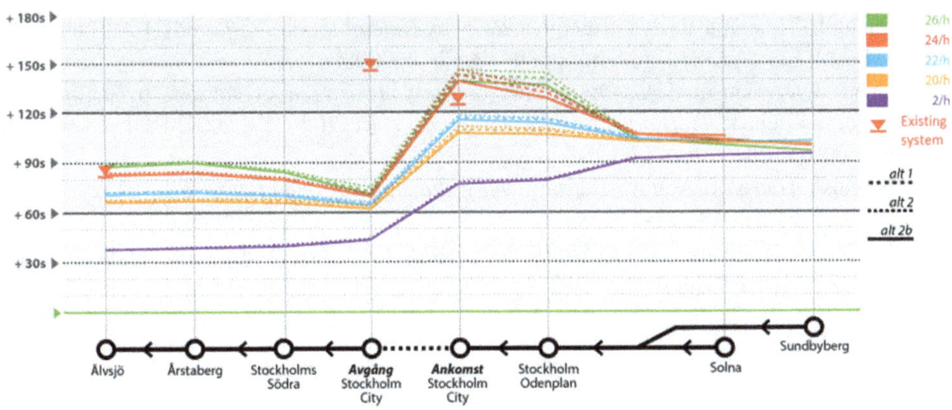

Figure 9: Simulation results for southbound traffic. The diagram indicates mean departure delays during rush hours. For Stockholm City both mean arrival and departure delays are presented.

The variance in arrival delay in Stockholm Odenplan is rather high and it results in train queues and delay propagation. The useful capacity for southbound traffic is therefore only 22 trains/hour, if mean arrival delay in Stockholm City is not allowed to exceed the current arrival delay in Stockholm Central station.

The differences between alternative door closing procedures are relatively small, about five s/train as a maximum. The capacity utilization has generally a greater impact on mean delays.

6 CONCLUSIONS

This paper explores how simulation can be used to determine useful capacity for a commuter line and how the effect of platform screen doors on capacity can be estimated through simulation.

Since the platform screen doors extend dwell times, the effect on capacity can be modelled through adjustments of the dwell time distribution in stations equipped with platform screen doors. The compilation and derivation of these distributions constitute a core activity in this simulation study, whereas the simulation itself follows ordinary standards.

The study indicates that the platform screen doors reduce the useful capacity, since they extend the dwell time by 12 s (mean value). The main goal, to operate 24 trains/hour/direction on the City line, is difficult to achieve, especially in the southbound direction where the bottleneck station is located right before the four-track station where scheduled dwell time can be extended in order to allow trains to recover delays.

The useful capacity might be even lower, since no failure rates for neither the screen doors nor other parts of the infrastructure or vehicles have been taken into account in the simulation studies. A straight comparison to empirical delay data from the current system is therefore not easily performed.

REFERENCES

[1] Lindfeldt, O., *Railway Operation Analysis. Evaluation of Quality, Infrastructure and Timetable on Single and Double-Track Lines with Analytical Models and Simulation.* Royal Institute of Technology: Stockholm. Thesis (PhD), 2010.

[2] Sipilä, H., *Simulation of Rail Traffic: Methods for Timetable Construction, Delay Modeling and Infrastructure Evaluation.* Royal Institute of Technology: Stockholm. Thesis (PhD), 2015.

[3] Pachl, J., Timetable design principles. *Railway Timetable and Traffic*, eds. I.A. Hansen & J. Pachl, Eurailpress: Hamburg, pp. 9–42, Ch. 2, 2008.

[4] Yuan, J. & Hansen I.A., Optimizing capacity utilization of stations by estimating knock-on delays. *Transportation Research Part B* **41**, 202–217, 2007. DOI: 10.1016/j.trb.2006.02.004.

[5] Swedish Transport Administration. *Delay statistics for commuter trains in Stockholm 2013-01-01 – 2013-12-31.*

[6] Radtke, A., *Software Tools to Model the Railway Operation*, Habilitation at the University of Hannover, Hannover, 2005.

www.ingramcontent.com/pod-product-compliance
Lightning Source LLC
LaVergne TN
LVHW062222200926
848484LV00006B/1457